The
First Science
and the Generic Code

Douglas J Huntington Moore

Published by Parmenidean Press

Hardback ISBN: 978-0-646-57127-0
Softback ISBN: 978-0-9873163-0-1
Parmenidean Press
http://www.parmenideanpress.net
info@parmenideanpress.net

First Edition
Rev. 3

To the memory of my parents
Thelma and Arthur.

Acknowledgments

This project started thirty five years ago when I first left the academic profession, this time for private study in Paris. I had the unique chance to spend the first six of my twelve Paris years, completely free to study and pursue whatever interests I chose. It was there that the foundational concepts of the present book were formed. During that time, I had the pleasure of attending many courses at the various universities of Paris covering philosophy, comparative religion, linguistics, and anything to do with signs. This included the stimulating lectures of Michel Foucault, the most erudite speaker I have ever encountered, François Châtelet, lectures in functional linguistics by Henri Martinet, the lectures of Algirdas Julien Greimas' in semiotics.

On returning to academia, I worked with my doctoral student Fei Lui to whom I am indebted for the many insights into logic using logical programming based on negative facts. I thank the Israeli entrepreneur Yaki Dunietz for supporting my multi-paradigm computer language project and who enticed me to quit academia yet once again. A deep understanding of the multi-paradigm paradigm has turned out to be crucial in writing this book.

Most of the intellectual influences shaping this book have come from the thinkers of antiquity ranging from the pre-Socratics to the Stoics, the ideas of the latter playing a pivotal role. Adding to this are the influences of Leibniz, Kant, Fichte, Hegel, and Marx. Marx, in particular, has been a constant influence for this project right from the beginning as the original aim of the project before going to Paris was "to make Marxism scientific." The influence of Charles Sanders Peirce has also been a constant theme in this project right from the beginning. I have also been strongly influenced by Vedanta Philosophy.

In modern times, the author is indebted to Iain McGilchrist's book on his parallel narratives concerning bilateral specialisation of the brain and the history of ideas. This gave me the confidence to be more forthright about the parallels between fundamental epistemological structure and the biological brain. In my search for the fundamental geometric interpretation of reality I was lead to the pioneering work in Geometric Algebra of David Hestenes of Arizona, and the physicists Stephen Gull, Anthony Lasenby, Chris Doran, *et al* at Cambridge.

This book might not have seen the light of day without the constant encouragement of my good friend Paul Marson over the past twenty years. As he did for my 1991 book, he has also provided much appreciated proof reading assistance. I thank my mathematician friends Andrew Macfarlane and James Juniper, for providing helpful criticisms on some of the more technical matters covered in the book. However, it goes without saying that any faults, omissions, or errors are entirely of my own making.

Last but not least, I must thank my wife Lyn and members of my family, both in Australia and in France who have had to put up with this preoccupation over these many years but have, nevertheless, supported me in my quest.

Contents

1

Prologue

It seems almost ridiculous, while every other science is continually advancing, that in this, which pretends to be Wisdom incarnate, for whose oracle every one inquires, we should constantly move round the same spot, without gaining a single step. (Kant, 1783)

Constantly moving around on the same spot, without gaining a single step, the quest for this ageless science continues. What was true in Kant's time is even more acutely evident today. Aristotle described it as the first of all the sciences. Those that followed named it metaphysics. This unborn science, destined to provide knowledge of the First Cause and even God, according to Aristotle, still stands there as an empty derelict monument to man's ineffectiveness at comprehending the deepest of the realities. It has driven many to despair and beyond into the depths of depression, and even madness.

Many have ventured out on this quest. Some are known for their attempts, others languish in anonymity along the way. The Scottish philosopher, David Hume, in his valiant quest locked himself away for six years before he suffered a nervous breakdown. He eventually settled for a sceptical response, advocating that science be condemned to empiricism. Several centuries later, William Rowan Hamilton, a fellow Scot, was leading a reasonably normal life as a very accomplished mathematician when something happened that would change his life forever. He discovered quaternions – or perhaps it was that the quaternions discovered him. He dropped all other interests and spent the final twenty years of his life on the slippery slopes of his newly acquired obsession. As has been appreciated by others, quaternions can be a beguilingly dangerous place to spend one's time. E. T. Bell in his book *Men of Mathematics* recounts:

The state of his papers testified to the domestic difficulties under which the last third of his life had been lived: innumerable dinner plates with the remains of desiccated, unviolated chops were found buried in the mountainous piles of papers, and dishes enough to supply a large household were dug out from the confusion.

During his last period Hamilton lived as a recluse, ignoring the meals shoved at him as he worked, obsessed by the dream that the last tremendous effort of his magnificent genius would immortalize both himself and his beloved Ireland ...

That is what happens on this kind of quest as one moves around and around on the same spot, not for a day or more, but for decades. Only those who have engaged on this quest can ever know the full depths of human emotion involved.

The great Isaac Newton also fell victim to the slippery slopes of the first of the first science. For him, the science would prove and demonstrate Absolute Knowledge, leading even to scientific knowledge of God. He spent over half his professional life on the quest, yet never published a single page on the subject. His instrument was that offered by the most advanced technology of his time: the crucible and flame of alchemy, that ancient science stretching back to the Babylonians.

Newton provided his own English translation of the mythical *Emerald Tablet* of alchemy. From the opening line of Newton's translation:

Tis true without lying, certain & most true

one can see that alchemy was not to be wholly an empirical science. Unlike the empirical sciences, this science is never approximate. It is simply true, certain, most true. Through alchemy, Newton's aim was nothing less than to unravel the code of the Cosmos embracing even a scientific knowledge of God.

John Maynard Keynes, who had in possession Newton's personal papers, made the following observation.

Why do I call him a magician? Because he looked on the whole universe and all that is in it as a riddle, as a secret which could be read by applying pure thought to certain evidence, certain mystic clues which God had laid about the world to allow a sort of philosopher's treasure hunt to the esoteric brotherhood. He believed that these clues could be found partly in the evidence of the heavens and the constitution of the elements. (and that is what gives the false suggestion of his being an experimental natural philosopher), but also partly in certain papers and traditions handed down by the brethren in an unbroken chain back to

the original cryptic revelation in Babylonia. He regarded the universe as
a cryptogram set by the Almighty... (Keynes, 1956)

Then there was Ferdinand de Saussure, a gifted graduate of the Sorbonne, full of promise, who started looking into understanding reality and language in terms of a possible Theory of Signs. He quietly slipped into obscurity for the rest of his life, totally engrossed in his research whilst teaching to a small stream of students at the Sorbonne. He never wrote a book and would have been forgotten if not for his former students who crafted the influential *Cours de Linguistique Générale* from the lecture notes they had taken during his courses.

On the other side of the Atlantic, Charles Sanders Peirce was engaged in a similar quest. He also never wrote a book on his philosophy, but left us his collection of published and unpublished papers. In addition to his philosophical writings, Peirce made important contributions to logic. Of particular interest are his attempts at a Theory of Signs and his use of non-linguistic tools such as his logical graphs that provided the initial inspiration, many years back, to this book.

The stakes are high, the anguish palpable. The routes are littered with the corpses of bygone warriors. So many have bravely embarked on this age-old quest for a goal that may not even exist. Starting from God knows where and leading to only where God knows, and even that is in doubt, they all return to where they started from, if indeed they return at all.

This book wallows in such anguish and is not for the faint hearted. Just as in a thriller movie, we have our backs to the wall and then the front wall starts moving in on us. There is no way back. There is no way forward. What can be done in such an impasse? This is the ultimate Mission Impossible, but pitched on an epic cosmological, ontological stage. This is not the time for fine-spun arguments. Dear reader, the author needs your continued patience and empathy to get us through this terrifying ordeal.

Discursive Technology

Some people talk too much. Other people write too much. When it comes to writing, this can certainly be said concerning the First Science. For over two thousand years the subject has been pored over, pondered upon and subsequently written about from almost every conceivable angle, yet no progress has been made. It would thus appear to be the height of impertinence and sheer folly, for us to declare that where all others have failed, we can possibly make any substantial progress.

Before even contemplating such an audacious project, we should first take a deep breath, and then start wondering actually, *why* so many writings over the centuries have only reported failure. Perhaps the writers just weren't smart enough. Considering that the writers concerned include such illustrious figures as Aristotle, Newton, Leibnitz, Descartes, Kant, right through to Charles Sanders Peirce, Whitehead, Wittgenstein and beyond, just to name a few, one should take yet another deep breath. Clearly, we should prudently close the book on that line of thinking. An entirely different line of thinking is called for.

The line of thinking that we advocate is not to up the ante in the probity and incisiveness of the writing, but to see *writing itself* as the obstacle to progress. To be more precise, the obstacle is seen to be in the form of the written symbol. In this context, writing is understood as a particular technology for communication and explanation, a particular *discursive technology*. The argument is that the discursive technology represented by the written word is limited and totally inadequate for expressing the essential nature of things. The technology of the Written Word is incapable of expressing the fundamentals of the First Science, the science expressing the deepest of the deep fundamentals of reality. The limitation of natural language in expressing principles of the profound is not a new idea. However, we will endeavour to give it a new twist.

Written Word technology was preceded by the discursive technology of the spoken word; the Oral Tradition founded on myth and ritual. Just as the technology of the Written Word in all its guises eventually supplanted the Oral Tradition, the unchallenged supremacy of the Written Word currently finds itself in decline. A new discursive paradigm is emerging. This will involve a new way of expressing knowledge and a new way of interpreting the world.

In order to comprehend the essence of the sweeping changes that are about us, it is important not to jump to hasty conclusions. One might see the changes in terms of their concrete technological manifestations in the form of communication and information technology. However, is this any more insightful than five thousand years ago in ancient Babylonia, when the written word was just starting to make its mark? Then, the concrete technological manifestation of the new emerging paradigm was in the form of the clay tablet. This was what it was all about, the exciting world of clay tablets..

The computer in our day is as the clay tablet of five thousand years ago, just as revolutionary, just as primitive. We have a long way to go before we encounter the more mature manifestations of the paradigm.

There are many revolutions upon us, information revolutions, computer revolutions, internet revolutions, and virtual world revolutions. All of these revolutions have one thing in common which distinguish the overall phenom-enon from the currently receding world that has been so dominated by the

Written Word. Instead of the static, immobile, passive symbol written on a piece of parchment, the new revolutions are based on the *dynamic symbol*, a symbol that is active and reactive. The symbol jumps off the page and starts participating in the real world.

The Scribe was the professional literate in the early days of the Written Word, charged with the daily task of producing the Written Word on demand. In these very early days of the new order, the new professional literate is the Developer. The Developer manually punches out software code on demand. There are many similar correspondences between emerging phenomenon in the new discursive technology and that of the written, just as there were between the written and the oral. In order to arrive at a useful perspective on current reality, such comparisons are instructive, but will not be pursued any further here at the low level.

At the higher level, the Written Word brought about innovations that were unachievable in the Oral Tradition, notably literature, mathematics, the sciences and the world religions based on sacred texts. All of this replaced oral storytelling, the myths and the sacrificial rites aimed at the pleasing of the gods. What concerns us in this book is what the new era holds in store as we move beyond the limitations of the Written Word to completely new ways of explaining and understanding reality.

In this book we end up reverse engineering a code inherently capable of coding, regulating, and explaining, not everything, but simply anything - an important distinction as we shall see. The code codes anything, or to be more precise, anything whatsoever. This is the underlying code that permeates reality. We will start to understand reality by its code, a code which is shared, not just by the animate but also the so-called inanimate - in fact by any entity that exists whatsoever, without limitation. The foundation of the sciences, the foundation of mathematics, and even the foundations of theology will be capable of being articulated in this code, and much else besides. One can call it the generic code, the genetic code, the cosmic code or simply God's code. It doesn't really matter. Pure reason alone leads to this code, and there is only one and the same code despite its many means of expression. This applies be it the organic, inorganic, or cosmic, be it the structure of matter or the structure of consciousness. There are no exceptions either to substance or to scale. There is only one such code, as indeed there must be.

This said the context is set. To say any more at this stage would be to spoil the story or perhaps risk being carted away in a big yellow taxi. As you can see, the anguish is already setting in.

Looking Forward

Ironically, the new science and philosophy that will dominate the twenty-first century may not be based on spectacular extrapolations of the present day sciences. Instead, as we argue in this book, the scientific and cultural revolution before us will be a revolution in the true sense, involving a return to past ideas. Ancient ideas dating from pre-Socratic times, such as the four element theory, the mysterious role of the masculine and the feminine and the active and passive principles, and in particular, the Stoic logic of Chrysippus, are all dragged screaming across the millennia into the present age. Ancient ideas are reinterpreted from a modern viewpoint. A new synthesis emerges and the beginnings of the new unifying science are developed.

Central stage to this science is a generic algebra capable of proscribing and describing *any entity whatsoever* - the *generic* organism. The algebra is based on a four-letter alphabet and has been around literally from the very beginning of time. In the biological arena, it appears as the genetic code. At the cosmological level, it becomes the *generic* code. An important aspect of this code is that it is based on a quite unusual construct, at least for modern times. It is based on gender.

Across the pages of this work, one starts to see the nature of this new science. It will perhaps be called *generic* science. Understanding this science requires learning to think differently from the moderns. Instead of "left hemisphere" dominated thinking, one must relearn to think like the ancients and allow right hemisphere dominant thinking to spring to the fore. The ultimate aim is to develop a right side science as an alternative and compliment to the traditional left side sciences.

As one Australian Prime Minister ruefully observed, "Life wasn't meant to be easy." The book starts with this dilemma. The author finds himself in a hopeless situation. He is confronted with the Kantian question: How can he construct a fundamental science based on pure reason alone? This is not just a question of interest to ontologists and epistemologists. It is also of central concern to even the smallest microbe that aspires to make its way in this world. This microbe is just as obsessed as anyone concerning the big question confronting us all. How can it *be*? How can *anything* be? How does life start? How does *anything* start? Answering these questions certainly does not look easy. Our story starts from this morass. From there, even the author was surprised by the eventual path taken. The project took on a much more noble and ambitious dimension than he ever envisaged. Quite amazing really.

We hope you enjoy accompanying us on this voyage.

Douglas J. Huntington Moore 2012

2
Lines in the Sand

Democritus and Heraclitus were two philosophers, of whom the first, finding human condition ridiculous and vain, never appeared abroad but with a jeering and laughing countenance; whereas Heraclitus commiserating that same condition of ours, appeared always with a sorrowful look, and tears in his eyes (Montaigne, 1580)

Why do some philosophers laugh and others cry? This imagery was employed in a very ancient illustration of two poles in philosophical thought dating back to several thousand years ago. Democritus was placed on one side of the divide and Heraclitus on the other. Democritus, who became virtually canonised as the laughing philosopher, was an atomist and reductionist, a precursor to that of present day sciences, and analytical philosophy. On the other side was Heraclitus, canonised as the crying philosopher. This is the side of brooding, troublesome musings, and contorted forms of reasoning. Heraclitus is famous for the immortal statement "You can't step into the same river twice."

There are many perspectives and angles of viewing this radical conceptual dichotomy. Some see it as a cultural division running across the university campus, with the sciences on one side and the humanities on the other. The sciences scoff at the humanities, whereas the humanities tear at each other's throats.

Formalising this age-old division between the laughing scoffers and the melancholic, paranoid depressives has often been seen described as the Demarcation Problem and has received considerable attention across the ages. In Cold War times, the communists built a militarised line in the sand in Berlin in order to make the demarcation more clear-cut. The paranoid depressives were kept safe on the inside insulated from the hedonistic scoffers on the

outside. In the West, Karl Popper countered with an epistemological version
of the Berlin Wall, based on a demarcation between empirical science on one
side and all the misfit non-sciences on the other. If your theory can be falsified
it is scientific, explained Popper. If it can't then it is unscientific. Marxism, of
course, was relegated to the side of the unscientific misfits.

For Charles Sanders Peirce, the two rival philosophical camps of his time
were Critical Philosophy and the Philosophy of Common-Sense. He remarked
that these two opposed ways of thinking were "at internecine war, impacifica-
ble." Bertrand Russel saw the dichotomy as a difference between atomism and
monism. The multiplicity inherent in atomism was already an empirical given
and so rationally obvious that there was no need to be agonizing over it. All
that it needed was for it to be simply and analytically sifted and dissected. The
other world of the monism involves the universe as a whole in a single
complex entity. Thus, it could be said that on the one side the universe starts
with a Big Bang, for the other it starts with the Big Birth, perhaps reoccurring.
The science of a world of objects on one side faces off to the allusive quest to
understand the enigmatic, introspective subject on the other, obsessed at not
only being itself but just being. Somehow.

The dichotomy we are talking about here is between two points of view,
two kinds of knowledge. One is often called objective knowledge, the other
subjective knowledge. They could be called object knowledge, knowledge of
object and subject knowledge, knowledge of subject. The justification is that
the terms objective and subjective seem to imply that the objective is more
exact than the wishy-washy subjective which is riddled with the whim of the
subject, destined to upset any well-planned experiment. A persistent view
dating from ancient times was that the opposite is the case. Sometimes this
view has been motivated by blind faith or religious fervour. However, a more
reasoned and painstaking approach will show that the seemingly murky world
of the subjective will reveal a breathtaking precision and accuracy surpassing
anything that a traveller from the traditional sciences could possibly dream.

This note of optimism does not dispel that fact that the task ahead of us is
not an easy one. We are to enter into an epistemological and ontological
hellhole. That is the bad news. The good news is that we eventually find a way
out of the morass.

Recipe versus Necessity

There are two ways of explaining something. One approach is to attempt
to describe it. The other approach is to reach an understanding of the tools
necessary to build it. One approach is the recipe. The other is based on
necessity: if it can't be built then it can't exist. It is the latter constructionist
approach that interests us here. Rather than attempt to describe reality, we

content ourselves with coming to terms with the essential apparatus needed to construct it. Rather than describing using some language, we concentrate on understanding reality *as a language*. Instead of trying to describe the thing in hand, we concentrate on a deeper understanding, an understanding of what is necessary to describe literally *anything*. The essential point here is that there is only one language, one code that can do the job.

Epistemology is traditionally defined as the science of knowledge. Our aim is to determine the underlying algebra, calculus or code by which the knowledge of knowledge may be expressed; this will be the calculus, the Logos, the language, the code with the necessary capacity to accomplish a rather daunting task, to lay the framework for what Kant called, the "Organon of pure reason."

Our approach will be far different from anything that Kant would have imagined. Kant argued for a new science in the name of metaphysics, writing that:

> *There is no single book to which you can point as you do to Euclid, and say: This is Metaphysics; here you may find the noblest objects of this science, the knowledge of a highest Being, and of a future existence, proved from principles of pure reason. (Kant, 1783)*

Our approach to the problem will involve reverse engineering a code. We will call this code the G code. What the G stands for is left up to the reader. Some might call it God's code because of the theological implications. Some might opt for the Genetic code due to structural parallels between the reverse engineered version of the code and the biological version. The author prefers the term Generic code because in the reverse engineering process, the initial problem to be solved involves how to provide a mechanism for describing *any entity whatsoever*. This *whatsoever* whatever is just one way of referring to the generic, hence the name Generic code.

All of this will takes some effort and demands some ingenuity to explain, as will be seen. At present, we are making a preliminary excursion into epistemology, the science of knowledge. Our initial excursions are designed to grease the mind, perhaps making it more susceptible to new ideas.

The Irony of Science

The type of knowledge that is of interest to our project is what we call *right side knowledge,* the kind of knowledge ultimately founded on pure reason alone. The only fly in the ointment is that the principles of pure reason seem to be, if not unknowable, at least totally unknown to us at this juncture.

Kant himself attacked the problem with what he described as his "fine spun arguments" which he admitted to be beyond the ken of the common folk. Indeed, his arguments continue to confound even the less common folk.

Unfortunately, Kant did not resolve his central problem. Perhaps one could say that his arguments were not fine spun enough. A greater degree of fine spinning might have delivered the goods. On the other hand, one could also argue that the problem may reside in the nature of the fine-spun argument itself. Such arguments claim objective neutrality. They display an impressive dexterity in logical rationality. However, in their obsession with technicalities, these fine-spun arguments often cloud over the stark truth that the essential subject in hand embodies trivialities bordering on the absurd. The argument somehow losses contact with reality and floats off into abstract nonsense.

So what can one do? What weapons can we bring to bear in order to defend ourselves from rationality drifting innocuously off into the patently absurd? The weapon is not abstract reasoning, as that is what gets us into the patently absurd pickle in the first place. One powerful weapon, mentioned by Aristotle, is something anathema to abstract reasoning. It is called *irony*. We note that left side scientific discourse is notable for its total lack of irony. Left side sciences and mathematics can be said to be irony free zones.

Certain strands of a society, particularly in the USA, are notable for being totally impervious to irony. This does not mean that such citizens cannot understand an ironical proposition; it just means that they understand it in a singular particular way. They understand it literally, so to speak. Other strands of the same society often seem to bathe in irony on a daily basis. They understand irony as a double bladed sword.

The logician and writer, Alexandre Zinoviev, in his writings on the former Soviet Union brought new dimensions to irony. Zinoviev writes in an ironic writing style and so it can be said that he described the Soviet Union ironically. Many writers employ irony in their works. Irony in these cases can be said to be a literary style of the writer. Where Zinoviev becomes interesting is that he describes a society where the people are objectively engaged in leading ironic lifestyles. Irony becomes a fundamental mode of being. More than that, he describes a whole society, which seems to employ irony as an underlying, systemic, organisational principle. Many have made the mistake of thinking that Zinoviev was like Solzhenitsyn, a ferocious critic of the Soviet Union. The Kremlin also thought so, and that's why they banished Zinoviev from Russia. Such is the nature of irony.

Aristotle discerned two kinds of irony. One was black irony and the other he called good irony. In most cases where Zinoviev is dealing with the Soviet Union, the irony is black or slides into sarcasm. But sometimes it is good, very good irony.

Zinoviev is important in helping us to grapple with the Kantian question. Kant asks what knowledge can be achieved when all experience and *a priori* knowledge is taken away. Zinoviev saw the Soviet Union as an ideal laboratory for grappling with a like question. The problem to be resolved was to understand the laws of human nature. To Zinoviev, the Soviet Union provided an ideal laboratory for studying the laws of human nature. In the Soviet Union laboratory, all market forces and *a priori* civilising influences had been taken away. Consequently, it was here that the fundamental laws of human nature could be studied in ideal laboratory conditions. In his quest, Zinoviev reveals that answers to questions like the Kantian problem have something fundamentally to do with irony.

With this in mind, we will now turn to look at the incredibly sophisticated epistemological machinery of modern day empirical science. We are of course joking. Many practitioners of present day science openly declare that they have no need for philosophy. Epistemology is for those that practice epistemology, not for those that practice science. It would thus appear that if we intend to engage with these practitioners of the modern art of science, reason alone would not suffice. When linear reason fails, an alternative is called for; enter irony. Our following attempt at irony probably falls a bit flat and comes out more as sarcasm. Pure irony is essentially a right side way of thinking. When projected onto left side thinking it loses potency and turns to sarcasm, or perhaps tongue in cheek, as witnessed by what follows. Nevertheless, it is worth a try. In order to get our story across, we will exploit any angle we can. We are the desperados in this plot.

Horse Kicks

Traditional left side sciences deal with conditional knowledge. Such knowledge involves knowledge of an entity in a given context. Conditional knowledge includes common sense kinds of knowledge, empirical scientific knowledge, and formal mathematical knowledge. Empirical sciences put an emphasis on precise and repeatable contexts for experimentation more than the actual entity of study. In formal mathematics, the context takes the form of particular presumptive axioms. The axioms provide the determined context for each particular kind of mathematics.

The concept of conditional knowledge may seem rather opaque and probably needs some freshening up with an everyday example. Conditional knowledge is the baggage one brings along to any given situation. One could call it *a priori* knowledge. For example, consider the example of a horse standing in front of you. That is the situation. Now most people know that it is not a good idea to stand behind a horse. That is an example of *a priori* knowledge.

This knowledge can be imparted to someone ignorant of this dangerous practice in a number of ways. This may become particularly pertinent if the person in question is actually standing behind an actual horse. One way of imparting the knowledge is to inform the person concerned that you know of a number of people who were kicked when standing behind a horse. This requires an application of empiricism and inductive reasoning.

Another way is to point out certain anatomical details of the hindquarters of the horse and its consequent capability to impart serious bodily injury by kicking. This is an example requiring deductive and predictive reasoning.

Finally, it is probably the mathematician who might get the most galvanised response by casually pointing out that he had read somewhere that the probability of being kicked by a horse satisfies the statistics of a Poisson distribution.

All three examples demonstrate the effectiveness and usefulness of conditional knowledge. The examples also clearly illustrate that the context, notably someone standing behind a horse, is always antecedent to the object of study, which, in this case, is the horse kick.

The small sciences of our day, in order to be judged as sciences, are various examples of the empiricism of horse kick rationality. This is not to devalue the importance of the empirical methodology. Empiricism can provide an important birthplace for a science in its primitive stages. Later, as it develops, it moves on to playing a loftier role in the academy of ideas. This is to say that there is a pecking order among the sciences. With the passage of time, the more fundamental sciences, the noble sciences, rise out of the ontological mud of empiricism and start to articulate a universality that begins to transcend the conditionality imposed by laboratory conditions. Conditionality does not need to be man-made.

Thus, noble ideas can rise from the mud, the mud of empiricism. However, as we shall see later, they can arise from above and trickle down from the heavens, so to speak.

...

The nice thing about conditional knowledge is that you know where it is coming from. Undoubtedly, there remain many hidden surprises and beauties in this domain but no worrisome, troubling, or really tortuous gyrations of the mind are required. This is, after all, the land of the laughing philosopher.

The same thing cannot be said for the other side. This is the domain of unconditional knowledge where roams, shrouded in myth, that most moody, melancholic depressive of all time, Heraclitus. With conditional knowledge, one always has something at the start to work with; a sensation, experience of some kind, a handful of assumptions, or even just an opinion could be enough to provide a tractable context. In the limit, perhaps even a bad joke might be better than nothing at least to get the ball rolling. However, with the unconditional you have nothing at all. You've got to start with something, but here you have sweet nothing, and it isn't even sweet and even that is uncertain There seems to be no traction at all for theory generation. It is enough to make a grown man cry.

The Kantian Question

In the Critique of Pure Reason, Emanuel Kant entered into this doleful epistemological hellhole posing the famous question: "What can we hope to achieve with reason, when all the material and assistance of experience is taken away?" It is the central question addressed in this book. How can you create knowledge working from virtually nothing but reason? Kant penned this question over two centuries ago. The question remains unanswered to this day. The question deserves to be raised once more and with even more urgency as Kant did in his time.

This raises the additional question of why does the problem remain unsolved? Many of the Germanic philosophers that followed Kant claimed to have had answers but then only in the form of tentative, intuitive and very informal glimpses. Hegel was the most successful. He read elementary and generic forms of reality into and from the historic movement of society. He combined metaphysics with an empirical historicism. The Hegelian perspective has held considerable influence right into the twentieth century, but has stagnated and lost potency in recent times.

Western philosophy presents itself today as a large soup of philosophical schools of thought, some quite decadent like the "post-modern." To each school doctrine there is, more often than not, a corresponding anti-doctrine. A characteristic of Western philosophy has been the tradition of different thinkers staking out their turf by taking sides along these doctrinal fault lines. Faced with two schools trumpeting totally opposed views, is it not so that one must be right and the other wrong? However, back in ancient times, Heraclitus did not seem to see any problem with this conflict of the opposites. To him, the whole Cosmos was a furnace of oppositions.

...

Basically, it is just a fact of life that at the deeper level, reality and the knowledge of reality tend to demand being understood in terms of doublets of ideas with one opposed to the other in some way. Aristotle developed his own way of coming to terms with the matter in the form of his famous Square of Oppositions. In general though, his basic *modus operandi* was to resolve oppositions by searching for some kind of "golden mean."

In modern times, Aristotle's Square of Oppositions construct has inspired the development of semiotics. Aristotle's square of oppositions becomes rebadged as the semiotic square. The basic idea of the semiotic square is that any particular view of reality as a whole can be analysed in terms of the semiotic square. The writings of Algirdas Julien Greimas (Greimas, 1991) provide many examples. In an early work (Moore, 1992); the author outlined his own primitive approach to the semiotic square, at the time.

Ferdinand de Saussure and Charles Sanders Peirce are considered the founders of semiotics in its modern guises. Charles Sanders Peirce interpreted semiotics as a Theory of Signs where the sign was essentially non-linguistic in nature. This contrasts with the approach adopted by de Saussure, which was essentially linguistic. Ferdinand de Saussure is more the founder of general linguistics rather than semiotics. He advocated that the relation between the sign as signifier and what it signified was completely arbitrary, a typical dictum of traditional linguistics and quite foreign to pure semiotics. The arbitrariness of the sign is also at odds with the long historic tradition dating from Plato, Aristotle, and particularly the Stoics, who followed the more semiotic tradition.

From our point of view, linguistics is a left side science of the same kind as all the other traditional sciences. It is on the left side that we find an immense variety of languages both natural and artificial. Such languages lend themselves to study using the traditional tools of abstraction, generalisation, empirical measurement, and mathematical modelling. On the other hand, we argue that semiotics is an integral component of our embryonic right side science, the science of the second kind. This semiotic breed of science does not treat the sign as a signifier of something "out there" as does linguistics. Pure semiotics is not concerned with knowing and describing "this and that." Rather, the objective is to know and describe something that simply just *is*.

This might give the impression that right side science must be inward looking. Traditional left side science looks outwards; right side science looks inwards. Hegel sometimes gave that impression, stating once that when he wanted to do philosophy, he studied his own ego. This evokes the image of one studying one's own navel.

Right side science may have introspective overtones, but this is not its characterisation. The relationship between left and right side science does not lie along any object/subject symmetry. If left side science can be characterised as a science of objects, it does *not* follow that right side science is merely the science of subject. The knowledge of the Self that *is*, does not merely involve knowledge of Self as subject. Self involves both object and subject. Even when Self is considered as object, the subject is always present. This contrasts with the left side perspective where in the case of the thing as object, no subject is present. Left side sciences specialise in what has been called the "view from nowhere," the "God's eye view," the view of the "detached observer." In right side science, the view is from the perspective of the subject or the object, depending on context. The Spectator never quits the Spectacle and the Spectacle never deserts the Spectator. There cannot be one without the other. In right side science, there is no desperately isolated Cartesian *moi pensant* left sitting alone on a rock.

Like the biological left hemisphere, left side sciences are intricately involved with languages. However, the treatment of the symbolic is rather skin deep as they only seem to be able to tackle language from a syntactical viewpoint: they are strong on syntax but weak on semantics. It makes one think of the golden droid C3PO in the Star Wars epic. C3PO looks good and sounds good. He is the master of over sixteen million means of communication. He knows all the protocols, all the customs, and is a highly valued asset when it comes down to etiquette. However, when the going gets tough he has to rely on his partner, the diminutive astromech droid, R2D2. R2D2 cannot speak any language at all except machine language.

Dramatic comrades like C3PO and R2D2 abound across the art forms. Don Quixote profusely provides a running commentary on his illustrious deeds of chivalry while Sancho just tags along with his feet never far from the ground. In the comics, Mandrake the Magician, with clever words and a snap of his fingers, baffles everyone with his illusions while dark and silent Luther provides logistical support from the shadows. The common theme is of a couple with the flamboyant character playing the role of a dexterous manipulator of symbols, the symbols having only the most tenuous attachment to the real world. On the opposite side, the other character is rooted in a simpler and simplifying world with the symbol anchored in the semantics of the real despite being strangely mute, or nearly so.

Therefore, it seems that, in both its biological and epistemological forms, left side thinking hogs the limelight and gets the good press. No better is this dramatized than in the writings of Karl Popper. For the sciences, using his falsifiable paradigm, Popper gives the bouquet to the traditional science as being scientific, whilst any other paradigm must necessarily be unscientific. In his book (Eccles, et al., 1984) with Eccles, he takes a comparable position on the relative values of the two hemispheres of the brain. Only the left side is conscious. The right hemisphere is relegated to playing the role of the "minor brain" and is basically devoid of consciousness, and by implication, devoid of intelligence.

As we have mentioned above, right science has a vocation for semiotics rather than linguistics. If there is any language at play on the right side turf, it will be the language of semiotics, not the language of the popular press and the masses. Moreover, the language will be unique as there is only one System of Signs. This is the central tenet of our work and the most dramatic. There can be only one such science, one such System of Signs. This is echoed as central tenet of the ancient science studied by Newton. The tenet, as stated on the emerald tablet, bears repeating:

Tis true without lying, certain & most true

In brief, the revolutionary, but ancient science we are looking for differs profoundly from its traditional science left side sibling. The right side science is certain and true, most true. It is a doubt free zone. The only problem is getting to understand it!

The task before us it to show that there is a unique, universal right side language that plays the role of uniting all scientific knowledge. We claim that this universal code can be reverse engineered. Moreover, the reverse engineered code will turn out to have exactly the same fundamental structure as the genetic code. Just like R2D2 who only speaks machine language, right side science only speaks in its own machine language; the machine language which is common to all life, language in the form of the genetic code. Any living thing whatsoever speaks and organises itself in this language. The other dramatic message of this book is that even the so called inanimate also speaks this language. This is why we will refer to it as the *generic code*. The question of what is living and what is not starts to take on another allure.

Old Science, New Awakenings

The question keeps teasing us: What kind of science can we achieve without resorting to experience? Science based on empiricism is a reasonably modern innovation and hardly practiced by the ancients. Rather than bottom up reasoning, the ancients seemed to have preferred a top down approach. Top down kinds of science sprung up spontaneously in the early civilisations. The same kind of basic structure seems to have kept raising its head in the process. A common recurring theme is the four-element theory of matter. For the Greeks the four Elements, air, earth, water, and fire, are attributed to Empedocles of the fifth century BCE.

Empedocles argued that the Elements were not gods, but rather got their power from the gods. There was a debate over which of the four prevailed over the other three. Heraclitus, like Empedocles, claimed it was fire. He associated fire with Zeus. Later, the Stoics also adopted this point of view in their cosmology, where Zeus was the only entity that survived the eternally recurring conflagrations. For Anaximenes, air was a god that "comes to be and is without measure, infinite and always in motion." (Cicero)

There is an important point to note here. Right from the beginning, the four building blocks were, more often than not, seen as a Three-plus-One structure: *one entity is determined in terms of a triad of others.*

Carl Jung saw this structure as an archetype for the gods:

Triads of gods appear very early, at the primitive level. The archaic triads in the religions of antiquity and of the East are too numerous to be mentioned here. Arrangement in triads is an archetype in the history of religion, which in all probability formed the basis of the Christian Trinity. (Jung)

Further to the East, in India, a four-element theory also flourished based on air, earth, water, and fire. Subtle thinkers like Sankara sometimes added ether as an ephemeral fifth element.

The godhead of Hinduism follows the Three-plus-One archetype where the Hindu trinity is made up of Shiva, Vishnu, and Brahma. Whilst the Christian trinity determines a living god with terrestrial manifestations, the Hindu trinity is transcendental. The Hindu trinity represent the three different ways of understanding the ultimate reality, where ultimate reality as the ultimate Self, reality as singularity corresponds to the allusive Brahman.

A Buffalo's Hoof for Buttocks

It was the VI century BCE. Nascent Hinduism was emerging from ancient Vedism. The first three Upanishads had already been written. It was in that time that Siddhārtha Gautama was born and raised as a Prince in a small kingdom in what is now known as Nepal. He eventually gave up his royal heritage and set out on a path to obtain enlightenment. He sought spiritual guidance from teachers and spiritual leaders down in the south of India. Legend has it that he attained advanced spiritual states, one being to control the seventh dhyana - the sphere of nothing. He then went on to control the eighth dhyana, the sphere of neither-perception-nor-perception. Still not satisfied, over a course of six years he embarked on an ascetic life, living in filthy poverty and eating one grain of rice a day. He pushed himself to the brink of death and is reported as saying:

Because of so little nourishment, all my limbs became like some withered creepers with knotted joints; my buttocks like a buffalo's hoof; my back-bone protruding like a string of balls; my ribs like rafters of a dilapidated shed; the pupils of my eyes appeared sunk deep in their sockets as water appears shining at the bottom of a deep well…."

At this point, Gautama realised the futility of such extreme ascetic practices and returned to meditation only.

The Buddhist Copernican Revolution

Gautama eventually obtained enlightenment, but radically different from what his original direction would have indicated. Under a fabled banyan tree he started to contemplate a complete doctrinal break from the ancient Brahman dominated Vedic religion. The notion of a supreme unqualifiable entity above all others, free from all determinations was a lofty notion but in reality, there was no such entity. In fact there could not be. How can the unqualifiable be qualified as supreme as this is a contradiction in terms? The Brahman starts to take on the dimensions of an oxymoron. A new doctrine was emerging in the Gautama's mind where no entity could be taken to be superior to any other. This doctrine eventually became known as *anatta*, the Doctrine of Non-Self.

Legend has it that under the Banyan tree he started to describe a new way of living that he called the Middle Way. He also espoused the basic doctrine of Buddhism in the form of the Four Noble Truths. In so doing, he had discovered a new godhead for a new religion; but in this case, it was a godhead without any transcendent or omnipotent god or gods.

The doctrine declares that no entity can be taken as superior to all others. No entity can be taken as the centre of the Cosmos. Buddhism was born as a Copernican revolution.

Sidenote:

A basic tenet of this book is that religion, and particularly the four major religions, are based on the same underlying reasoning, and logic even, of the right side science that we are researching. In this respect, we are at complete odds with the fundamentalist atheists that abound today. The popular writings of Richard Dawkins (Dawkins, 2008) are a prime example. Dawkins characterises religion as simply a "delusion" and offers his concocted alternative, Evangelical Darwinism. Dawkins joins Karl Marx, who infamously declared that religion was the "opiate of the people," a ruse to keep the masses happy while they are being exploited by the ruling classes. Just as for Dawkins, this kind of attack does little to advance knowledge and understanding of religion. Their agenda is not to encourage people to abandon a stultifying dogmatic creed, but rather to swap one dogmatism for another. Dawkins' Evangelical Darwinism leads to a modern version of Epicureanism whilst Marxism has already almost run its course and is in need a severe makeover.

Although not a fervent preoccupation of this book, the religious dimension cannot be ignored. Instead of being a target for attack and ridicule, religion, its literature, history, and its art, can serve as a very rich source of informal, potentially scientific knowledge, at least for the right side science we are developing. In this respect, Buddhism is at the forefront. Buddhism is the most difficult to understand of the World religions as it is devoid of a fundamental role for a deity of any kind, As a result, this doesn't leave much for the enquiring rational mind to latch on to. In the final score, Buddhist theology becomes the science of Emptiness and so becomes excruciatingly delicate to formalise.

The Buddhist paradigm involves a negation of the ancient Vedic paradigm based on the supremacy of an all-pervading transcendental Brahman. The negation is not a single, simple negation but, to use a heavily misused Hegelian term, it involves a *double negation*. The first negation totally negated that there can be any absolute hierarchy of beings. There are no absolutes. The Vedas claim the absolute supremacy of the Brahman. There is no transcendental reality. The only absolute truth here is that the Vedas must be wrong and so must be abandoned. However, this simple negation does not characterise the Buddhist paradigm. The first negation merely negates the possibility that any being, God or like creature, can be absolutely superior to any other being. In its perspective, this negates the existence of a transcendent as well as any omnipotent kind of deity as in Christianity. However, it does not negate the possibility of an imminent type of deity, a deity present with other beings but not over-towering. To eliminate that possibility, a second negation is required.

In what follows, only the first negation has been discussed and so the argument is over simplified. No attempt has been made to make corrections. Better to ride with the error than to introduce yet more errors. The discussion at this juncture is only exploratory anyway as we presently lack the tools for a more incisive kind of precision. We continue blithely onwards.

Buddhism and the Godless Godhead

The Four Noble Truths discovered by Siddhārtha Gautama provide the Three-plus-One semiotic structure underpinning the worldview of Buddhism. As such, they effectively define the Buddhist "godhead," where the first Noble Truth captures the supreme truth of the doctrine whilst the other three Truths articulate the triad of determinations by which the first truth is manifested and can be comprehended. In the other world religions the supreme truth, the One in the Three-plus-One formula, is seen as the ultimate expression of God. In Buddhism, the One is the ultimate expression that there is *no* supreme One: the Buddhist godhead has the same form as other religions, but differs from them all by being virtually godless relative to the other religions.

The First Noble Truth declares that there is absolutely no permanence in the universe, everything changes, and everything eventually perishes. Clearly, such a Noble Truth does not leave much of a career opportunity for gods claiming to be immortal. However, this does not mean that Buddhism is atheist. Deities, demi-gods, and deva of all kinds are tolerated as long as they do not claim to be of the imperishable supreme variety. In brief, deities are permitted in Buddhism but are excluded from the doctrine's godless godhead. Only the First Noble Truth and its accompanying triad of truths, enjoy the status of being supreme and eternal.

The triad of Noble Truths declares the three categories of how the supreme truth can be approached. The Second Truth declares that the non-persistence of everything leads to cravings for permanence and that this is the source for all suffering. We crave to retain what we have and crave to have what we have lost and wish to have. The Third Truth declares that the cure for suffering is to eliminate cravings. The Fourth Truth, which is called the Middle Way, prescribes a way of living, which provides the means to eliminate cravings and thus avoid suffering.

The Counter-Copernican Revolution

The doctrine discovered by Siddhārtha Gautama became known as Buddhism. Buddhism rapidly expanded throughout the Indian subcontinent to become the dominant religion in India for a thousand years.

Meanwhile, Hinduism was in epistemological trouble. Siddhārtha Gautama had exposed a deep structural flaw that had been glossed over for thousands of years going back to the earliest Vedic times. Now, with the emergence of writing, and a heightened climate for critical analysis, this flaw could not be ignored.

The problem is central and confronts any budding theological engineer aspiring to construct a World religion. The religion must be based on a Three-plus-One semiotic structure. It will be explained later why this should be the case, but for the moment it suffices to know that that is common practice in theological circles. The central problem is that the *One* part of the Three-plus-One structure must demand a supreme entity that is impervious to disqualification. (For the moment, we choose not to consider entities of the Jehovah and Allah variety, as they can be a bit tricky in this respect.) The best place to start is with the Vedic Brahman, as it is simply the entity the most immune from disqualification possible. Now a possible disqualification of the Brahman is to say that it does not exist. Fortunately, the Vedic Brahman has been engineered to withstand such an ontological onslaught because, as is well known, it is so devoid of qualification that it cannot be determined whether it exists or does not exist; it is beyond such trifling existential details. This means that it is true that it does *not* exist, but equally true that it *does* exist. Going the other way, it is false that it does not exist but equally false, that it does exist. The same argument applies to any property that may be attributed to it. Neither in the positive or negative can it be said that this Brahman is eminent, transcendent, constructive, destructive, expansive, contractive, smaller than your thumb, bigger than the universe, or any other qualification that comes to mind.

Clearly, this Brahman is a quite incredible entity. It stands out and above all others and so surely merits being entitled *Brahman*, the most supreme being of all. However, there is a problem. By considering the Brahman as the Supreme entity, the pinnacle of freedom from qualification, the adept had qualified the unqualifiable. The supreme unqualified being becomes a walking oxymoron. There cannot be any Supreme Being. Siddhārtha Gautama took his body, mind, and soul to the brink of the non-qualified and come back with the message that there is nothing there. There was no Imperishable One, but only a multiplicity of multiplicity. Everything changes.

Over the next millennium, Buddhism flourished while Hinduism had to refine its thinking. The process started to mature in the seventh century with Gaudapada building upon the Mahayana Buddhist concept of *shunyata,* or "emptiness." Gaudapada developed *advaita*, the philosophy of non-dualism. (Adv11). Sankara followed on to refine the doctrine and founded the Advaita Vedanta school of Indian philosophy.

Advaita Vedanta is first and foremost philosophical, rather than theological and theist. The philosophy rescues the ancient Vedic Brahman from attack from Buddhism by lifting the argument up to a higher level. Buddhism claims that there is no Supreme Being and, in particular, no Brahman. Advaita Vedanta has no alternative but to accept this argument as being literally true. There is not and cannot be a literal Supreme Being. To overcome this apparent impasse, the philosophy introduces Nirguna Brahman, that is so totally lacking in qualification that it cannot be said to be literally a Supreme Being. Moreover, it cannot be said that it is not. Sankara concludes that Nirguna Brahman must be totally beyond comprehension.

Nirguna Brahman is clearly out of range of any criticism that Buddhism might level at it, but the victory seems rather vacuous. How can one comprehend the totally incomprehensible?

In order to comprehend Brahman, the finiteness and limitations of the human mind must be content with a lower realm of understanding. In this lower realm, Brahman appears as Saguna Brahman, which, unlike Nirguna Brahman, is endowed with determined characteristics. Thus, in principle, Saguna Brahman is knowable and is called *Ishvara*. The determined characteristics of Ishvara are those of the empirical reality in which we live. Although these characteristics, attributes, and properties may appear real, according to Advaita, they are mere illusion produced by a mechanism called *Maya*. The only thing that is real is Nirguna Brahman, which is the only objectively true reality. Advaita Vedanta, philosophically speaking, is monistic; but rather than declaring that Nirguna Brahman is the One, as does ancient Vedism, it must be qualified by what it is not. It is non-dual. The role of the Maya mechanism is to maintain non-duality of the only fundamentally real reality, Nirguna Brahman.

The big question now looms. What is Maya? The first impression might be that Maya is all smoke and mirrors. After all, the principle role of Maya seems to be to create illusions. However, this would be like saying that the principle role of a motorcycle is to create noise. A more rational answer would be to say that the role of Maya is to maintain system integrity. System integrity, of course, is synonymous with maintaining the non-duality of Nirguna Brahman. In this context, Nirguna Brahman starts to take on the guise of being the *system invariant*.

What is at play here is a very profound form of relativity. In relativistic physics, it is the speed of light, a cosmological constant that is invariant for all inertial reference frames as explained in the Special Theory of Relativity. Not maintaining the invariance of the speed of light is tantamount to violating causality, allowing effects to precede their cause. Thus, relativity theory in physics can be seen as an essential aspect of maintaining the fundamental integrity of our universe. The profound relativity implied in Maya mechanism of Advaita Vedanta would embrace modern relativity theories, but go much further and deeper.

...

Here is not the place to attempt a treatise on Advaita Vedanta philosophy. Our intention is merely to indicate how the philosophy moved from the Vedic literalism of Brahman to a non-literal way of thinking. Instead of talking literally, the philosophy reaches for a lower degree of qualification. This can be achieved by changing the attention from the privileged and mythically supreme Brahman to a radically different entity. The supreme entity of ancient times morphs into *any entity whatsoever* which plays the role of an entity in its own right. It is this being-cum-entity that takes centre stage. The whole Cosmos gyrates around it. The centre of the Cosmos is *any entity whatsoever.* Here we start to see the Maya mechanism coming into play. If any determined entity, any determined being, were to enforce itself as being more important than any other then duality would be the outcome. A duality always exists where there is a ruler and the ruled. It is only in the case of the ontological democracy of *any entity whatsoever* as the centre of the Cosmos that such dualities can be avoided.

The Advaita Principle of Non-Duality is a subtle monism. It applies to a pair of entities that are deemed indistinguishably different. One entity is without qualification and the other qualified. At the most profound level we find the couple formed by the Nirguna Brahman and the more qualified Saguna Brahman. However, qualification is not an absolute: it is relative. A better-known couple than the Nirguna and Saguna Brahman is that between the individual soul ātman, and the cosmic soul, simply referred to as Brahman. The qualification of this Brahman would be that it is comprehensible by the ātman as being unqualified. These two entities form a non-dualist monism.

It is thus that anybody, any ātman, can declare that it is the absolute centre of the universe. It is in this kind of way that Advaita Vedanta philosophy can be viewed as a counter Copernican revolution, albeit an ironic one. You are indeed the centre of the Cosmos. The whole Cosmos gyrates around you, or at least your ātman.

As for the Maya mechanism, it is not smoke and mirrors but relativistic affects, some of which are already known in physics, and the involvement of a complex interplay of considerations, which remain unknown at this stage. The Maya mechanism may still be obscure but its role, as outlined in the Advaita Vedanta, has not changed: maintain a non-dualist reality. This is applicable to the Cosmos to maintain system integrity. It is also applicable to any entity whatsoever, including ourselves, in order to maintain system integrity.

3

Semiotic Adventures

There are two takes on reality, a left side way of thinking, and a right side way of thinking. This applies to the biological brain and to our "epistemological brain," the main thrust of this book. The left side of our epistemological brain deals with the traditional sciences. These bottom-up sciences work from experience, empirical data, learnt rules, axioms, prevailing opinions, official dogma, and so forth. On the right side, the science features a top-down mode of thinking which primarily concerns acting as a unifying and unified system. This right side science must unify not only mathematics, geometry, physics and cosmology, but also spiritual, moral, and aesthetic values.

Whereas left side sciences involve labelling things and manipulating labelled things, right side science works with oppositions and dichotomies. A preliminary excursion into right side science requires becoming familiar with the fundamental fault lines that mark out the epistemological landscape. On left side of the fault line is the aggregating, atomistic, labelling technology of the traditional sciences. Our task is to construct a right side science that must somehow express a science of oppositions.

Dealing with oppositions has tormented thinkers across the ages, not only Heraclitus. After spending a great deal of their life in the quest, these thinkers tend to settle for some variant of the semiotic square as a means of managing and expressing oppositions. The more ambitious of them, such as Kant and Hegel, can even employ a semiotic square of semiotic squares. In this section, we develop an informal introduction to semiotics and the semiotic square. Beforehand, however, there is the matter of whether semiotics really belongs to left side or right side science.

In a classic essay, John Locke (Locke, 1689) wrote that there are three kinds of science. The first two of his threefold classification correspond to what we have been calling left and right side science. The contingent sciences are on one side and the science of "things in themselves" on the other. He added a third kind of science dealing with the communication of one or the other of the first two kinds of knowledge according to the "doctrine of signs"; in other words, a semiotic science. Locke's view of semiotics is of a science that bridges both the first and second kinds of science and is intimately involved with signs. Such a triadic division of knowledge has appeal. However, the picture can get complicated as Locke's triadic epistemological structure for the sciences can claim to be a semiotic structure in its own right. With this in mind, we must take the wider view. We must study the *whole*.

We refer henceforth to Locke's doctrine of signs as semiotics. However, rather than interpreting it as a third kind of science, we take the stance that there are two different kinds of semiotics, one belonging to left side science and one belonging to right science.

The left side variant could be called applied semiotics. This was the semiotics pioneered by Ferdinand de Saussure in his abstract approach to signs in linguistics and the social sciences. Ferdinand de Saussure's work is more in the domain of General Linguistics than semiotics as a universal System of Signs. He tackled the problem of the *general* theory of signs, not the universal.

Charles Sanders Peirce attempted to develop semiotics as a pure System of Signs aimed at providing the overarching meta-reasoning for all the sciences. Peirce was one of the main inspirations for this project in its early days. Peirce believed that a System of Signs could be achieved with mathematical rigour, envisaging an abstract, deductive science of signification. In this respect, his theories were destined to remain as fragments of left side science. The great unifying science of signification he sought *cannot* be abstract as, by its very nature, abstraction imposes a dualism. Abstraction is always detached from the real world it fancies to explain from on high, hence the dualism. Right side science is the alternative to abstraction. It provides the generic, non-dualist solution. Ironically, the exponents of non-abstract science were none other than Peirce's nemesis, the ancient Stoics, thinkers he scorned as being crass.

What Does Semiotics Study?

Traditional sciences are characterised by the objects they study. Botany studies flora, biology studies fauna, axiomatic mathematics studies abstractions determined by axioms. These are all left side sciences. Semiotics bridges both traditional left side science and right side science. It studies the structure of wholes. In its applied form, the semiotic structure of a whole can be considered as an abstraction for the understanding of things that left side sciences study. In this context semiotics is a left side science. However, in right side science all abstraction must be avoided. There is no "God's eye" view of the object of study, the perennial source of duality and abstraction. The approach must be "non-dual." One could be forgiven for saying that it sounds very abstract. It all comes down to how to avoid the "God's eye" view.

Profound, but not abstract, is our reply. If the approach is to be profound, then it must be simple and simplifying. The subject must not be separate from the picture, but part of the picture. In fact, the subject must make up one half of the picture. The other half of the picture consists of object. Thus, in addition to considering the left-right takes on reality, there is an additional generic specificity, the subject-object dichotomy. This defines the front-back dichotomy of our world of knowledge. Place the subject in the frontal lobe section and the object in the rear. We end up with the semiotic square or, if you like, the semiotic architecture of our metaphorical brain.

The semiotic square describes the architecture of a whole, *any* whole. A whole is reality discerned from a particular point of view, the point of view of subject. The semiotic square includes subject as half of its structure. It is both Spectator and Spectacle on the one piece of paper.

Briefly, the semiotic square is a quasi-formal way of understanding wholes. This should become the first point to get across for anyone teaching semiotics. All of the nuances and ramifications can be initially glossed over. The next step in the learning process involves carrying out semiotic analysis of wholes through practical examples. By applying this quasi-formal tool, the student gains a practical introduction to understanding qualitative structure. We provide some examples in the next section. The point to note is that, unlike traditional forms of philosophy that claim to teach the qualitative, the semiotic approach lends itself to doing practical drills.

Education in the left side sciences, to its credit, is full of drills. The same approach is possible in philosophy if it is taught as a left side science. Analytic Philosophy provides such an example and offers training in the use of the analytical tools of logic. Graduates thus obtain skills that go beyond a prowess in spinning words. The only negative aspect of such an education is that the graduates are prone to display symptoms of right hemisphere deficit syndrome.

Currently, in many Western universities, the right side version of philosophy is rather curiously called Continental Philosophy. In this domain, it appears that the only drills offered to students are those of writing voluminous essays on European thinkers, either on the past masters or on the modern boutique, mainly French philosophers. Aping the left side disciplines, Continental Philosophy is noted for its extreme degree of specialisation. It seems that the only thinkers interested in actually accomplishing a unification of philosophy and the sciences have long since been dead. In many universities, the masters of antiquity are not considered very "Continental" and so often receive only scant attention.

In Australia, and possibly the US, Continental Philosophy is thought of as an "over there" philosophy. As some compensation, the US does have the legacy of the locally grown Charles Sanders Peirce. Any Australian philosopher of note will tend to be found at the extreme end of Analytic philosophy. They spend a lot of energy contemplating what it would be like to be a brain in a bottle, little realising that the experience can easily be duplicated by merely going along to a lecture on the subject.

A Semiotic Analysis Example

It is now time to look at some examples of semiotic analysis based on the semiotic square. The first example comes from business and illustrates the non-abstract nature of this way of thinking. The second example is from the other end of the spectrum and looks at theology.

Whilst left side thinking requires a form of deductive reasoning involving labelled or named entities, right side thinking dispenses with labels and drops the emphasis on syntactical structure. Right side thinking expresses itself in terms of oppositions. The reasoning process becomes *dialectical*. A basic understanding of semiotics provides an introduction into dialectical reasoning.

Dialectical reasoning in its simple form commences with an opposition between two apparently contradictory viewpoints, sometimes simplistically described as the thesis-antithesis divide. A prime example is the left side, right side dichotomy that divides traditional scientific conditioned bottom-up knowledge from the misunderstood unconditioned top-down knowledge domain.

A simplistic interpretation of dialectical reasoning takes the form of the thesis-antithesis-synthesis triad, often attributed to Hegel and his dialectics. Hegel never used these terms in his own work and it is doubtful that he ever would have. In this triad, each of the three terms only plays the role of labels. As such, the triad belongs more to left side reasoning. Such structure is sequential and more concerned with linear syntax structure than semantics. The sequential triad construct contrasts with semiotic structure, which is primarily geometric and a thus synchronic.

Geometric semantic structure leads to semiotics. Instead of the triad, a second orthogonal dichotomy is projected on the first dichotomy. The result is a semiotic square. The subject and object provides the second dichotomy. Subject provides the "frontal lobes" of the semiotic square, object bringing up the rear. This semiotic square can play the role of a metaphorical brain that projects its architecture onto the world it wishes to understand. Conversely, the world demands this kind of brain architecture in order to be understood.

The practical example in the next section should help to demystify this construct and its use as an aid to understanding. In the initial instance, the approach might appear a bit simplistic. Most of us have a hunger for the noble and dignified. That comes later after at least some mastery of the facile.

Kiyosaki's Cash flow Semiotic Square

E	B
S	I

- **E** for Employee
- **S** for Self-Employed
- **I** for Inventor
- **B** for Business Owner

Figure 1 Kiyosaki's semiotic cash flow quadrant.

In the market of "Get Rich" books, the book (Kiyosaki, 1998) of Kiyosaki is a classic. Kiyosaki had two fathers. His biological father was well educated but a financial disaster waiting to happen. His stepdad had no tertiary education but was a rich, self-made man. Kiyosaki makes no mention of semiotics nor undoubtedly did his stepdad. The topic was not semiotics, but establishing a personal cash flow. The stepdad looked at the problem as a whole and came up with the four cardinal modes of establishing cash flow. The construct then serves the purpose of being able to talk and reason holistically and coherently about the question in hand, notably how to become financially independent.

The geometry of Kiyosaki's cash flow semiotic square agrees nicely with that already outlined in our work, even satisfying the same polarity conventions. It is worthwhile making the connection. Firstly, there is the left side, right side dichotomy. On the left side we have Employee (E) and Self-employed (S) cash flows. These two entities are both contingent on something else, the basic characteristic of left side entities. The employee is dependent on being employed by an employer, the source of his cash flow. Self-employed survives thanks to his clients; no clients means no self-employment. Clients are the source of his cash flow. Both Employee and Self-employed rely on an Other.

On the right side, we find Inventor (I) and Business owner (B). These entities are free of *a priori* conditioning. They are lone wolves. Inventor has only himself to rely on. Inventor, more often than not, can also include the unemployed. Both have similar cash flow characteristics and are obsessed by innovation. Thus, Inventor includes working on a patent, and for the unemployed variant, working on an innovative curriculum vitae or an even more innovative scheme to get welfare payments. As for the Business owner, he only relies on cash flow from what he owns, not from something else.

That completes the left-right dichotomy; now for the front and back split of the cash flow world. According to our semiotic square, this involves the dichotomy between subject at the front lobe end and object at the back. In Kiyosaki's square, this makes E and B as subject and S and I as object. Projecting the subject-object interpretation onto Kiyosaki's square is quite interesting.

One way of looking at it is to take Karl Marx's perspective. Instead of talking about cash flow, Marx would talk about the flow of surplus value. In this perspective the poor old Self-employed and the Inventor are not involved in the flow of surplus value, they are too caught up with things. Thus, they disappear from consideration. We are left with Employee as proletariat, the source of surplus value who confronts, face to face, the Business owner as capitalist, the accumulator of surplus value. The innovation of new product by I and the maintenance of existing product by S do not enter into the surplus value equation. Innovation and service industries are ignored in the classical Marxist framework. They also fall outside the realm of exploitation. It is hard to exploit your plumber, even harder to exploit the inventor as the person about to invent the replacement for your iPhone. To sum up from a Marxist perspective, by the surplus value argument definition, only subjects can exploit or be exploited. Thus, E and B are subjects. All the rest, like innovation I and all the service industry cash flow S people, come along for the ride as mere objects. In order to explain the difference between subject and object we have had to resort to classical Marxism. Subjects engage in class struggle, self-employed and innovation objects do not. This explains the front-back dichotomy of Kiyosaki's semiotic square from a Marxist perspective.

Here, we are using the semiotic square as a source of educative entertainment and somehow we are learning something. Also, keep in mind that semiotic squares should not be used as a *static* classification device. A real world business example is instructive in this regard.

The author himself has been splattered all over Kiyosaki's cash flow semiotic square at one time or another, but his best example comes from a seminar given by the CEO of a large successful start-up. The seminar opened up with the speaker walking into the room with a trolley full of a great variety of lettuce. He promptly proceeded to throw lettuces at the audience. After being hit twice by two different kinds of lettuce, the author started to take notice.

The speaker recounted the story of his family start-up. It makes ideal semiotic square material. The family started off as independent vegetable farmers and so commenced as cash flow S. Times were hard and they were being ripped off by the supermarkets. A few of the family member moved over to I and started innovating, as Inventors do. They discovered that a lettuce is not an industrial product and so not a good way of making money. However, properly selected and packaged it became a salad pack, and a salad pack *is* an industrial product. The family was about to become important players in the packaged fresh fruit and vegetable industry. They raised venture capital for R&D, marketing, and promotion. Like most venture capital exercises, this does not move the recipient from the ranks of Self-employed and Inventor to the exalted ranks of the Business owner B, smoking a cigar. It is venture capital that becomes most of B, while the family moves into the ranks of employees of the company owned by B. As the dust settled, the family was left owning only 13 per cent of the company, and it was quite a large family. This became particularly vexing as the company, after all the trials and tribulations, eventually became extremely successful, being worth many hundreds of millions of dollars on the stock exchange. Fortunately, one family member was thinking straight at the beginning and had put in a clawback clause in their contract with the venture capital firm. The family could buy out the venture capital firm as long as they paid ten times more than the money invested, within a certain time limit. This was a very large amount of money, over 60 million dollars. However, since the company was worth many times that amount the family had no trouble arranging a management takeover and everyone lived happily ever after. It just shows what can happen when you move around on the semiotic square.

Kiyosaki's books illustrate the difference in thinking between his highly educated, abstract-thinking poor dad and his uneducated, but smart, rich dad. The author's favourite way of summing up the situation is with the observation:

> *Most people who successfully complete university studies end up working for someone that did not.*

The author is being deliberately provocative. However, the present state of world affairs demands such sorely needed provocation. Left side, abstract thinking increasingly dominates university education, even in the humanities. However, such a way of thinking leads to a high rate of failure in the real world, and particularly in the business world. Apparently Kiyosaki's clever, university educated dad was a dedicated left side thinker and so had all the ingredients for becoming a flop in the world of business. His other dad had learnt business acumen by avoiding abstraction, shunning opinion, calling a spade a spade, and relied more on right side thinking. He also employed squared shaped thinking where all the cards were placed on the table at the same time.

In an attempt to overcome this left side phenomenon, and cash in on business demand, many universities have set up their own version of the famed MBA, the Master of Business Administration. Other university degrees are oriented to produce future employees for the universities themselves, government, and corporations. In contrast, the MBA provides some kind of remedial epistemological therapy for graduates that aspire to run these enterprises. To some extent, the MBA can be successful. An important part of the MBA is real world experience in a wide number of business case studies, rather than offering abstract *theories* about business. Often these case studies involve the student moving to another country to carry out the work. One can also detect some attempt to experiment in modes of thinking other than the abstract and analytical modes. The MBA can dabble in the synthetic and the lateral. The author is unaware of MBAs that do drills in semiotic analysis, thus taking a leaf out of Kiyosaki's book, but would not be surprised to see such cases.

In the next section, we take a deeper look at the difference between left and right side rationality.

The Semiotic Product

This section is about multiplication. In the large sense, multiplication brings two things together to make a third. In the case of numbers, this leads to simple arithmetic. In the case of two inebriated men at a bar, it can lead to a bar room brawl. The ancients, both in the West and the East, were interested in bringing two principles together, one masculine and one feminine. The multiplication of these two principles created the Cosmos. We will visit the ontological and epistemological roles of gender later. For the moment, we are interested in multiplying together two different ways of thinking, two different takes on reality. In mathematics, there are so many different kinds of multiplication that it can be very overwhelming. We are particularly interested in the role of multiplication in geometry. There is one kind of geometry that is pertinent. It is called Geometric Algebra (GA). We will consider it in more detail in other sections of our work. GA only has one fundamental product called the geometric product and is closely intertwined with quadratic forms. In this section, we avoid any technicalities, and our algebraic product does not explicitly lead to a quadratic. It leads to another kind of energy powerhouse, the brain. The brain considered is the epistemological brain. This brain is constructed by the product of two different kinds of rationality. We like the product idea so much; we play God and apply the operation to itself. This section is a prelude to things to come.

We have been repeating many times the mantra that there are two takes on reality. From an epistemological viewpoint, this can be seen as two kinds of knowledge, knowledge conditioned by the contingent and knowledge conditioned by the demands of pure reason. We gloss over all of the nuances involved and simply refer to them as conditional and unconditional knowledge. Conditional knowledge corresponds to all of our traditional sciences, including axiomatic mathematics. The big question lies on the unconditional side of knowledge. The only science possible for this kind of knowledge is Aristotle's First Philosophy, what we will call the First Science. The kinds of thinking involved in these two forms of knowledge, we refer to as left side and right side knowledge, the inference being that the fundamental epistemological dichotomy is mirrored in the architecture of mind, and correspondingly of its implementation as brain.

One common approach to a dichotomy is to interpret it in the thesis-antithesis format leading to some kind of synthesis, which is supposed to resolve the inherent opposition. Some might even claim that this constitutes dialectics. Semiotics offers an alternative approach. Rather than attempt to explain how Nature or Mind might function sequentially as a temporal or logical process, the emphasis is placed on the geometric spatial structure. We concentrate on the *shape* of knowledge and the corresponding shape of Mind. Appealing to poetic license, we interpret the corresponding structure as the *epistemological mind*. We may also employ other terms that help us to understand the notion, such as the *semiotic brain*, and sometimes the *metaphorical brain*.

Initially, this metaphorical brain architecture is characterised by a great split down the middle, dividing it into a left and a right hemisphere. This structure carves up scientific knowledge into two camps. Traditional scientific knowledge, and its characteristic way of thinking, is the affair of the left hemisphere. On the other hand, the mysterious yet to be understood, holistic, unifying science, and its way of thinking, belong to the right hemisphere.

The left-right dichotomy expresses the first fundamental principle of consciousness. There can be no consciousness without something to be conscious about. The fundamental dichotomy at play here is between the object of consciousness and the consciousness of subject. Without any hesitation, we place the object on the left and the subject on the right side of the two hemisphere epistemological brain. What we have here is a basic formula that conscious being can be expressed as am opposition between the subject and its world as a conglomerate of objects. it can also be thought of as the "sum" of subject and objects.

Thus

Being = objects + subject   (1a)

This simple formula illustrates the left side view of the traditional sciences: any being, as a whole, is nothing more than the sum of the parts. The subject is the organism as a whole, and the objects are the parts. Set Theory in mathematics is a good example. From a Set Theoretic perspective, the predominant mathematical entity, the subject, is the set. Sets are constituted of objects called elements. Of course, the elements themselves can possibly be considered as sets in their own right, just as the sets can themselves be considered as elements of other sets. Bertrand Russel considered this kind of hierarchy of entities in terms of his Theory of Types. According to the theory, a set was of a different type to the elements it contained. A set of sets was thus of a different type to a set, and so on. The main claim to fame of Russell's Theory of Types is that it provides a simple way of avoiding logical paradoxes such as Russell's Barber Paradox. However, this is of little interest to our project.

The Opposition of Generic Being

The right side view interprets the relationship as an opposition between two contraries. The opposition can be intuitively expressed in the form:

Being = {objects, subject}   (1b)

Whilst the left side approach deals with the mereological sum of the two sides of the opposition, the right side approach treats the opposition as a fundamental dichotomy. The dichotomy is not an absolute dichotomy as what is subject and what is object depends on context.

Just as in the left side perspective, subject and object are of different type. However, in the right side perspective there is no hierarchy of types. The types don't nest as do Russell's type system they combine geometrically. In this respect, the right side expression (1b) is allows a geometrical interpretation. The objects are on one side of the dividing line and the subject is on the other. Which side is which? We adopt the convention that the objects are on the left side and the subject on the right. There is nothing fundamental about this convention. Handedness is determined relative to each individual. However, like all convention, once adopted, to avoid all future controversy and paradoxes, the convention should be adhered to, at least as for the individual concerned.

The opposition expressed in (1b) leads to cognitively dividing objects from subject in terms of a geometric opposition between a left and a right side typing of entities. The entities on either side of the oppositions are *generic* entities. The notion of left side and right side are also generic notions. At the level of individuals, what is right side for one individual may be left side for another.

Further on in this work, we will replace that object-subject dichotomy with a more fundamental and more generic concept, but that requires an understanding of *ontological gender*. For the moment, the object-subject dichotomy will suffice. The subject and its kingdom of objects make up a whole. This construct is truly universal as it applies anywhere and everywhere.

Sidenote

Many thinkers make the mistake of confusing the general with the universal. The general is an abstract construct and is limited to a closed, conditioned reality. Within its confines, a general law applies to everything. By contrast, the universal is not based on abstraction but is a *generic* construct. The universal does not apply to everything in some confine, as does the general. The universal, as a generic construct, is free of confines. It is an open system construct. As such, it applies *anywhere*. In brief, the general applies to everything (within a confine), whilst the universal applies everywhere (but not to everything. From a generic point of view, 'everything' is an abstraction that does not even exist).

To be pedantic and more strictly correct, the reference to 'everywhere' should always be read as a reference to 'anywhere'.

The two "parts" in expression (1b) form integral moments of a unified whole. They are inseparable, as one is needed to determine the other. The expression possesses what C. S. Peirce called "vagueness." In fact there is nothing more vague than this simple expression. The expression literally applies to anywhere you can think of. Wherever the consciousness, any consciousness, directs itself, expression (1b) jumps to the fore. Peirce remarked that vagueness should be a focus for formalisation. How do we formalise such a construct so reeking of vagueness?

The vagueness comes about due to the highly undetermined nature of the two entities involved. Essentially, consciousness demands the absolute separation of the object of consciousness from the agent of consciousness. It remains that the object of consciousness is so undetermined that it drowns in Peirce's sea of vagueness. One must agree that the validity of expression (1b) is inescapable. The concept expressed is absolutely universal as the object of consciousness can be literally *anything*. On the scale of vagueness, to be conscious of *anything* would be hard to beat. In order to reduce the level of vagueness, the natural direction would be move from the consciousness of anything to at least the consciousness of *something*.

We are thus led to two expressions of consciousness, one where the object is *anything,* and the second where the object is *something.* Thinking about this can make the head start to spin, but it's not that complicated. The *something* referred to here can be *anything*, but it is not. It is more determined than the allusive *anything* it is *something.* The two things are different, so are the two subjects for that matter. The two kinds of Being are also different. However, both expressions must be valid in the same moment. Each of the two expressions determines an opposition. The whole secret of the underlying right side formal system involves oppositions and the oppositions of oppositions. This is what dialectical thinking is all about. Left side thinking is analytical and is expressed rhetorically in order to convince, Right side thinking is dialectical and based on oppositions in order to understand. The question now is to understand how the dialectical notion of two oppositions opposed to each other.

Aristotle provided the germ of the answer but gave few details. He proposed a new science that was different to all others. Traditional sciences all deal with a determined genus of a thing that could be decomposed into species of things and particulars. He proposed a science of a thing that had no determined genus whatsoever. The science became known as metaphysics. Developing such a science has proved so allusive that no progress has ever been made.

A science of a thing, totally lacking in genus; here we see Peirce's 'vagueness' appearing in a new light. According to Aristotle, metaphysics is the science of an entity so vague that it is totally devoid of genus. The study of such an entity was to study Being itself, the science of pure ontology. In *Metaphysics*, Aristotle grappled with this mysterious entity-without-genus. All of the entities that the traditional sciences study fall under the umbrella of some kind of genus. Aristotle eventually settled for naming this mysterious entity of study for metaphysics in a particular way. He called the object of study for this first of the sciences, the study of being *qua* being. The exact meaning of this term, being **qua** being, has been a subject of constant debate over the millennia. Some say that ontology should be simply the study of being, with the term being '**qua** being' taken as unnecessary.

To avoid sliding into a long interminable debate, we simply present our interpretation as follows. The entity of study, notably pure being, has none other than itself for its genus. Thus being is both an entity and a genus. The genus itself is an entity in its own right. There are thus two entities in play here. The first entity is an entity that *has* a genus. The second entity is that genus. Unlike the ordinary entities of the traditional science, the entity and its genus are not separable, nor even distinguishable from each other. They from two modes of a common unity.

The unity of a determined entity with its less determined counterpart has long been interpreted from a religious perspective as the unity of the individual with his god. A most obvious example is the non-dualist doctrine of Hindu Advaita philosophy. There the individual. soul is the ātman and the cosmic soul is Braham. They two are different but non-dual. The entity becomes indistinguishable from its genus.

However, such lofty theological matters don't immediately concern us here. All the same, it is always interesting when one can imagine God in one's mathematics. Of course, only right side mathematics has that vocation.

In both cases, an opposition between subject and object is involved. A fundamental principle at play is that subject and object are themselves different. In this way, the principle of being is inseparable from the principle of consciousness. Both principles demand a separation of subject from its other. We don't delve into whether this difference is the result of a conscious effort of a knower or a characteristic of the known. Whether Mind moulds the world or the world moulds Mind, is a question we want to avoid, rather than attempt to answer in one way or the other.

Accordingly, we can consider that the terms 'being' and 'consciousness of being' can be interchanged. Similarly, the terms 'object' and 'subject' can be interchanged with the terms 'consciousness of object' and 'consciousness of subject.'

The Opposition of Individual Being

We can start the ball rolling by saying that there are two forms for ex-pressing being as an opposition between subject and object. We represent the first opposition geometrically as a geometric left-right opposition in (1b). This expression is valid for any being whatsoever. It involves the minimal aspect of consciousness of Self or Subject and the consciousness of what is not subject, the object part of the equation. This expression is the universal principle of consciousness. That is why we have referred to it as *generic*. This form of consciousness has a very high degree of "vagueness" about it, to use Peirce's term.

Consciousness Incompatibility Principle

In addition to the requirement that consciousness demands the difference between subject and subject, there is also the matter of attention. The fundamental principle at play here is that only one pole of the object-subject can be the centre of attention at the same time. We state this as the *Consciousness Incompatibility Principle*. As we shall see later, when we reinterpret all of this in the context of ontological gender, the inability principle plays a dominant role in the overall structural integrity of organisms, be they animate or even inanimate.

The next form of being applies, not to universal, generic being, but to individual being. We geometrically represent this opposition as a front-back polarity and express it algebraically as

$$\text{Individual Being} = \begin{Bmatrix} \text{subject} \\ \text{objects} \end{Bmatrix} \tag{2}$$

where the 'subjects' term is at the front and the 'object' term is at the back of the front-back geometric representation of this opposition.

What now interests us is that Aristotle's *qua* functor can be interpreted algebraically. We will interpret the *qua* operator as a kind of *semiotic product*. In this respect, substituting the individual and generic forms from the above two expressions we get that

$$\text{Being } qua \text{ Being} = \begin{Bmatrix} \text{subject} \\ \text{objects} \end{Bmatrix} qua \{\text{objects} \quad \text{subject}\} \tag{3}$$

Interpreting Aristotle's '*qua*' operator as an '*as*' operation, we get the two by two tensor:

$$\text{Being } qua \text{ Being} = \begin{Bmatrix} \text{subject as objects} & \text{subject as subject} \\ \text{objects as objects} & \text{objects as subject} \end{Bmatrix} \tag{4}$$

In Figure 2, the four-element tensor is interpreted as the four quarters of our epistemological brain. The basic requirement is that the organism be conscious of its Self and the Self of others. As the diagram illustrates, there are four possibilities. All four possibilities are of equal value. All four are present in consciousness and provide the common ground.

Each of the four regions of this epistemological brain has been categories for its speciality. The science of consciousness divides into four components, as does the epistemological brain that it must study. We have labelled the left frontal lobe as *Cognition*, the science of studying subjects as object. Cognitive science, seen in this perspective, becomes a traditional left side science. This is the realm that Hegel would refer to as the "empirical Ego." Cognitive science is the study of the empirical self. The right frontal lobe of our epistemological brain corresponds to the science of Consciousness pure. Hegel would describe this as the realm of the true Ego. This science must be an integral part of right side science. Up until now, right side science, the Frist Science, has not seen the light of day. The task of this work is to make a significant contribution in this regard. In its purest form, this First Science has no determined objects whatsoever. It is a pure science of Subject, a pure science of the conscious Self.

		The Epistemological Brain
Consciousness of Objects	**Consciousness of Subject**	**Cognition** ǀ **Consciousness**
Empirical awareness	Non-empirical awareness (of self)	Subject as Objects ǀ Subject as Subject (no objects)

X   →

Cognition of Subject

Non-sensory apprehension
_____

Cognition of Objects

Sensory apprehension

Cognition ǀ Consciousness

Subject as Objects ǀ Subject as Subject (no objects)

Objects as Objects (no subjects) ǀ Objects as Subject

Sensation   **Perception**

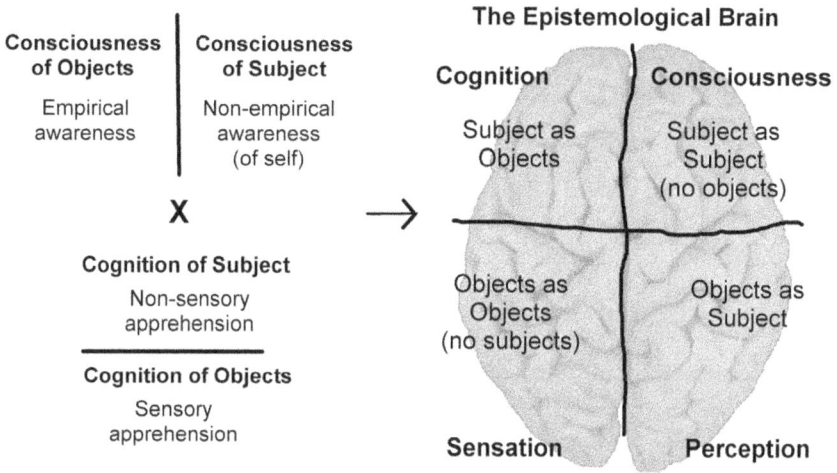

Figure 2 Superimposing via an informal "semiotic product" the two fundamental dichotomies of knowledge leads to an epistemological square. Only right side science is aware of the full significance of this opposition of oppositions and so has the potential for explaining consciousness.

We have categorised the left back area of the epistemological as Sensation. In this region, there are no determined subjects whatsoever, nothing but pure objects, a Buddhist paradise. This is a realm of unhindered sensual objects devoid of any determined subject to sense them. This was a realm explored by David Hume. In this realm, there is no determined knowledge or even direct reference to the Self. The Self appears as a bundle of "lively" sensations. The science of Sensation was touched upon by Hume where he described a world of pure sensations autonomously growing into rational forms. This must be a left side science but of a distinctly different kind than the traditional Western sciences. As will be explained later, such a science will be free of any traditional logic and totally free of abstraction.

On the other right hand back side, we see the realm of perceptions. A science of perceptions must form an integral part of right side science. The ancient Stoics seemed to have had a lot to say about such a science.

...

In constructing this semiotic square, our approach has been highly informal and lacking in any kind of rigour. However, we do end up with an embryonic cognitive form of mind that is not totally ridiculous. Also we have touched on something that will turn out to be very fundamental, an algebraic notion of product, albeit devoid of any mathematical precision. In later development and in the appendices, we will relate this product to the geometric product of the geometry pioneered by the mathematicians Grassmann, Hamilton, and Clifford of the nineteenth century. Can this qualitative, informal approach of semiotics eventually reach maturity and match up with a new kind of geometric algebra based on geometric products? We have a long road to hoe, but that is one of the objectives of this work.

Epistemological Brain and the Sciences

When interpreted from an epistemological perspective, the resulting semiotic square illustrates the four fundamental kinds of scientific knowledge. On the left side, the traditional sciences of object correspond to the physical sciences. Axiomatic mathematics, which has the study of axiomatically defined abstract objects as its vocation would also belong here. The traditional left side science of subject would correspond to the social sciences and would include empirical, behaviourist psychology. In passing, one could consider the possibility of an axiomatic mathematics of subject. Perhaps mathematical Category Theory would fit into this slot, as it is an attempt to provide an abstract overview of all mathematics. It is the closest that the axiomatic can come to the notion of mathematical subjects rather than mathematical objects. However, we will not labour over that point.

There are a myriad of traditional sciences and the number keeps on growing. This is a characteristic of left side science: there are many of them. There is an explosion of knowledge into an ever-increasing number of specialisations. Each science is fundamentally atomist in its philosophy and the overall epistemology ends up as an atomism of sciences. We then come to the question of what kind of sciences should occupy the right side slots of the semiotic square. The first thing to note is that the monism of the right replaces the atomism that dominates the left side. There is only *one* right side science. Right side science has for its vocation, the unifying of knowledge. There can only be one such science. Right side science, as the science with the role of unifying reality science, is the *science of unified reality*. The one right side science must still answer two questions and provide a science of object and a science of subject, but in a different manner from the left side science. On the left side, knowledge splits irrevocably into the sciences and the humanities. On the right side of the epistemological divide, object and subject present themselves as two sides of the one coin. As a preliminary attempt, the epistemological square shown in the diagram below has knowledge-of-Self in the frontal lobe and Knowledge-of-Object in the rear.

Sidenote:

We use the term *monism* loosely here. No rigorous science or philosophy can ever be a true monism, but rather an expression of *non-duality*.

Right side knowledge of self can be interpreted in many ways. It can include knowledge of the personal self and knowledge of the impersonal, generic self. It includes knowledge of the personal god and knowledge of the impersonal, transcendental god. Unlike the left side knowledge obsession with labelling everything, right side knowledge dispenses with labels. Instead of trying to put *this* and *that* into individually labelled, cardboard boxes, right side knowledge is always relative: it knows *this* relative to *that*. As such, right side knowledge can accomplish something to which the left side mode of thinking is totally oblivious: the right side can know the cardboard box as well as what it has come to contain. Right side knowledge works as a doubled edged sword. Even what is *container* and what is *contained* becomes relative.

Left Side Obsession with Truth Value

Left side reasoning can be formalised in terms of the various kinds of logic: propositional calculus, predicate calculus, modal logic and so on. All of these different species of logic come under the common genus of *symbolic logic*. Here we see the first obsession of left side reasoning, the obsession with symbols and the manipulation of symbols. Symbols form the basic atomic material of symbolic logic. It is a truism to say that, if one takes away the symbols and there can be no symbolic logic. Sequences of symbols lead to logical expressions: reasoning takes on a linguistic character. At this point, the second obsession of left side reasoning comes into the fray; attaching truth values to logical expressions. A logical expression is either true or false. Of course the expression must be meaningful as a well-formed-formula (WFF). Thus, the meta-logical expression "the expression X is a WFF," can itself be true or false. Left side reasoning embraces the law of the excluded middle, and so militantly ignores any possible truth-value other than true or false: the logic is systemically binary valued.

Left side reasoning, as mirrored in axiomatic formulations, involves non-constructionist formalisms. As is well known, constructionist formalisms require reasoning that breaks with the law of the excluded middle. A third truth-value of "unknown" is required. To cater for this situation, three valued, and even unlimited multi-valued logics have been introduced into formal left side reasoning. However, this is just a smokescreen. Ultimately, at a lower level than the three-valued proposition P, there can be another proposition Q that ultimately states, "P has the truth value of *undefined*" where, once again, Q is either true or false. This inbuilt, systemic obsession with the truth-value of symbolic expressions and the ultimate binary valued nature of the true or false, is a primary characteristic of left side reasoning. (In passing, right side reasoning is not based on the true and false but rather on the truth, an important nuance promoted by the ancient Stoics.)

Left side thinking relies on sequential, deductive forms of reasoning dominated by a binary valued truth system. Pioneered by George Boole in the mid-nineteenth century, the symbolic logic mode of reasoning became increasingly formalised. Extended to the higher order predicate calculus logic it ends up underpinning all present day sciences and left side schools of philosophy such as Analytic Philosophy. In The West, it has become the dominant mode of formal reasoning in our time. However, this unchallenged supremacy may be short lived.

Left and Right Reasoning in the Biological Brain.

Symbolic logic is an example of abstract reasoning, the speciality of left side thinking. A prime consideration of this book is to attempt to avoid the fatal attraction of abstract thought, a primary characteristic of left side thinking. This can be difficult at times, due to the profound questions that we have to confront. The psychiatrist Iain has been a source of inspiration for the author in drawing parallels between philosophy and the biological brain. (McGilchrist, 2009). Consider what he has to say concerning the difference in reasoning between the two hemispheres of the biological brain. First, there is his basic thesis, namely:

> My thesis is that the hemispheres have complementary but conflicting tasks to fulfil, and need to maintain a high degree of mutual ignorance. At the same time, they need to co-operate. How is this achieved, and what is their working relationship like?

Here, McGilchrist is talking about the biological brain. He is also implicitly talking about what we have been calling the epistemological brain. In his chapter entitled *The Triumph of the Left Hemisphere*, he ponders over what the world would look like, if the left hemisphere of the biological brain were dominant.

> ... the world would change into something quite different. And we can say fairly clearly what that would be like: it would be relatively mechanical, an assemblage of more or less disconnected 'parts'; it would be relatively abstract and disembodied; relatively distanced from fellow-feeling; given to explicitness; utilitarian in ethics ; over-confident of its own take on reality, and lacking insight into its problems ... the neuropsychological evidence is that these are all aspects of the left hemisphere world as compared with the right.

Once again, McGilchrist is talking about the biological brain; however, by attending a seminar in any one of the sciences there is a good chance you could come out with the same impressions and feeling the same sentiments. It looks as if the left hemisphere has taken over in the scientific world.

The two biological hemispheres harbour two different ways of thinking. McGilchrist describes repeatable experiments carried out by Deglin and Kinsbourne that clearly show this difference. How does each hemisphere process a syllogism?

Take the following example of a syllogism with a false premise:

1. Major premise: all monkeys climb trees;
2. Minor premise: the porcupine is a monkey;
3. Implied conclusion: the porcupine climbs trees.

Subjects with one or other of their hemispheres disabled, or both intact were asked, "Do porcupines climb tree?"

As Deglin and Kinsbourne demonstrated, each hemisphere has its own way of approaching this question. At the outset of their experiment, when the intact individual is asked 'Does the porcupine climb trees?' she replies (using, of course, both hemispheres): 'It does not climb, the porcupine runs on the ground; it's prickly, it's not a monkey'. (Annoyingly, there are in fact porcupines that do climb trees, but it seems that the Russian subjects, and their investigators, were unaware of this, and therefore for the purposes of the experiment it must be assumed that porcupines are not arboreal.) During experimental temporary hemisphere inactivations, the left hemisphere of the very same individual (with the right hemisphere inactivated) replies that the conclusion is true: 'the porcupine climbs trees since it is a monkey'. When the experimenter asks, 'But is the porcupine a monkey?' she replies that she knows it is not. When the syllogism is presented again, however, she is a little nonplussed, but replies in the affirmative, since 'That's what is written on the card'. When the right hemisphere of the same individual (with the left hemisphere inactivated) is asked if the syllogism is true, she replies: 'How can it climb trees — it's not a monkey, it's wrong here!' If the experimenter points out that the conclusion must follow from the premises stated, she replies indignantly 'But the porcupine is not a monkey!'

Deglin and Kinsbourne's experiment can be repeated with syllogisms ranging across many different subjects and the result will be the same. The experiment clearly illustrates the fragilies of left side reasoning based on what is essentially a simple form of symbolic logic.

Symbolic logic is very easy to teach and master, even though the resulting apparatus has no grounding in reality. However, despite many philosophical objections, it has to be granted that left side forms of reasoning underpin the most successful scientific ventures to date. Left side reasoning underpins present day science. This raises the question: What is the corresponding Organon for *right side* rationality?

Right side form of thinking was much more dominant in ancient times. In pre-Socratic times, we find Heraclitus, who exclaimed that reality could only be understood in terms of oppositions. To every proposition, there was an equally valid second proposition in total contradiction with the first. Heraclitus's picture of reality consisted of a ferment of oppositions. His philosophy lacked discipline and gave a picture of reality dominated by irrational anarchy. Even with Kant, several thousand years later, the picture had not progressed significantly. One way Kant treated the oppositions was in the form of his four antinomies. There might be many oppositions in metaphysics, but he claimed his four antinomies were the most important. Kant put the spotlight on the possibility of a new, noble science and that somehow it would be fundamentally involved with rationality based on oppositions, in some way a form of dialectical reasoning. However, he provided little guidance as to how such a science of oppositions could be organised. He also presented his versions of the Categories (hastily prepared, according to Hegel). The categories were arranged in a four by four structure, vaguely indicating that deep reality could be understood in terms of organised semiotic square-like structures.

The purpose of this book is to show how Heraclitus's Cosmos of apparently confusing oppositions *can* be presented in a rational manner. Empedocles made the first step with his science of the Four Letters. Our project must formalise this ancient way of seeing the world.

The first step towards understanding the science is to achieve an elementary familiarity with handling oppositions. This involves practical exercises in semiotic analysis where the basic tool is the semiotic square. We have already considered one practical example, Kiyosaki's semiotic cash flow quadrant. A number of other examples follow below.

Teaching semiotic analysis should be an essential part of the school curriculum, and at an early age. The approach is simple, simplifying, but can reach into the most profound areas. This elementary training in right side thinking can help to counteract the excesses of left side rationality emphasised in present day education. To be successful, students must work across many problem domains. Unlike left side thinking, specialisation is not a characteristic of right side thinking.

Fundamentally, reality can only be understood in terms of oppositions. Semiotic analysis introduces such an understanding. At the basic level, two oppositions, themselves opposed to each other, form the semiotic square, a kind of semiotic binomial, a semiotic quadratic, as we shall see later. The next task is to start to understand the semiotic square in more detail. The next example of semiotic analysis will illustrate how left side thinking, obsessed as it is by truth-values, is based on *belief*. On the other hand, we will come face to face with right side thinking with the startling realisation that it doesn't believe in anything. In the process, we discover that the right side belief-free zone nourishes a much more reliable form of truth.

The Semiotics of Belief and Faith

Our present intention is to look at a semiotic analysis of religion. Before doing so, we briefly explore a few aspects of religion and in particular the relationship between religion and science.

Traditional Sciences form a Belief System

The left side sciences bathe in abstraction. This great bubble of floating rationality works from propositions, which have truth-values. As such, each proposition expresses a *belief*. The left side subject believes in propositions that have 'true' truth-values, and disbelieves propositions, which are deemed to have 'false' truth-values. A considerable source of angst for the left hemisphere is figuring out what propositions to believe in and which to disbelieve. The source of angst comes from the fact that the whole logical side of the apparatus is suspended mid-air in a world of abstraction. This abstract bubble of rationality has no logically expressible relationship to the non-abstract world, whatever that might be. Any coherence with the real is incidental and not part of the formalisation. The only critical faculty available to left side reasoning is the demand for internal logical coherence of its belief system.

The end result is that left side rationality can be very sharp for detecting the most subtle logical irregularities, contradictions, and variances within the current prevailing belief system. This is the strong point of the reasoning. The weak point is that the resulting belief system can creep so far away from reality that it becomes quite whacky; fundamentalist religious belief systems and political belief systems can even become very dangerous and destructive. Modern sciences, exploiting the analytical clarity of left side rationality, try to avoid creeping insanity by searching for sanity in consensus. By demanding peer review, and repeatable experiments, whacky science becomes must less likely, but not impossible nevertheless.

The corpus of knowledge making up present day sciences appears as a gigantic belief system. Karl Popper cottoned on to this fact by providing his well-known criterion for a belief system. A belief system is one where no proposition in the system is absolutely and definitively true. For Popper, a belief system was one where every proposition is provisionally true, but may be "falsifiable." For this to be possible, all propositions must enjoy the rational status of possessing a truth-value: hence, providing the possibility of being either true or false. Popper effectively declared that modern science, according to his falsifiability criterion, was fundamentally a belief system. He then went on to use the criterion in the reverse sense: If any pretender to scientific knowledge was *not* a belief system then it was "unscientific."

Traditional science is based on abstraction. A fundamental characteristic of abstract reasoning is that it does not demand whether objects exist or not. This is seen as its power. A favourite topic for abstract reasoning is the proposition "God exists." Is the proposition true or false? The same question can be asked about unicorns, electrons, and gravitons. Do they exist? According to the Law of the Excluded Middle, the answer must be true or false. The basic assumption of abstract reasoning is that existence is an attribute. Something existing or not existing is like something being soft or hard, being coloured blue or not blue. Existence is a mere attribute that some things have at a particular point in time. Unicorns will never have existence because they are fictional. Unicorns do not exist and never will exist. Socrates also does not exist, but for a different reason: he is dead. The Judeo-Christian god is an entity that possesses this existence attribute. God exists. In the form of his son, he even once existed in the flesh. What is more, he can return in the flesh at any time. The Judeo-Christian god is distinct from any other god by its existence attribute. Grace to this attribute, the citizen is faced with a stark choice. The citizen, being an abstract thinker, must respect the Law of the Excluded Middle. He can believe that god exists and so enter into the communion of believers. Alternatively, he can believe the contrary: God does not exist, he declares. He thus enters into the club of the Atheists. Theist or atheist? That is the question. It is in this way that the god-fearing believer and the god-hating atheist join hands in a common exploit. They are all people that believe that the god question is a reasonable question with a clear and precise answer. They are all creatures driven by belief. Of course there might be a third option, that of the agnostic. However, the agnostic must climb to even more tortuous heights and start musing over whether the Law of the Excluded Middle is valid or not, and why.

Not all people are creatures of belief. This is the case for followers of Allah and the Hindu gods. In the case of the secular Islamic world, for example, there are no atheists as there are in the secular Judeo-Christian world. It might seem paradoxical but no Muslim believes in Allah. Not even the most devout Muslim believes in Allah. This differs from the Christian doctrine, which is founded on belief. For Christianity, the proposition that "God exists" can be either true or false. To be a Christian, you must believe that the proposition is true. If you believe that the proposition is false then you are an Atheist. It is on this point that Christians and Atheists are at one. Both Christians and Atheists adhere to the primacy of a belief system. They merely disagree on the details concerning the particular truth value of a proposition.

A problem with the Christian doctrine is that it admits a God that is weak, a god whose very existence is conditional on belief. For Islam, the picture is radically different. Allah is not an object of belief as Allah is beyond the true and the false. Rather than being determined by the true, Allah is determined by truth. There is no room for belief and its negation. With Allah, belief is inconsequential, what matters is faith. Allah is determined by the faith of the individual. If you hold such faith then Allah is your god. If you are secular, not only do you have no god, you have no concept of god. There is no debate. There cannot be any debate between the faithful and the infidel, just a different state of being based on faith or the lack thereof. This difference between belief and faith can be difficult for Westerners to comprehend.

There is a huge difference between belief and faith. For example, someone can believe in fairies but it is difficult to imagine having faith in fairies. In Christianity, belief comes first and faith second. It is quite possible for a Christian to have a crisis of faith and so even lose the faith. Nevertheless, the Christian will still believe in God.

The Christian god can be qualified by a proposition that satisfies the Law of the Excluded Middle. The proposition "God exists" thus can be considered as a scientific hypothesis. This is where Popper steps in and adds an extra requirement for a proposition to be acceptable as a scientific hypothesis the proposition must be falsifiable. The consensus amongst both Judeo-Christian theists and atheists alike is that the proposition "God exists" is not falsifiable. There is no scientific experiment that could possibly refute the proposition. Thus, by Popper's criterion, the question of whether god exists or not cannot be covered by science. Once again, the theists and atheists usually concur on this conclusion, something that underlines the unanimity of theists and atheists in Judeo-Christian culture. Theists and atheists mutually agree on everything except the particular truth-value of a proposition.

Darwinism and Swerve Theory

At the time of writing there seems to be a battle raging between the Creationists and the New Darwinists. The conflict seems to be essentially a Midwest and Southern American affair, peculiar to those cultures. What is of concern in this section is the scientific status of Darwin's Theory of Evolution. Firstly, one should note that the basic epistemological foundation of the Theory of Evolution was due to the Epicureans of Ancient Greece and so preceded Darwin by several thousand years. Despite a similarity of their world scientific outlook, the Epicureans differed from the New Darwinians in their views on how to enjoy life. The New Darwinian advocates getting meaningful pleasure out of life by ranting on the Internet and in literary tracts that there is no God. Some get pleasure out of giving children science books for Christmas. The Epicureans took a different tack. Rather than pleasure being a mere by-product of certain kinds of scientific pursuit, they turned the pursuit of pleasure into the central object of science itself. They argued that one of the worst obstacles to leading a happy pleasurable life was fear of the gods. This lead to the Epicureans taking the theological position that the gods were distant from humans and totally uninterested in human affairs. Of particular interest was their scientific outlook. The Epicureans, although not empirically minded, held a similar philosophical outlook to the traditional sciences of our day. They were strict materialists, atomists, and determinists. The whole world was in the vice of a strict determinism of cause and effect. But, like modern physics, there was an exception to this draconian determinism. Epicurus called it the Swerve. Atoms moved and interacted with each other in a totally deterministic way, but every now and then, an atom would execute an imperceptible, totally random "swerve." Epicurus exploited this notion to develop his Swerve Theory of the universe. At the beginning of the cosmic cycle, the world is non-structured: all atoms were falling down in straight vertical lines, according to Epicurus. After an immensely long time, because of the accumulated random swerves of the otherwise deterministic atoms, the universe micro-swerved into the way it is today.

Amazingly, this picture is no different in principle from that of modern science. The random beginnings were a bit different, but the micro swerving into the world of today is essentially the same belief. Since Epicurus' time, Swerve Theory has come a long way. The random swerves of atoms have been confirmed and even quantified. Nowadays the Epicurean Swerve Theory is known as Heisenberg's Uncertainty Principle and is explained in the Wave Equation.

Swerve Theory has been applied to the biological realm, where it manifests itself as genetic mutations. Just as collections of atoms micro-swerved to produce the first single celled living creature, further random swerving eventually lead to the animals that we have become today. It's all a product of Epicurean Swerve Theory.

Darwinism also adds in the survival-of-the-fittest paradigm as an embellishment of Epicurean Swerve Theory. It is thus claimed that genetic swerving is not a completely random process, as some swerves are more successful than others are and hence are directed by success. The successful swerves then go to propagate other successful swerves. The end result is that only the fittest survive. In fact, the survival of the fittest paradigm is a huge red herring. It's about as meaningful as saying that the survivors of a car crash are the fittest compared to those that perished. The paradigm is a simple tautology. Who survive *are* the fittest; who are the fittest *are* the survivors. The semantics are the same, only the labels change. All up, the survival of the fittest paradigm adds nothing to elementary Epicurean Swerve Theory. To name the survivors as being the fittest is just a change of terminology. We are all the "fittest," we are all the last men standing; we are all the survivors of a trillion times a trillion Epicurean Swerves. That is the way we came to be the way we are today, believe it or not, says the Theory.

Dawkins and Epicure:

The idea that the world we live in is the end result of an infinitude of micro swerves, as Epicure postulated, appears absurd despite being quite a rational point of view. Equally, absurd appears to be the notion that we humans and all other life forms in the world today are the end result of countless random mutations. However, traditional left side sciences abound with such apparent absurdities. Take particle physics, for example. According to the physics, all matter is ultimately composed of point-like particles operating in a void. This is much the way that Epicure imagined it. However, his elementary particles, his atoms, had extent. In modern physics, all matter is explained in terms of quarks, leptons and bosons, none of which have extent. Although all of these views of reality appear absurd, it would be equally absurd to reject them. They are not wrong or false views of reality but the consequence at looking at reality through the prism of a certain epistemological paradigm, what we call the left side paradigm. We argue that in order to get a complete view of reality, a second take, a second paradigm, is necessary. This is the right side paradigm and it provides a holistic view of the world, as we shall see.

The zoologist Richard Dawkins also admits to the absurdity that we evolved through an endless process of random genetic accidents. He argues that there needs to be a more palatable explanation, and one that stacks up probabilistically. Of course, one possible explanation is that God directs the process, and not the laws of chance. Apparently, if we take Dawkins' word for it, this is the explanation of choice of the readers and writers of The Watchtower, and the banjo plucking pig farmers of the Appalachian Mountains. It is enough to galvanise a man of science into action. However, the rigour of the science Dawkins proposes is somewhat lacking in detail and rigour. There are two ways of explaining how the world is populated with life, declares Dawkins, one is the God Hypothesis, and the other is Natural Selection. He then goes on to explain the scientific basis of Natural Selection. Without going into details, his argument consists of three cardinal points.

Firstly, evolutionary change is not random:

Chance does not make sense.

Secondly, he explains the dynamic of the process not to be directed by God but by Natural Selection as this is the only alternative to chance:

Once again, intelligent design is not the proper alternative to chance.
Natural selection is not only a parsimonious, plausible, and elegant

solution; it is the only workable alternative to chance that has ever been suggested.

Thirdly, you can only really appreciate and understand these natural processes by having your "consciousness raised."

But perhaps you need to be steeped in natural selection, immersed in it, swim about in it, before you can truly appreciate its power.

Thus, Natural Selection involves more than an understanding at the intellectual level. Dawkins declares that he has been there, done that, and has come back to report to us that he saw absolutely no sign of God anywhere, just "natural forces" at work.

It is true that, in his professional work, Dawkins has revealed important insights into how genes interact, including certain algorithmic phenomena. He also advocates a notion that we develop in this book, the idea of genetic geometry. We call it generic geometric algebra, something applicable to even beyond the biological sphere.

The central critique that we have to offer is that he is endeavouring to explain "natural selection" from the perspective of the traditional left side scientific paradigm. Our view is that the only possibility of a rigorous explanation comes from operating a second take on reality, looking at the problem from the right side perspective. Certainly, a raised consciousness is required, as one has to change gears and move to a different epistemological camp to get to the truth. Our overall impression is that Dawkins is very close to the truth, but just on the wrong side of it.

Is Darwinism Scientific?

It would seem that Epicurean Swerve Theory and its modern biological successor in Darwinism are capable of being expressed in terms of a theory that people can believe or disbelieve. Thus, the theory could be taken as a traditional scientific hypothesis. However, there is no possible way to refute the hypothesis. That things change deterministically with a random component: this is hard to refute. This is what Karl Popper himself eventually recognised. By his falsification criteria, Darwinism was unscientific! Popper initially accepted this conclusion and only later tried to worm his way out of it. Refuting that we didn't just drift to where we are today and thus refute Darwinism is a task that even Popper can't convincingly achieve.

Where Darwinism wins prestige is the notion that the theory articulates a certain knowledge of the world. It describes change in terms of evolution. However, it only has descriptive not explanatory powers. It *describes* evolution. As we know, the evolutionary process goes in the face of what is predicted by the Second Law of Thermodynamics. According to the Second

Law, there should be a drift to increased entropy, an ineluctable drift towards thermodynamic death. However, this drift has a deterministic feature, its entropic direction. In the evolving world we live in, the opposite seems to be the case. Evolution leads to an apparent *decrease* in entropy, a steady rise in diversity rather than a steady fall. Darwinism describes this phenomenon, but does not explain it. The central difference between the thermic death principle of thermodynamics and the evolutionary explosion of life is the context. The Second Law and its death principle, only apply under the assumption of a closed system. It applies to any gas, for example, as long as it is in a bottle. Life evolves as a world where there are no bottles, just the world. Any entity is its own world.

All traditional sciences including mathematics, the left side science, all are sciences of closed world systems. Any attempt to apply any of these sciences, including traditional mathematics to the study the fundamentals of an open system will fail because, the system is closed before even taking the first step. To study the life of the entity outside the bottle, you have to start with no baggage whatsoever. Nothing *a priori* is permitted. This leads to the startling conclusion that, in order study the system not qualified by anything a priori, you have to solve the Kantian problem.

...

We are now coming to the end of this section with the basic understanding that science is based upon falsifiable belief. As for religion, it is based either on non-falsifiable belief or on faith, which is impervious to belief. In the third slot is Darwinism. It appears that Darwinism is somewhere in the domain of the Epicurean Swerve theory. Alternatively, it can be taken as a non-falsifiable belief that things, particularly living things, evolve and so are in the same boat as the religions. The New Darwinians seem to prefer the latter option and see it as a viable religion substitute, but still a religion nevertheless.

After our brief exploratory sojourn into science, belief, and faith, it is now time to carry out a further exercise in semiotic analysis. This time we will end up with a system based on belief on the left side and a system based on something else on the right. The right side system is based on *faith*. This will be an exercise in theology.

Right Side Reasoning is Based on Oppositions

Left side reasoning flows sequentially from premise to conclusion via intermediary steps. Every step in the process involves a proposition of some kind, be it simple, modal, a predicate or whatever. The reasoning involves a flow of logical steps. A mathematical proof, for example, presents itself as such a flow.

In the West, we have becomes so accustomed to this way of thinking that we refuse to countenance the possibility of any other kind of formalism of reason. Our task is to demonstrate the contrary. Orthogonal to the sequential reasoning of left side thinking is a non-sequential form of reasoning. We call it right side reason. Instead of thinking in terms of sequences of single units of rationality, right side reasoning takes the path of Heraclitus and expresses itself in terms of binary oppositions.

The atomism of left side logic can involve complex propositions being formed from the sequential concatenation of simple propositions. On the right side, the equivalent to the compound proposition is the compound opposition. There we find oppositions between oppositions. The simplest such compound structure is the semiotic square.

Non-trivial aspects of left side reasoning can be taught at an elementary age. The student can be introduced to the propositional calculus, including the concepts of negation, disjunction, and conjunction, for example. Venn diagrams can also be used as an aid to an intuitive understanding. Right side reasoning should also provide its own gentle introduction. The semiotic square is the first artifice along the road to understanding. Understanding this artifice requires many practical examples. At all times, it must be stressed that right side reasoning *cannot* be an abstraction. Abstractions always leave something out of the picture. In right side reasoning, the entity is investigated *as a whole*. The semiotic square, composed of two opposing oppositions encapsulates the whole. The whole can be thought of as Totality viewed from a particular perspective. The perspective itself also forms an integral part of the resulting whole. The semiotic square and its dialectic is a simple tool for understanding these concepts.

The Dialectic of the One and the Many

The semiotic square can be used to analyse a particular kind of whole. In this case, the basic oppositions are deconstructed from the whole. Deconstruction is the right side equivalent of analysis. The approach can also be constructive, starting from a primary opposition; a generic whole can be constructed. This latter approach will be illustrated here. First, we look for a fundamental opposition to work with.

There are many fundamental oppositions. The most fundamental of all oppositions is the masculine-feminine gender principle of ancient pre-Socratic times. This will be considered in later chapters. Here we choose a rather more tangible opposition, that between the Many and the One. We remark, in passing, that this Many-One opposition can be thought of as a simplistic version of the masculine-feminine opposition, and so paves the way towards understanding the more advanced concept.

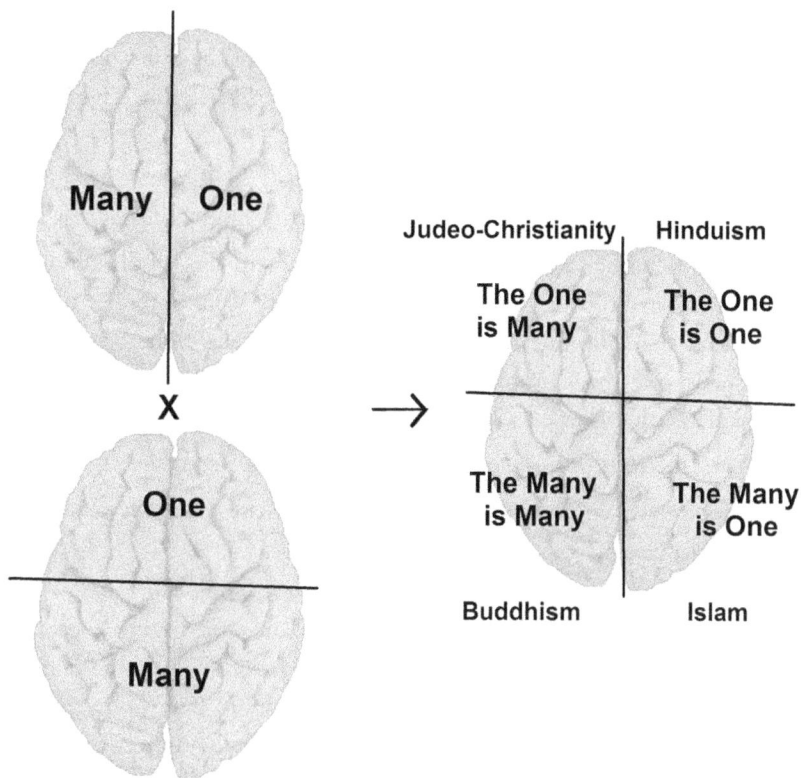

Figure 3 The Theological Brain. A semiotic square for the four takes on reality based on cardinality oppositions.

The cardinality, Many-One opposition can be considered as a dichotomy that divides reality into two. On the left side, there are the Many, on the right side there is the One. These two different realities can also be seen as the one and the same. The two realms are a consequence of the two possible takes on reality, the left side *atomist* take, and the right side *monist* take. One sees an exploding multiplicity; the other sees an all-embracing unity.

This Many-One opposition provides the left-right cut for our semiotic square. The next cut is the split into the front and back. Cutting a long story short, this cut can be made by reapplying the same Many-One opposition *to itself*. The One side goes on the front and the Many takes up the rear. The reasoning here is that any One can be many, any Many can be one and there can be many Many. Last of all, there must be one One. Having exhausted all possibilities, we end up with the semiotic square in Figure 3. This is the semiotic square of some undetermined entity when viewed as a whole, when viewed from the perspective of its cardinalities. The end result is that this semiotic square demonstrates not *two* different takes on reality, but *four*. There are four completely different ways to conceive reality. Such a mind-boggling vision has not gone unnoticed across the ages. Corralled by the emergence of the written word, formalised in canonised texts, different ethnic and geographical regions have coalesced around one or the other of these great doctrinal viewpoints. Each gave birth to its own world religion. There are four such world religions.

Godhead Semiotics

Left side sciences rely on abstraction, the "view from nowhere," the God's eye view. It could be said that right side science takes the "godhead view." It sees the world from four different vantage points. What interests us now is that each of the four different takes elaborated in the Many-One semiotic square spells out the central tenet of a world religion. The fundamental question of theology can be asked:

What is the relationship between the One and the Multiple?

As can be seen from the semiotic square, there are four answers. Each is discussed below.

The One is Multiple (Christianity)

The One is Multiple formula defines Providence, the giving, creator god. In Christianity, the Multiple corresponds to the individuals of the flock. The Multiple is showered on by the Benefactor on high. The loving god, the greatest giver, even sacrificed his own son for the redemption of the often egotistical individualists massing below. The One is Multiple doctrine emphasises free will and the temporality of the individual. Exercising free will is inextricably concerned with temporality. The Christian God is essentially temporal in nature. Christ is sometimes interpreted as the Lord of Universal History. The culture is very conscious of its history. God manifests through history argued Hegel. Hegel thought that this is why the Christian civilisation was superior to others. It had a sense of history. This is a temporal God. Being a *left* side religion, the emphasis is on *belief*. The pious should believe in this god. It is possible, but not advisable, to disbelieve.

The Multiple is One (Islam)

In the Islamic take on reality, the tables are turned. God does not provide for the Many, the masses of the Many must give to god. Islam is literally the doctrine of surrendering to god. The Christian individual gives way to the collectivist convictions of the peoples of Islam. Gone are the historico-temporal preoccupations of the first paradigm. Enter spatiality. If Christ is the Lord of Universal History, Allah is the Emperor of Space. It is the spatiality of Allah than unites the Multiple into Oneness. Allah is truly the Greatest because he is everywhere, with no exception. Mecca is the iconic centre of all spatiality, dictating the spatial direction of the praying, emphasising the spatial unity of all, and the spatial, all-embracing nature of Allah himself.

All the major religions involve right side reasoning. However, Christianity is on the left side of the right and so based on belief. Islam is a pure *right* side religion and so is not based on a belief system. No Muslim believes in Allah. "There is no god," declares the Koran. To believe in Allah would be to admit the possibility of not believing in him, a sacrilege. Situated on the right side of rationality, belief gives way to *faith*. Allah is the god that Muslims have faith in. "There is no god, except god," states the Koran. Lose your faith and you lose your god. You cannot disbelieve something of which you do not have any inkling. The only way to have any inkling of Allah is to have faith in Allah. Allah walks the tautological line.

The Multiple is Multiple (Buddhism)

The Multiple is Multiple paradigm underlies Buddhism. The One has no place in this worldview soaked in Multiplicity. Each religion has its favourite icon. The iconic Buddha statue, usually in bronze, says many things. It radiates wellbeing, but what is so captivating is that famous ironic smile. The author's strongest Buddhist image is of a lawn covered square in a Bangkok primary school. Each side of the square features a row of identical Buddha statues, all looking across to the other side; multiple to multiple from left to right, multiple to multiple from front to back. Student desks were also arrayed under the awnings around the square. What an incredibly idyllic place to start school!

This Multiplicity doctrine is a *left* side religion, like the Christian, and so is *belief based*. People believe in Buddhism. The problem is that, unlike its theist front side neighbour, there is not anything specific to believe in. All is multiplicity, there is no "One." This is the belief. Working this into a tractable religion requires the work of someone of considerable intellectual agility. Hence, the Buddha's ironic smile.

The One is One (Hinduism)

Advaita Vedanta is the core philosophical school of Hinduism and teaches the Principle of Non-Duality. Ignoring the fine print, ultimately only the One is real. Apparent multiplicity is subjective and illusory. Advaita Vedanta articulates a highly nuanced monism. The One is One does not mean that there is only the One. There are two One type entities, the individual soul (ātman) and the generic, universal soul (Brahman). These two kinds of soul are different. The One-is-One expresses that fact that they must be at one with each other, and so indistinguishable. This is expressed in the Doctrine of Non-Duality.

This Hindu doctrine is situated well and truly on the *right* side of the theological semiotic square and so, like Islam, is *not belief based*. According to Advaita Vedanta, the truth of the doctrine is obtainable via pure enlightenment. For Sankara, the founder of Advaita Vedanta, you do not believe in the Oneness of the One. By enlightenment, *you simply know.*

The Theological Square

The theological semiotic square provides a way of understanding the four world religions and how they relate to each other. Every religion has its own semiotic square and so there are semiotic squares "within" the semiotic square. Take Hinduism for example. The One in the One-is-One doctrine is the *Brahman*, the impersonal god that cannot be worshipped directly. The Brahman can be comprehended in the form of a triad of personal gods consisting of Shiva, Vishnu, and Brahma. This leads to the semiotic square in Figure 4 where Vishnu occupies the same front top slot as did the Christ god, with Shiva occupying the same back left slot as Allah. Brahma, who is rarely worshipped, takes up the Buddhist slot. Theologians warn that this is an error. The Christ god and Vishnu are quite different, as are Shiva and Allah.

The semiotic structure shown in Figure 3 corresponds to a very unconditioned reality. Figure 4 shows four semiotic squares, one for each entry in the first semiotic square. Each such entry corresponds to a qualifying context for the four more qualified structures. Each of these four semiotic squares corresponds to a *whole* and should be comprehended in isolation, as they are mutually exclusive. You can't be a Christian and a Muslim in the same instant. Each semiotic square is shown with a small accompanying square representing the qualifying context.

With experience from studying semiotic structures, and further reading of the material presented here, the student should be able to start providing detailed explanations concerning the mythical beliefs and imagery of the world religions. Students could be of a relatively young age, but we have no expertise on such matters. Students could respond to question like, why are Vishnu and his reincarnations such as Krishna all dark skinned and Shiva vivid white. Why does Vishnu have reincarnations and Shiva none?

Why can the Christian god have an offspring and Allah none? Compare the Islamic Trinity with the Christian Trinity. Why are the principle deities featuring in major religions all male? The four elements of Empedocles are personified by two goddesses and two gods. Discuss.

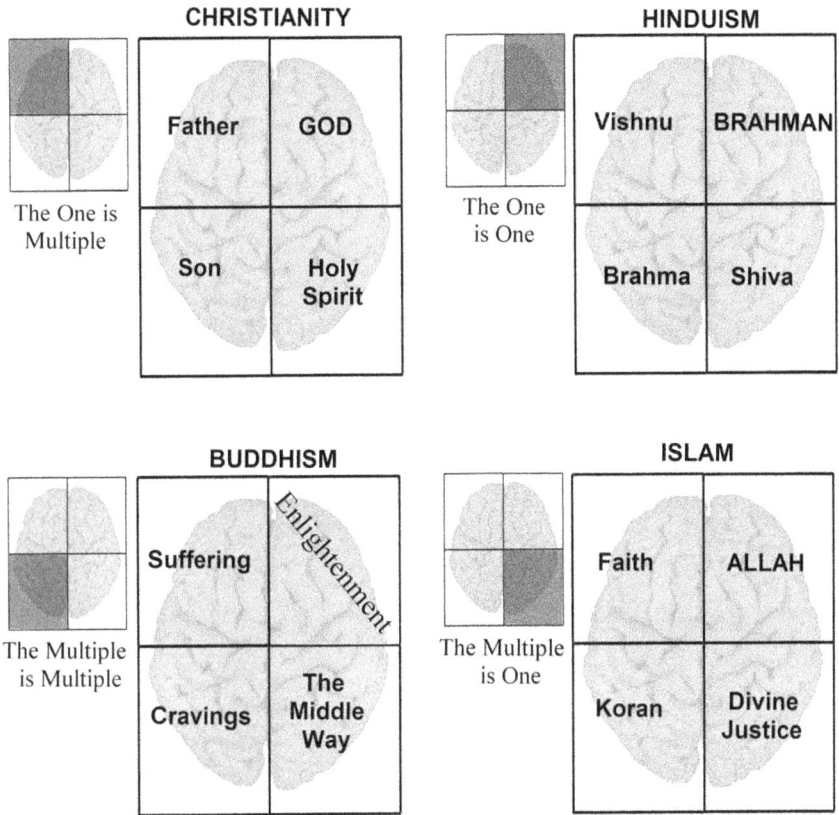

Figure 4 The godhead of each of the four world religions. Any well versed theology or divinity student should be able to provide some kind of reasonable response to these questions. It is hoped that the material provided by the author will expand the horizons somewhat. The author also hopes that the layman, together with the theoretical physicist might also profit from the experience.

4

Science with and without Subject

In the previous chapter, we have started exploring elementary cognitive or generic structures and their interplay with reality. The theological semiotic square has provided a way of understanding a holistic view of reality and that there are four fundamental "takes" on reality. These four worldview paradigms are mirrored in the four world religions. In this chapter, we start to explore the semiotics of the Subject in science.

Science without Subject

In present day scientific circles, it is generally thought that there is only one kind of science worthy of the name. These are the traditional empirical-based sciences that we have been referring to as left side sciences. Left side sciences are considered as objective, as they study objects under controlled conditions that attempt to remove any subjective factors. As such, left side science specialises in the domain of *objects* where any reference to or involvement of the subject has been eliminated. The operational paradigm here is *science without subject*. Such sciences end up in a single opposition between two different kinds of object. The objects involved are the *real* object and the *imaginary* object. Left side science is dominated by a duality between the real and the imaginary.

The duality between the real and the imaginary runs right throughout the science and appears in many different and sometimes surprising guises. In the first instance, it appears as the dichotomy between the object of a science and the theory of that object. The object is considered as belonging to the real, whilst the theory belongs to abstraction and hence is imaginary. This primary opposition resurfaces in Analytic Philosophy in another guise as the Mind-Body problem.

Classical physics provides the most dramatic expression of the Real-Imaginary duality. According to the classical physics doctrine, all entities have properties. In physical reality, only entities are deemed to exist whilst properties, not being entities, are deemed not to exist, In other words, entities are real whilst their properties are imaginary. This is an amazing situation, as the only way that classical physics can get to know the real components of reality, the entities, is via perceiving and measuring the attribute. Real world entities can only be known via their imaginary component. This is left side science at its purest and most intriguing. Not only does the science lack a subject, it has to be content with a second rate access to real objects. The whole system bathes in the imaginary. There might be two halves to the world, but we can never know more than one half, the imaginary half.

Sidenote

Much further on in this book, we come across the real and the imaginary when viewed from a right side perspective. However, no fundamental dichotomy is involved. The construct is of a relativistic nature. The real and imaginary form a whole and are determined relative to that whole..

Science with Subject

The opposing paradigm to the Science-without-Subject doctrine leads naturally to Science-with-Subject. This is where we find right side science; the paradigm that insists that Subject *must* be present with Object at all times, and treated on *equal terms*.

It is at this point, right at the beginning of the presentation, that we see the essential difference emerging between left and right side reasoning. The left side cuts the world into imaginary and real chunks, and only retains the chunk that it feels it can know. It can only know the imaginary. This left brained beast can only know what it feels. This creature lives in a half world that keeps on fracturing into finer and yet finer half worlds.

Contrary to the fractured atomistic half worldview of the left side, right side science specialises in always seeing the world in the form of *wholes*. The wholes are reminiscent of Leibnitz's monads "without windows." Each offers a holistic view of the universe. The left side takes refuge in the left side, satisfied with the half world of objects. Things it can easily understand. The right side paradigm must also make its home on the right side. However, unlike its tunnel vision sibling, it is conscious of a bigger world and its domain of influence extends over the whole world. It is even conscious of its sibling and acknowledges left side deftness in dealings requiring finer, crystalline aspects of reality. On the other hand, the left side is prone to believe that it is an only child, and often behaves accordingly. This is particularly notable for left side science.

The critical step to obtaining a holistic view is to introduce a *second* cut, orthogonal to the first. The first cut determined the dichotomy between *impersonal* subject on the right and its corresponding object on the left. The second cut determines the *personal* subject at the front and its corresponding object at the back. Instead of two takes on reality, the paradigm provides four takes. We have returned to the semiotic square structure.

Just as for the left side, the concept of the real and the imaginary emerge. This time the real becomes associated, not with the impersonal subject, but with the confluence of the impersonal subject *with* the personal. This corresponds to the front, right hand corner of the semiotic square. Now, the left side paradigm declares that the real can only be known via the imaginary side, via attributes, abstract theory, and so forth. The right side paradigm has its own take on this matter.

The real world is no longer that thing-out-there seen from the God's eye view from nowhere. The real becomes relativised to the personal subject. The real becomes the identity of the personal subject with *its* reality, the impersonal subject. This becomes an expression of Sankara's Principle of Non-Duality, expressing the formula the *One is One*. The impersonal reality and the reality of the personal become indistinguishable. This is real, all else is illusion, so declared Sankara's Advaita Vedanta. For consistency, we prefer the term imaginary rather than illusion.

Sidenote:

In passing, we note that the imaginary has a certain meaning in mathematics in terms of real and imaginary numbers. In Appendix B, this is shown to be no accident.

Then suddenly, we hit the goldmine. The left side, after cutting itself off from the real has to content itself with knowing the real via empirical attributes flowing from God know where and what. The right side paradigm does not need to go out there rummaging through the dustbin for leftover attributes in order to understand the real. The right side already *has* three attributes in immediate possession. This can be seen from the semiotic square produced by the two fundamental oppositions. Relative to the personal, real subject in the front right hand side of the square, there are three other boxes left over that provide an opportunity to get to know the allusive One-is-One entity. These three boxes can be thought of as three attributes of the real. They can also be thought of as imaginary.

Thus, in summary, left side knowledge exhibits essentially a single kind of fundamental dichotomy, that between Body and Mind, the real and the imaginary, the concrete and the abstract. Knowledge of Body, the real, the concrete, is via empirical attributes. Left side science is thus based on one single opposition, that between the real and the imaginary. We can say that it is based on *first order semantics*.

On the other hand, right side knowledge is based on two oppositions applied to each other. As such, we can say that it is based on *second order semantics*. Second order semantics is what semiotics is all about. There, we find a fourfold structure, where the real appears as relative to the subject and there are three *a priori* generic attributes, applicable to *any* determined subject.

The generic structure thus involves the real entity with three imaginary entities. This fourfold structure we interpret as the *epistemological brain*. In a moment of foolhardiness, we also interpret the biological brain to be organised along these same generic principles. Moreover, the right side of this structure is capable of comprehending the *whole* structure in these terms. The left side is anchored in an atomistic view of reality and is incapable of such an overview. The left side, be it epistemological or biological, by its very nature *cannot* be conscious of the whole.

Sidenote

Many years ago, the author's first glimmer of understanding came from some quite elementary mathematics. Very briefly, it concerns the nature of number. Most students are taught that there are two kinds of number, real numbers, and imaginary numbers. Imaginary numbers are necessary in order to provide a tractable way for handling the square root of negative numbers. Numbers made up of real and imaginary parts are called complex numbers. Most students, even engineers and scientists, go through their studies only knowing about complex numbers and that there is only one kind of imaginary number. This is left side mathematics at work.

There is a right side slant on imaginary numbers as presented in Appendix B. In this case, one gets hyper-complex numbers with one real and three kinds of imaginary number. Where do these complex numbers come from? The answer is simple, from solving quadratic equations. What we end up realising is that the semiotic square is just a semantic way of generically representing binomials. There is no mumbo jumbo here.

Since all of the geometries of the Clifford Algebras and Geometric Algebra are based on quadratic forms, the apparatus can be very powerful. In fact, this is all the geometry you need for mathematical physics and Engineering. All that is required is a better understanding of some of the nuances involved.

The Hard Problem

Philosopher David Chalmers remarks that the confidence in the traditional scientific method "comes from the progress on the easy problems." Over the past decade or so, Chalmers has argued that it is time to tackle what he famously calls the "Hard Problem," notably to develop a rigorous, scientific theory of consciousness. Chalmers' Hard Problem is "hard" to tackle because its requirements are antithetical to the very essence of the scientific method. The objectivity of the scientific method demands that only the object data be under consideration. All reference and interactions with the knowing subject must be eliminated. Thus, to turn the tables and make the knowing subject the object of scientific enquiry means that all data has disappeared. Thus, the problem of knowing the subject, this entity without data, becomes indeed a very "Hard Problem."

As we have sketched in this book, this kind of problem has a long history, going back to Aristotle who saw it as the problem of developing the First Science, which he called the First Philosophy. Kant raised the ante in his time, calling for a science that didn't rely on any *a priori* experience. Kant called such a science, metaphysics. In modern times, we now see it presented as the challenge of understanding consciousness badged as the "Hard Problem." Nothing has changed over the past few millennia. Whether it is called metaphysics or the Hard Problem, the problem still remains distinctly difficult.

Chalmers' Hard Problem nomenclature raises possible objections. The emphasis on consciousness, as the last man standing, implies that traditional science has victoriously swept all before it, conquering practically everything in its way and has finally come to the final and last frontier to be conquered. Charmers offers no critique of the scientific method except that when it comes to consciousness it doesn't work. This ignores the many foundational crises that riddle present day traditional science. What is needed is not just a science of consciousness but the noble, unifying science that Aristotle and so many others since have called for.

Having said this, we have no fundamental disagreement with what Chalmers has been saying. He is just presenting the scenario in terms of the measured language of Analytic Philosophy. He has considered all of the armaments and munitions at our disposal, inspected the terrain, and has reported back to base. Despite all the equipment we possess and may develop in the future, it appears starkly apparent that there is absolutely no way we can win the battle. Game over. Chalmers' message is clear. If you want to win the war, you will have to start from scratch. You need an entirely different scientific methodology.

This, of course, is precisely our message. In order to start getting traction we have illustrated our thinking by using the biological brain as a *metaphor* for the required epistemological framework to do the job, the *epistemological* brain. The traditional sciences are what we call the left side sciences and correspond to the left hemisphere of our epistemological brain: reductionist, analytic, abstract, and obsessed with raw data. To resolve the Hard Problem we need another kind of science, the unifying right side science, the one that mysteriously operates out of the right hemisphere of our conveniently confected epistemological brain. In employing this pseudo-biological terminology, we take the same convenient path as Chalmers and effectively rebadge the ancient metaphysics problem as an organisational problem of mind. One could be tempted to say that it is a brain problem. However. other than sounding a bit crass, the epistemological brain we are constructing is more based on a *metaphor* than sticky grey matter.

In order to resolve Chalmers' Hard Problem, we are faced with the challenge of developing right side science. Using the biological brain as a metaphor, this requires understanding right side reasoning, a totally different kind of reasoning from left side reductionism.

We have a fair idea of how linear reductionist left side reasoning works. The student can start off with elementary logic, truth tables, Venn diagrams, and so on as an introduction to symbolic logic. The abstract exercise can be combined with practical applications, so that at least some semblance of contact with the real world is inferred. This is all part of the Easy Problem.

What is the corresponding right side way of reasoning? We have already provided a preliminary response to this question in previous sections. Right side reasoning works with *oppositions*. The only way to understand something is in opposition to something else. In left side reasoning, it suffices to give a label to something in order to get a conceptual handle on it. What's more, as general linguist Ferdinand de Saussure pointed out, the label can be completely arbitrary. This is first order semantics in action. This is labelling technology at work. However, labelling technology does not work for right side reasoning. Arbitrary labelling is not allowed. Ferdinand de Saussure stayed clear of the Hard Problem and stayed at home on the left side, the easy side.

Unlike left side rationality, the semantics of symbols form an integral part of right side reasoning and do so in an incredible way. However, that most exciting and positively overwhelming part of the story must wait until the later part of this work. For the moment, we must work in a label free world.

Rather than say "Let A be such-and-such, consider A," our first examples were based on oppositions of cardinality, the opposition between One and Many. This is not the most fundamental opposition. It is too simplistic.

However, the One-Many opposition is useful for an introduction. We then introduced a *second* opposition, another version of the same One-Many opposition. The second opposition was opposed to the first. The first was assumed to apply spatially from left to right, the second from front to back, as shown in Figure 3.

Relative Relativity

Now here is the rub. Something has been cut into four with these left-right and front back cuts. However, *what* has been cut into four? Nothing is really being cut in this first application of the semiotic square. What is being established is simply a unique frame of reference from which to comprehend reality. We build this tennis court-like structure in the middle of the Cosmos and demand that the whole Cosmos gyrates around it. From this unique pedestal for viewing the world, we have a ready-made reference frame of what is left and right, as well as what is front and back. This is all set in the polarity convention shown in Figure 3. We have discovered the location and shape of the centre of the universe! In fact, it has the same shape as the centre of your universe.

Right side science must be simple and simplifying, whilst continuing to climb out of the trap of appearing simplistic. Granted, our square-shaped mind situated in the centre of the Cosmos might appear a little simplistic. However, the situation can be saved by this egotistical mind-sprite admitting that there might be other entities in the Cosmos that enjoy the same viewing rights as itself. In this less determined context, the centre of the Cosmos becomes not *that* entity but *any entity whatsoever*, the true centre of the universe. One might argue that maybe only one such entity has the necessary four-part brain to join in the fun. This would not be an impediment, provided the consciousness in question could imagine itself in the place of any one of the other mindless entities and would thus see that same thing as the mindless (that is, if it had a mind). However, even that requirement could be weakened because the single mind might lack the capability of imagining changing places with another. In that case, it would not matter, as long as the same result would have occurred if it had such a capability.

At this point, we pull the ripcord even though we have not finished the story. These little naïve adventures into right side reason can be like a voyage into insanity. The benefit for the student is probably to wean them off dependence on left side linear thinking and on to binomial thinking. It should be kept in mind that similar tortuous adventures can be entered into by, for example, simply explaining in words something like the clock paradox in the special theory of relativity. In applying the theory mathematically, the formal methodology works quite smoothly and effortlessly. Right side relativity, once endowed with its own formalism, a relativistic relativity, rather than the classical, should also be smoother and effortless.

It is time to look at some more practical examples of semiotic wholes.

Semiotic Square of Freud

The intention here is to provide a gentle introduction to right side science via practical example of the semiotic square. The approach is informal and intuitive at this stage. It may even be entertaining.

The semiotic square is an informal way of understanding wholes. A whole is Totality looked at from a particular perspective. Any thinker contemplating reality in a fundamental, non-abstract way is lead to semiotic squares of some kind. We have already seen this with the case of Kiyosaki, the less educated but "rich dad" who thought holistically about the rationale of generating cash flow. Kiyosaki thus sneaked into the ranks of the great philosopher's like Hegel. In fact, these ranks are full of autonomous autodidacts like Kiyosaki. Unlike Kiyosaki, Hegel was highly educated, but both these figures shared one thing: an aversion for abstract thought. Abstract thinking is left side thinking. Both Hegel and the entrepreneurial Kiyosaki emphasised *right* side thinking. They reasoned in terms of *wholes*. Wholes are not abstractions, as they include the subject. This is right side thinking. The abstract thinker gives way to the *generic* thinker, a much more powerful breed.

One can only think of Bertrand Russell, one of the greatest analytic, left side thinkers trying to understand Hegel, a great right side dominant thinker. Trying to understand the right side with left side thinking mindset is like trying to square the circle. The atomist trying to understand monism? That is really a Hard Problem, as Russell freely admitted.

With the exception of Kiyosaki's quadrant, the semiotic square examples have so far been concerned with the rarefied environment of the highly non-qualified. Understanding generic entities, which are practically devoid of any qualifying specificity, presents quite a challenge. A practical path to understanding is via the theological interpretation. This has many advantages, as the entities are often seen as divinities that naturally spring into life in one way or another. The figures may be adorned in myth and shrouded in icons, which can be confusing but also illuminating. At the same time, every myth, every icon, is the product of various people's deep intuitions and wisdom accumulated across the ages. The author believes that a desirable part of modern education should be to impart the skill of being able to read and interpret these icons, myths, legends, and sacred texts, not to mention radio and TV commercials. This is not to debunk them, but to revel in them. The necessary skills can come from even an elementary mastery of semiotic analysis of the kind presented here.

The whole examined in this section is more qualified than the theological variety. Instead of subject as the impersonal self, with all of its theological overtones, we are going to consider subject as *personal* self. We are going to consider the human mind from the perspective of psychology. What is the basic generic architecture of the psyche? Our response will be in the form of Freud's semiotic square interpreted from a viewpoint somewhat like that of Freud's student, Jung. Once again, we will start from scratch.

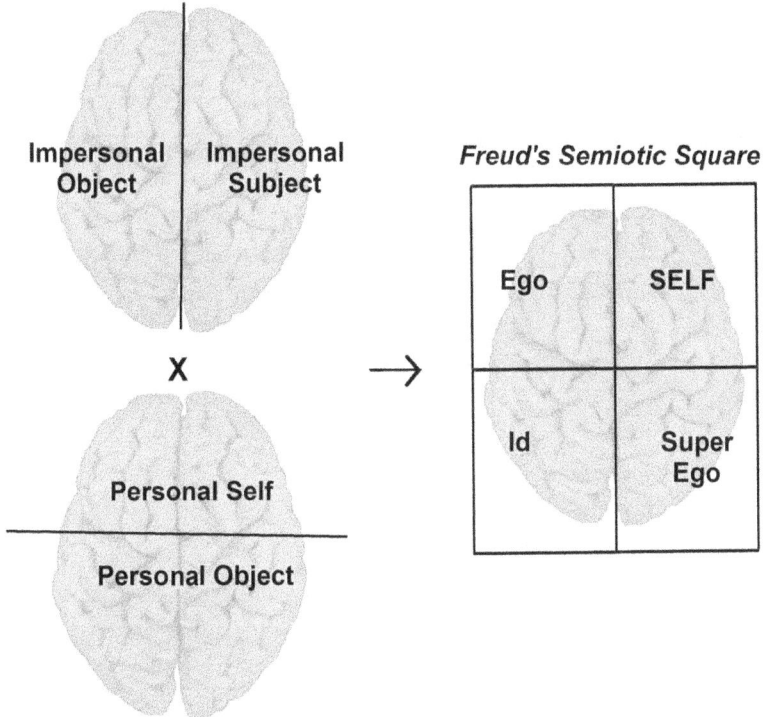

Figure 5 Freud's semiotic square of the personal Self in the form of the human Psychic Self.

We start with the left-right divide of reality as conceived by modern present day science. Modern science splits the Cosmos into two sides. On the left side can be found objects which are completely untainted by subjectivity of any kind. The Cosmos itself is sometimes referred to in hushed and hallowed tones as the Laboratory. In between the objectified objects on the left side and the other side of the laboratory is a glass wall. On the right side of the glass wall is the observing subject. This subject is not like any ordinary subject as he is the Supreme Scientist, beyond and above all other. The Scientist, represented iconically as dressed in an impeccable white dustcoat, a sure sign of divine objectivity, is completely fair, dispassionate, and unbiased in any way. This means that he is devoid of any determined viewpoint or favoured perspective. The Scientist is endowed with the unique ability of being able to see everything from literally nowhere. He has the God's-eye view. These characteristics form the essential ingredients for being the Supreme Scientist, Lord of the objective universe.

Ordinary, everyday scientists, who have to work for a living, aspire to emulate the Supreme Scientist and obtain his God's-eye view. Frustratingly, they never quite achieve their objective. Some scientists are so impressed that they take on the Supreme Scientist as their personal god. Like George Berkeley, they believe that you cannot have a Spectacle without an omnipotent Spectator; and that even applies to the lonely tree on a hill spectacle. Other scientists are completely unaware of or refuse to embrace the existence of any scientists cleverer than themselves. These are the atheists who spend all their time on the Left Side.

Once again, we have made a literary excursion into the realm of the great left-right dichotomy. It paves the way for looking at the great left-right cleavage of the biological brain, used as a metaphor for understanding the personal Self.

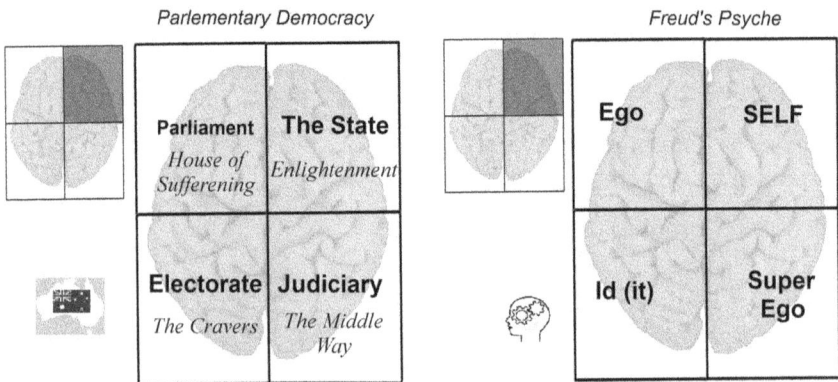

Figure 6 Two semiotic squares with a generically common ground but different figures.

The relationship between the personal Self and the impersonal Self has been a long discussed topic. Advaita Vedanta provides the most elaborate accounts, where the impersonal Self (Brahman) and the personal self (ātman) are considered as non-dual. Advaita non-duality is a more subtle way of saying that they are indistinguishable but different. For our immediate needs, it suffices to say that they share the same structure with the personal Self, simply being a more determined version of the impersonal Self. The extra determination comes about via introducing a second cut of reality, with the personal Self occupying the front; and the rest of reality, the personalised objects relative to the personal Self, taking up the rear, as shown in Figure 5. As can be seen, the semiotic square illustrates the human psyche, the Self, which can be understood in terms of the triad of psychoanalytic terms: the Ego, the Id (the "It") and the Super Ego.

Carl Jung, one of Freud's students, claimed that the right hemisphere of the brain was the "religious" side, an observation that has already become tantalising more evident. The Freudian semiotic square is essentially, what Jung described as a Three-plus-One structure. He remarked that these structures underlie most religions. This is in agreement with the theological semiotic structures that we have examined in previous sections. We wish to push the envelope further and demonstrate how *any structure*, treated holistically, will present in a Three-plus-One format. Thus, instead of explaining Freud's' Ego, Id and Super Ego in strictly psychoanalytic terms, we will take a more *generic* approach. In the next section, we add in the semiotic square for a holistic system that is easy to understand, a political democracy.

Political Psychology and Psychological Politics

The material in this section is probably better suited for discussion in a tutorial situation with a small group of students. It involves an exercise in lateral thinking across several semiotic squares. The importance here is to have some fun as well as perhaps getting a deeper understanding, without actually learning anything in particular. Our fascination is in the *generic shape of knowledge* and less in specific content.

Figure 6 shows two semiotic squares, the Freudian square and one for parliamentary democracy. To avoid any diplomatic incidents, the democratic square has been grounded in Australian democracy, hence the flag. Freud's square has been grounded in the psyche of a person of undetermined nationality. We will now spend a few moments explaining the democratic square as a subterfuge for explaining Freud: the author knows only a little about Freudian psychology. Like most people, he knows more about democracy and particularly how it works in his home country.

Side note:

As the author started to fill in the details of the left side of the square, as reported below, he inexorably slid into a mode of thinking that he can only describe as Zinovievian (but without the talent!). The world starts to take on a *Yawning Heights* (Zinoviev, 1979) character. Despite having read most of Alexandre Zinoviev, he is not really an influence, but represents rather a syndrome. It is a kind of mental disposition. Describing left side reality from a right side perspective seems to be the catalyst.

One way of explaining the democratic square is to exploit a few Buddhist insights. This turns our subterfuge into a double subterfuge, but it can shorten a long story. Besides, everyone likes Buddhism.

Take Parliament for instance. From a Buddhist perspective, Parliament can be thought of as the house of suffering. All suffering ends up here. The house is full of suffering because of the craving. Craving stems from the Cravers down below in the bottom left side slot. The role of Parliament is to try to appease the Cravers, which presents a perennially difficult problem; hence, the suffering and angst.

Parliamentarians publically refer to the Cravers as Voters, which gives the impression that somehow the Cravers control Parliament by voting for it. Nothing could be further from the truth. Voting is compulsory in Australia. The main reason people vote is to avoid a fine and so have more money to spend on their cravings. However, sometimes they will vote for a Parliamentarian who seems to identify with their particular craving. In private, Parliamentarians refer to the Cravers as the "It." The word "It" might refer to the Electorate, but more commonly, the word is used *generically*. Those clever enough to translate the word into German and creatively back into English might end up referring to the Cravers as the Id.

The Id is a teeming mass of opposing desires. Down-river irrigators confront up-river cotton farmers. Talk back radio Shock Jocks inflame the airways, railing against the boat people arriving on shore. Indigenous people writhe in the consequences stemming from when the forbears of the Shock Jocks arrived in boats on what used to be their shores. Greenies battle against loggers. Every complex, every syndrome imaginable will be found here amongst the craving Id.

That completes this little section on the left side of the Freudian psyche, written in Zinovievian mode. Coming over to the right side of the Freud square, the desire to write in Zinovievian mode vanishes. Actually, it feels a little bleak on this side as all we have is the Self in the frontal lobe and a thing called the Super Ego equipped with some powerful jurisdictional and moralising capability. There also seems to be some law enforcement capability as well. Super Ego seems to be full of lawyers and law enforcement officers.

Although we could pursue this topic at length, that is not on the agenda. So far, we gained some experience in semiotic analysis and had some fun along the way. The author has used these informal semiotic forms of analysis over many years in his profession developing software systems and computer languages.

5

Gender

The author first became interested in gender when living in France and learning French. In the strict language context, gender is only a grammatical category and may appear of little importance. It is interesting to note though that all Latin languages feature grammatical gender. Moreover, French seems to be the most "gender centric." For example, unlike Italian and Spanish and other Latin languages, French demands that even the names of countries must have gender.

This has posed problems in the past. For a while, there was some confusion over the gender of Canada: is it masculine or feminine? At that time, about two thirds of the population judged it as masculine and one third thought it was feminine. The final arbiter was Voltaire when he disparagingly described "le Canada" as being just a few acres of snow. That clinched it. The emerging Grand Robert dictionary entered Canada as masculine, which it has been ever since.

Right from those early days, the author has been fascinated by gender. What is gender? What is it for? He even founded a Venture Capital funded software company called Gender Systems Pty. Ltd, which unfortunately was did not survive. In the wake of the collapse of Gender Systems, the author wrote an exploratory book trying to grasp the deeper fundamentals of gender. The editor gave the book (Moore, 1992) a horrible title, but gender was essentially the thrust of it. La Trobe University historian, Don Ferrel, who proofed the book, noted the author's persistent interest in gender and quizzically asked whether he had attended a boys school or not. A bit taken aback, the author sheepishly had to admit that he had.

A few years after writing the book, the author made his first breakthrough in understanding and formalising gender. He came up with a construct that was not only compatible with its use in natural language, but also compatible with its usage in the sciences and philosophy of ancient times. The essence of the insight is summarised in the side note definition of gender below.

And so it has come to pass that the main, unifying plank of this present work is gender. The gender principle is posed as the organisational principle of the Cosmos and anything associated with it. In the pages that follow, we will endeavour to gently introduced the reader to this simplest but most profound of concepts. So far, the nearest we have got to the dialectics of gender is the dialectics of the One and the Many. The One-Many opposition is an opposition based on relative cardinality but it can serve as a tractable, easily understood introduction to gender. In this case, the One can intuitively be associated with the masculine and the Many with the feminine. However, gender is much more profound than that, as we shall see,

This chapter is a first, informal introduction to gender as a generic organisational principle mixed in with some further semiotic examples. The more formal treatment of gender has been relegated to the appendices. There we find gender playing a foundational role in geometry and even the theory of number.

The Other Science

Left side reasoning relies on a linear, sequential, punctual form of rationality. This has become the standard and generally accepted form of reasoning in science and mathematics. Nowadays, few professionally educated people would countenance the possibility of a science based on a completely different form of reasoning. Indeed, if such a science were to be proposed, the common belief is that it would be characterised by being fuzzy, woolly, mystical, irrational, and stereotypically New Age. The author *does* propose an alternative form of reasoning which is quite "orthogonal" to left side reasoning and he intends to refute any such characterisations. He claims that ultimately, at its core, the new science will be *more* rigorous that left side science. This is because the science will not rely on the vagaries of assumed *a priori* knowledge, be it by measurement, hypothesis, established opinion, or clever hunches. Instead, the science must rely on reason alone.

So far we have illustrated elementary forms of right side reasoning by getting acquainted with thinking in terms of oppositions rather than just labelling things and then manipulating the resulting symbols. It is now time to step the reasoning up a rung.

So, right side reasoning deals in wholes. The left side deals in fine detail. Let us now look at the semiotic square more closely. Our simple way of understanding the square is in the form of the One-Multiple oppositions as in Figure 3. The alternative to left side reasoning is naturally called right side reasoning and we infer that has something to do with right hemisphere biological brain function. However, linking the alternative form of rationality to the biological brain may be considered as an ambit claim at this stage of development. The veracity of our argument does not rely on such an association. However, the argument would gain potency if this were to turn out to be the case. For reasons that will become apparent with time, we will assume that the ambit claim is valid anyway. If we have to stick our neck out, then we may as well do it courageously.

The author has argued that traditional left side rationality is based on only one fundamental opposition, that between abstraction and the real. This naturally culminates in the "view from nowhere," the God's-eye-view form of objectivity popularly called the Scientific Method. The Scientific Method excludes one half of reality, the subject side, and only considers the half world of objects.

Right side rationality goes the other way. Rather than excluding the subject, the subject must be present at all times. It keeps both subject and object present at all times. However, the subject of the Scientific Method is only the impersonal subject, the formalisation of the view-from-nowhere. In addition to embracing the impersonal subject as partnered with the impersonal "objective" object, the right side demands that the *personal* subject also be present. This means that, instead of being based on just one opposition like left side science, right side science needs a *second* opposition. The second opposition is orthogonal to the first and is intuitively formalised in the form of the semiotic square discussed earlier. The result is fascinating.

The right side presents two kinds of subject, the personal and impersonal and defines the real as that which coincides with the personal *and* the impersonal, the conjuncture between the classical science "view from nowhere" of the impersonal subject and the "my view" of the personal subject. This gives one "real" slot and three "non-real" slots that we refer to as imaginary. The real can only be properly known via its three imaginary partners. The front-right lobe of the resulting square provides the slot for the conscious subject entity and there are three imaginary entities. This is yet another version of Jung's Three-plus-One structure that can be discernible in all the important theological configurations. We have also seen an instance of it in Freud's version of the semiotic square for the human psyche. Many more versions will become apparent as our incredible story unfolds.

So far, our gentle introduction to right side science has been a very intuitive and qualitative affair. On the left side of the classroom, the student has already been introduced to some elementary symbolic logic. It's time to catch up. On the right side we can teach about symbols too, different kinds of symbols of course, and we don't need as many as those that travel in the left lane.

Elementary Gender and its Simple Compounds

It is now time for the first introduction to the most fundamental and profound concept of right side science. It is a binary construct that we technically call *gender*. Gender has two sides to it, a masculine, and a feminine side. Here, we only consider gender from an elementary point of view. Later we will work towards a more fundamental understanding and will call it *ontological gender*. However, for the moment we are only going to consider it from a rather Pythagorean point of view, the point of view of cardinality. Nevertheless, just to get the mind to start working, we include here, as a side note, a more profound definition of gender. We come back to this definition later on in the book.

Sidenote: Gender Definition

As will be discussed later, the fundamental approach to gender starts with the notion of the totally unqualified entity, completely lacking in any specificity whatsoever. Such an entity is defined as being of *feminine* gender. The entity of pure feminine gender possesses the attribute of total lack of determined specificity. Thus, the feminine has an attribute, albeit nonspecific. This attribute must be an entity in its own right. It will be considered to be of *masculine* gender. Thus, two entities with only one attribute between them. The feminine **has** an attribute. The masculine **is** that attribute. This expresses a very profound concept and takes some time to get one's head around. It is for this reason we make the simplification of assimilating the gender dichotomy to the easily understood, simplistic One-Many opposition. Unspecific cardinality is not an attribute, whilst the cardinality One is. The Many **has** the attribute of oneness, the One **is** that attribute. The feminine **has**, the masculine **is**.

Gender comes into play in the relationship between what can be considered a singularity and hence have cardinality One, and that which is *not* a singularity and hence has cardinality Many. The first kind of entity will be considered to be of masculine gender, the second of feminine gender. In introducing the cardinality interpretation, it must be stressed that we are not talking about *absolute* quantification. The cardinalities are *relative*. An entity is One, and hence masculine, relative to something which is *not One* and hence the feminine Many.

From now on, rather than talk about the One and the Many, we will talk about the masculine and the feminine. An entity that is of masculine gender will be labelled or typed M and the feminine labelled or typed F. A more correct statement would be to say that the entity of masculine gender can be used as the masculine label and the entity of feminine gender can be used as the feminine label. In this game labels are not only made of the same stuff that they label, but *they are what they label.* I *am* my name. My name *is* me. We will not labour over this point, but it should be kept in mind that there is no arbitrary relationship between the signified and the signifier in this domain. Such arbitrariness is only permitted in the sciences of the left side.

The interesting thing about right side science is that that's it. Two symbols are all that you require to construct a code capable of describing and specifying any entity whatsoever in a rational universe. More complicated things than the elementary One and Many can be described by concatenations of M and F letters. The first compound entities are made up of binary combinations of M and F. There are of course four of them and they lead to the semiotic square shown in

The Four Letters of the Generic Code

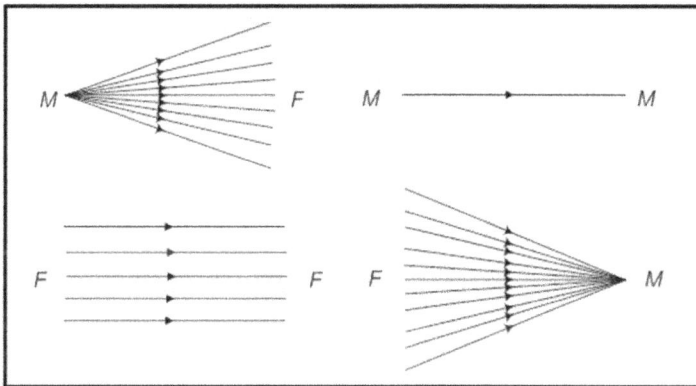

Figure 7 Semiotic square with the four generic types of structure in terms of elementary compound gender, MF, FF, FM and MM.

An informal, intuitive illustration of the four possible generic structures ia shown in Figure 7. These elementary structures are necessary to describe and/or construct a coherently rational reality. Like all of right side science, the idea is simple, simplifying, but subtly profound.

The four elementary compound genders MF, FF, FM, and MM can be interpreted in many ways, depending on context. They can be interpreted as compound types. They can also be considered as entities in their own right, in which case they pose as the "four elements."

Figure 8 shows a useful schematic representation of the four gender typed bases. We will later refer to these as Chrysippus diagrams. The diagrams also indicate an embryonic algebra. At the finest level, the qualifications are all in terms of gender, a relativistic typing system. The general corresponds to MF, the particular to FF, the universal to FM and the singular to MM. There is no need for external, traditional style empirical attributes.

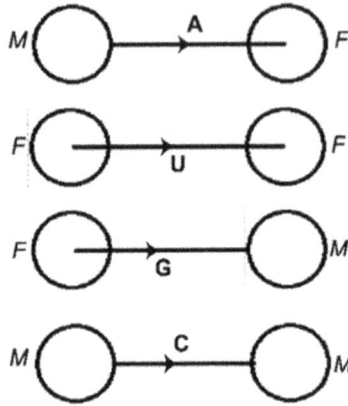

Figure 8 Iconic representation of the four bases and compound binary gender and the four letters A, U, G, and C.

The author has introduced a shorthand terminology where the four binary compound gender terms have been replaced by four single letters. It is at this point that it might appear that the author has lost his mind. Rather than invent his own lettering scheme, he has borrowed that of another general, universal, particular algebraic four-letter scheme for describing and organising generic entities. The four letters, of course, are A, U, G, and C used to denote the four bases of the genetic code. At this point, the reader can merely assume that any structural resemblances with the genetic code would be sheer coincidence.

Base Pairing

There are some readily noticeable structural resemblances between the generic code and our gender-based structure. In the double helix structure of DNA, the base A pairs with G and the base U pairs with C. In our case, if we assume that masculine M pairs with feminine F, which seems reasonable, we then get FM pairing with MF and MM pairing with FF. In other words, the bases A, G, U and C, have to pair as stated. Base pairing in the genetic has an underlying gender pairing. However, we are getting ahead of ourselves.

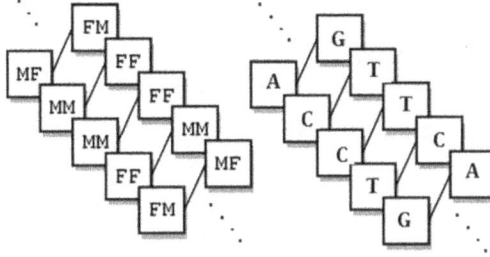

Figure 9 The RNA pairing of bases in the double helix of the generic code is claimed to have an ontological basis in gender. This diagram uses the DNA coding instead of the RNA coding. The U is replaced by a T.

Univocity

Traditional left side science and left side thinking have obvious and well-known proven strengths. One role of this book is to point out some of their terrible failings and how a radically alternative right side science can remedy the situation. However, right side thinking also has its peculiar traits and limitations. Some have said that to work in this domain of Kant and Hegel, the first thing to go out the window is common sense. It certainly takes some getting used to. However, an uncanny aspect that is particularly hard to accommodate is the absence of scale. Not only is there a lack of specificity in scale but a lack of specificity in general. The French philosopher Felix Deleuze noted this in his studies of Spinoza and incorporated the concept into his personal philosophy. He called it *univocité*, Even our little examples with the semiotic square have brought this absence of specificity to the fore. One minute we are looking at a semiotic square of how to get rich, and then we're looking at the Cosmos as a whole, followed by the Freudian Psyche and Parliamentary Democracy all in the one breath. It seemed that we didn't even have to change tablecloths. It was all done with the one semiotic square. We then end up with semiotic squares articulating base structure of the genetic code. This is univocity in action. Using this technology, we can say so many things in the one voice. That is what univocity means. In our case, rather than the one voice, we use the one semiotic square.

Of course, there is nothing mysterious about univocity. Univocity is simply another word for describing the generic. Speaking and thinking generically, we can say the same thing that is applicable to a vast number of possibilities. One could say that univocity is the generic equivalent of abstraction. Abstraction speaks in generalities. The generic expresses the univocity of being. The generic expresses universality... One voice articulates the essence of being and is applicable to any being whatsoever. Ultimately, univocity is revealed in the generic/genetic code applicable right across the kingdom of beings. Univocity ultimately means one code codes all. The code is generic and speaks in universals. The unrestrained universality and univocity of right side science is possibly one of its most confronting features.

In what follows, we try to present some of the essentials of this science without any mathematics or even anti-mathematics. A more thorough and more formalised account can be found further on and in the appendices.

The Start Codon and the Generic Ground

The semiotic square representing the *whole* provided a common rational ground, a common launching pad for the analysis. The structure of the common launching pad can be sketched out as in the semiotic square shown below. The square represents a generic whole and illustrates that *any* whole can simultaneously be looked at from a general, a particular, a universal, *and a* singular viewpoint. The square does not represent a spectacle; that would be a left side way of interpreting the representation, the spectacle-without-spectator, the object-without-subject. Right side reasoning demands that the *subject is always present.*

Relative to the one spectator, there is only one spectacle. In the One-plus-Three structure, the spectator, the subject is the "One." The One is the real part. The "three" is composed of the three relative attributes. That forms the imaginary part, the subjective part. Time and time again, example after practical example yields the same result. Three attributes labelled A, U, and G form the ground attributes for the whole. This applies to religions, for example the Christian Trinity, as Carl Jung consistently observed. These three synthetic attributes, A, U, and G, can be thought of as the generic general, particular, and universal aspects of the whole. It applies to Freud's Ego, Id, and Super Ego triad making up the human Self. Readers can construct their own versions of this semiotic launching pad for analysing their favourite wholes. Each time, the ubiquitous three modes A, U, G keep raising their heads.

MF: General *A*	MM: Singular *C*
FF: Particular *U*	FM: Universal *G*

Figure 10 One interpretation of the semiotic square is that of four generic types, the general, particular, universal and singular.. The types can be designated by four letters. The four letter A, U, G, and C habe been chosen as shorthand for MF, FF, FM, and MM respectively.

Here, we have the embryonic beginnings of a code, a code for coding *anything whatsoever*, as long as it forms a holistic aspect of an autonomous organism based on fundamental reason. We associate this code with Nature's code for any singular animate subject, the genetic code. In genetics, the three bases AUG form the start codon that, on a messenger RNA molecule, marks where protein synthesis begins. In positions other than at the start, the AUG codon codes an amino acid just like other codons. AUG in this case will code methionine, but the biochemistry is not of central interest as it is purely the implementation technology and does little to explain genetic code semantics. Only semiotic structure can do that.

A central plank of right side science is that it is *not* limited to scale; it is also unlimited by application domain. Generic structure is generic structure no matter what the problem domain and what the implementation substrate. The illustrative semiotic analysis cases considered so far do not *prove* this assertion, but they do illustrate it. The rationalisation comes from overall systemic coherency of reality in its ensemble, something that we have barely touched on so far.

In order to get some kind of handle of the role of AUG as the start codon in the generic code, we should try to look at other problem domains. Anyone with some background in physics would find that the semiotic diagram in Figure 8 looks a bit familiar. The cone of arrows for MF is evocative. Could this be interpreted as a cone of timelike arrows used in the customary explanation of relativistic spacetime? In addition, the arrows in the FM cone, are they spacelike lines? For that matter, what about the bundle of parallel lines? Are these optical lines? This line of thought is expanded in later sections and in the Appendices.

In this case and other cases, the AUG structure keeps on cropping up as one kind of underpinning generic ground for a more detailed overlay. The details of the overlay may change, but the ground stays as a constant.

The Singular and its Three Flavours

The author has been privately carrying out semiotic analysis exercises over the past twenty years or more, both in his profession and in philosophical and linguistic interests. His basic conclusion is that right side rationality is based on a common, generic code. Later, we will come back to the cones and lines illustrated in Figure 8 and add some life into them.

Present day Western education places a lot of emphasis on developing highly focused analytic thinking. Analytic thinking deconstructs the world into a fine granular mosaic of ever-increasing detail. In constructionist mode, analytic thinking ends up producing abstract theoretical models. The big picture for the linear, analytic thinker is the *model*, and it stops there.

Of course, we have been referring to this mode of thinking as left side. What interests us is right side. The biological brain provides the metaphor. The left is generally considered to specialise in an analytic, abstract, atomistic and language oriented mode of thinking. According to Iain McGilchrist, one role played by the right hemisphere is that of "bullshit detector," a talent sorely lacking in the raw data driven, data obsessed left side. Perhaps, the left side is unaware that the source of most bullshit in the world today comes from data.

The right side is considered creative and oriented towards the big picture. The left hemisphere is skilled in manipulating the symbolic entities of language, but is blind to linguistic emphasis and intonation, nuance, ambiguity, irony, empathy, and the metaphor. It is the role of the right hemisphere to provide these missing ingredients in the mix. The strongest point of the left side is its rule driven, focussed precision. The weak point is that left side processing leads to tunnel vision.

The left side can home in and fix its attention on anything. The problem is that, left to its own devices, it will resort to its default attention directing mechanism, the pursuit of anything producing pleasure. This is not a value judgment, but a physically demonstrable fact. Dopamine transmission is more extensive in the left hemisphere. What's more, the effect of dopamine in the brain is asymmetrically distributed with the left hemisphere being positively biased and the right hemisphere negatively biased with dopamine producing an inhibition effect. (Heller, 1990) (McGilchrist, 2009) Whilst there are higher levels of dopamine on the left, there is more of the neurotransmitter norepinephrine on the right. Norepinephrine stimulates the awareness of the environment.

6

The Four Letters

The farmyard hen's left-brain is connected to its right eye, the right brain to the left. There is no partial sharing of retina connections across the hemispheres as in humans. When foraging for that elusive grain of wheat, the hen will use its left brained right eye to focus in and seek it out. For awareness of its environment, it relies on its right brained left eye. The hen will check out possible threats, menaces and escape routes with the right brained eye, even if this requires cocking the head around to the other side. The opposite applies for finding the noodle in the haystack, in which case it will use its left brained right eye. The farmyard hen carries around two worldviews, one in each side of its cranium, two worlds, and one hen reality. The two worldviews are interchangeable by just a cock of the head. Is this farmyard chook more epistemologically aware than present day science?

For the author, some of the subject matter covered in this book provokes a personal feeling of profound wonder, particularly when it comes down to the Code. However, if we are not careful, this wonder can be lost in the technical detail. This can particularly be the case if the reader focuses in with left side analytic reasoning where so often, the finger points to the stars and the narrowed mind merely stares incredulously at the finger. The subject matter can only be fully understood from a right side perspective, the awareness side of consciousness. Without awareness, there can be no wonder and no deep comprehension; the sharpness of focus can be a distraction.

To bring this subject onto centre stage as a full-blown science requires a different kind of mathematics, some details of which are sketched out in the appendices. In the meantime, we can make do with some elementary apparatus accessible to any mind curious enough to go along with the flow. The kind of reasoning is not traditional reductionist, analytic reasoning of the ordinary sciences. Left side reasoning is an "open loop" form of thinking that necessarily involves labels and meaningless symbols. Allocating arbitrary names and symbols to things is a shaky start in the search for the deeper truths.

We have been referring to this label kind of thinking as left side. What interests us is right side thinking. Right side reasoning is *closed* loop thinking, where concepts are expressed in the form of oppositions and oppositions between oppositions. The oppositions express tensions and tensions within tensions, which constantly require attention. Eventually there may be a good fit and something resembling truth is achieved. However, this truth may not be long lasting. There will be a process of continual shuffling this way and that to relieve the tensions between oppositions and oppositions of oppositions. The central principle at play is to relieve violations of First Classness. Some might call this dialectical reasoning, but no one has yet succeeded in formalising such reasoning. This is one of our objectives. The basic ideal of such reasoning is that everything is determined and understood in reference to something else.

The opposition, the dichotomy, is the means to expressing such references. The semantics of *object* is lost without a present *subject* and so this leads to the fundamental opposition between subject and object, each giving meaning to the other. The opposition between subject and object articulates a dichotomy. If this dichotomy is absolute, FC will be violated. The whole process is dictated by non-duality, a core requirement of FC. In the final analysis, this non-duality expresses itself by the processes of mind being indistinguishable to physical processes. A generic form of absolute relativity acts as the guarantee.

We paint the picture with broad brushstrokes. However, even before the brushstrokes there comes the canvas. The canvas has four corners and is sufficient for an artist to paint a whole picture. So far, we have looked at a number of wholes and found that, as a whole, they can be painted on a four-cornered canvas. In previous sections, we saw that the "rich dad" Kiyosaki painted his cash flow quadrant on his four-cornered canvas. We saw that Freud accomplishes the same thing for the architecture of the psyche. We discussed the functioning of Freud's mechanism by talking about another semiotic canvas, parliamentary democracy; to demonstrate that by talking about one thing you can be really talking about another, a favourite pastime of artistic expression. As an attempt at some dangerous semiotic acrobatics, we talked obliquely about the personal psyche in terms of the political psyche, with a dose of Buddhist philosophical semiotics thrown in. It appears that we have stumbled on the universal language of the artist, a language that can talk across the board. Rather than just describe the scene, we can describe the canvas, the common ground for any painting. It also provides the elements for a common universal language that can operate across the board, a language that speaks in one Voice.

Oppositions of Oppositions

The canvas can be understood in the form of a semiotic square that encapsulates the two kinds of subject with the two corresponding worlds of objects. This semiotic structure is based on the opposition of two oppositions, a kind of "semiotic binomial." The first opposition, termed the left-right opposition, was seen as that between the impersonal subject on the right and the impersonalised objects on the left. Empirical scientists dream of this dichotomy where a pure, dispassionate, non-entangled subject surveys a non-disturbed world of objective objects. Such a subject has the highly sought after "view from nowhere," the God's eye view, the holy grail of empirical science, the unachievable dream. This was the first understanding of what constitutes a whole. The whole presents as an amorphous mass of objects together with the necessary, but totally undetermined, impersonal subject   t.

In order not to be stranded in the domain of the unachievable dream, a second kind of subject must enter the scene. This leads to a *second* opposition that we referred to as the second dichotomous cut across the canvas, the front-back opposition. In the frontal lobes resides the epistemological domain of the personal subject. The rear is the epistemological domain of the other side of the whole, all that is *not* personal subject. This is the personalised object domain. The result is a canvas cut up into four regions. These regions are not spatial divisions. One could say that they represent epistemological regions describing the four aspects of a whole, *any* whole. This is ground zero. We have considered a number of examples already that share ground zero. The content has changed, but the ground has been constant throughout.

A natural question is to ascertain where ground zero is located. It all depends on where the personal subject is located, and that can be literally anywhere. The exception is where you are the personal subject. Obviously, in that case the personal subject will be located at where you are at. Everyone possesses his own ground zero. It is usually located somewhere in the region between the ears and behind the eyes. This is your own personal canvas for picturing the universe. Functioning correctly, it will be aligned with the impersonal version. It is split into left and right sides that in turn are split into front and back. This, in itself can be an immense source of wonder. However, we have not yet finished with the technicalities.

The generic ground for any entity taken as a whole can be understood in terms of the semiotic square. The square is generated from an opposition applied to itself, the generic binomial. We have already interpreted this opposition in a number of ways. There was the opposition between subject and object. Another version was the opposition between the One and the Multiple. The most fundamental version of the opposition is that conveyed by ontological gender, the opposition between the masculine and the feminine. Gender will be revisited in more detail and precision later. Here we simply consider the masculine-feminine opposition as involving a more generic opposition than the cardinality opposition between the determined One and undetermined Many. Gender is not limited to cardinality and goes right across the board from the quantitative to the qualitative. It is really gender that speaks with one Voice. In all cases, the masculine appears as the determined singularity, that which is determined *as* singularity. The masculine is the only certainty in the equation.

The feminine, on the other hand, is a totally unknown quantity. The best way to understand the feminine, albeit from the masculine viewpoint, is that it is a total wildcard. And this is the key. There is nothing wrong about knowing nothing about something, as long as that knowledge is just, as long as you *know* that you don't know, then you know something. Here we find the Socratic confession of ignorance as the lynchpin of a whole algebra of the Cosmos! The only thing that Socrates knew with certitude was that he knew nothing with certitudes, and of that, he could bet the bank on. The purest and deepest knowledge is founded on the purest and deepest ignorance. The ignorance is encapsulated and carried along in the feminine wildcard. Encapsulated in the masculine is the absolute certitude of knowledge that this wild card *is* a wild card. The singularity of absolute certitude meets absolute uncertainty. This is the ultimate Principle of Uncertainty. What's more, it provides the two letters capable of coding the whole Cosmos, any Cosmos. One letter of this generic code is a total wildcard; the other codes a singularity and certainty absolutely proportional to the ignorance coded by the wildcard. The two letters are spelt M and F respectively. Is not ignorance a wonderful source of wonder?

The Four Letters of Antiquity

The above material will be revisited at a more leisurely pace in later sections. What we wish to retain here is the notion of a two-lettered generic alphabet. Intuitively we can say that these letters are M for masculine and F for feminine. These letters have semantic implications. The two letters have meaning as has been explained above. For example, the feminine F is the wildcard and is totally devoid of determined meaning, which, when you think about it, is really loaded in meaning. In a recent seminar given by the author, a young lady in the audience was taking notes and wrote down the letter F and then the word "wildcard" followed by a string of exclamation marks. So F seems to have meaning of some kind!

In passing, as a question to the audience, which would you rather be: a complete mystery even to your closest friends or an open book to the whole world? Did you know that the whole Cosmos is made out of these two kinds of beings?

The physics of pre-Socratic times and later the physics of the Stoics were founded on the theory of the four elements, sometimes called the Four Letters. According to the Stoics, two of the four elements were masculine and two feminine. The Stoics were not innovators in this domain and seemed to have just adopted the older versions of the science from previous generations with little modification. In addition to the masculine feminine opposition, the Stoics also include a second opposition based on the Active and the Passive principles. The way we interpret it, the gender opposition is the primary *impersonal* opposition and fits the left-right polarity convention. The Active Passive opposition can be interpreted also as a gender opposition like the first. However, this time it involves the *personal* version, the one corresponding to the front-back polarity convention. The Active corresponds to the personal masculine (the personal singular subject) and the Passive to the personal feminine (the personal non-singular). The four ancient elements, similar to those mentioned in other cultures such as those on the Indian subcontinent, were air, earth, water, and fire.

Stoic Qualia	Pure Gender Algebra	Element
masculine active	MM	Fire
masculine passive	MF	Air
feminine active	FM	Water
feminine passive	FF	Earth

Figure 11 Table showing the four ancient elements, the Stoic qualia for the elements and the pure generic gender algebra version.

Figure 11 shows the four elements together with the Stoic qualia and the pure gender versions. Heraclitus associated Fire, the doubly singular MM element, with Zeus. Note that Earth, the doubly non-qualified FF element is a kind of "double wildcard." As a substance, Earth would have to be interpreted as devoid of any specificity whatsoever. It is neither expansive like air or contractive like water. It is pure "stuff." In this F and M algebra, the F can be replaced with a question mark. The other three elements do possess specificity, but only relative to subject. Water with the specificity FM has for its only specificity the singularity of the personal subject. Air with the specificity MF has for its only specificity the singularity of the impersonal subject. Fire, on the other hand, being MM enjoys both the specificity of both personal and impersonal subject.

In the light of the above, it does not take too much imagination to realise that this ancient way of reasoning about the substantiality of reality is nontrivial and, in fact, very profound. Keep in mind that this is not *abstract* thinking that is involved here; it is thinking of a different kind, what we call *generic* thinking. For several millennia, this brand of thinking made up the dominant scientific view. This generic kind of science has been totally eclipsed by the dominance of the abstract sciences of the last few centuries. The generic science perspective has fallen in such disarray that it has become a source of ridicule. "Four elements! Everyone knows nowadays that there are at least 96 elements." The thinking of thousands of years of the greatest minds of the times has become an object of scorn and derision. It is time to reverse the tables.

The physics of the ancient world was not based on empirical left side thinking, but rather on an intuitive version of an embryonic right side science. In later sections, we will endeavour to reconstruct the ancient science and move it to a more rigorous and potentially formal footing. Of fundamental importance is the concept of gender, the most fundamental of any ontological principle. At present, we are content with an intuitive understanding of the concept.

Figure 11 shows how gender coding can be used to provide the elementary algebraic expression of the ancient four elements. The table includes an additional column that describes how the same gender coding codes the genetic code. It is a relatively simple exercise to determine the exact match between the genetic code and the gender coding. Suffice to say that there are so many constraints to the puzzle that only one combination stands out. We do not go into these details here.

The Genetic Code viewed Left and Right Side

The gender code mapping to the four bases C, A, U, and G of the genetic code, as shown in Figure 12, is incomprehensible from a left side science perspective. Implicit in the gender coding is a right side science of language. Before going down that track, it is worthwhile considering the genetic code as seen from the viewpoint of traditional left side science.

The left side linguistic theory of the genetic code is quite elementary and predictable. Basically, the genetic code is seen as a simple transcription language. This is in accordance with the standard left side concept of the binary relationship between the signified and the signifier. The sign and the signifier are assumed to be the one and the same. Language thus presents in the standard way outlined by Ferdinand de Saussure: first as a sequence of signifiers and secondly as a sequence of entities signified. The relationship between the signifier and the signified is considered as completely arbitrary. In human everyday languages, this means that the actual sounds, the phonemes, are devoid of any meaning. For example, the three phonemes making up the words C-A-T are considered arbitrary and have no meaning. It is only the morpheme CAT that signifies something. Three arbitrary markers, taken together have come to signify a cat.

The left side view of the genetic code is along the same lines. The signifiers in this case are the four bases A,U,G, and C. True to the left side paradigm, these bases are considered to have no meaning, they are just markers. The equivalent to a morpheme in the case of the genetic code is the codon. A codon is a triplet of any combination of the four bases, three signifiers per morpheme. The codon is considered to have meaning because it signifies something. As we know, in most cases it signifies one of the twenty basic amino acids that make up proteins. Three of them, the "stop" codons, act as punctuation marks signifying the end of a genetic sequence. One of them, the AUG "start" codon, also acts as a punctuation mark, signifying the start of a genetic sequence. If situated in the middle of a genetic sequence, it signifies an amino acid, methionine in this case.

We pause for a moment and give left side science its due. To arrive at the present day understanding of the genetic code is an incredible achievement and well worthy of a Nobel Prize or two. However, after the dust has settled, there comes a time when deeper questions come to the surface. Some of these questions are quite simple. Why do all creatures, ranging from the smallest microbe, the smallest streak of slime, right up to humans, all use practically the very same code and the very same coding? Even very specifically, why do they all use the very same start codon? If all life were a product of evolution, then surely the genetic code *itself* would also be a product of evolution. However, the evidence points to it as never changing and never have being in a state of change. Right at the beginning of life, there was the Code. Why didn't the genetic code evolve? Where is the survival of the fittest code? Where does the genetic code come from? Why this particular code and this particular coding? Does this code *precede* life?

Even more importantly, we ask the question as to whether the genetic code can be *reverse* engineered. This is the problem addressed in this book: determine how to reverse engineer the genetic code. Our approach will be to attempt to reverse engineer a *generic* code which is capable of coding, not just the visibly animate, but *anything whatsoever* in a rational reality.

Asking the above questions, posing them to the left side dominant scientific thinker, inevitably results in being confronted by incredulity or shear blankness. From the left side perspective, the situation is in hand. The genetic code has been "cracked." How the genetic coding transcribes the building planks of life has been revealed. All of these ontological questions of where it came from and why it works the way it does, lie outside the scope of science. The role of science is to describe, not to explain.

Gender Code	Ancient Code	Genetic Code
MM	Fire	C
MF	Air	A
FF	Earth	U
FM	Water	G

Figure 12 The ancient four-letter code can be understood in terms of the gender code. So can the generic code.

The Full and the Half Paradigm

There are two takes on reality. One is a full take and the other a half take. Left side science is based on the half take. It appears that the biological brain is similarly inclined. Thus, before investigating the scientific ramifications and avoiding any abstract musings, we look at the personality traits and competencies of the human brain when operating on a single hemisphere. What is the difference between the take of the left brain operating alone, from the take of the right brain acting alone?

When only the left hemisphere is effectively operational, the subject suffers from "hemi-neglect," as McGilchrist explains:

Since the left visual field, and the perceptions of the left ear, are more available to the right hemisphere, and by the same token the right visual field, and the perceptions of the right ear to the left hemisphere, one would expect, and indeed one finds, a gradient of attention from left to right, or right to left, across the experiential world for either hemisphere. But these gradients are not symmetrical: there is a fundamental asymmetry of concern about the whole picture. The right hemisphere is concerned with the whole of the world as available to the senses, whether what it receives comes from the left or the right; it delivers to us a single complete world of experience. The left hemisphere seems to be concerned narrowly with the right half of space and the right half of the body — one part, the part it uses.

In split-brain patients, for example, the right hemisphere attends to the entire visual field, but the left hemisphere only to the right. This refusal

of the left hemisphere to acknowledge the left half of the world ac-
counts for the fascinating phenomenon of 'hemi-neglect' following a
right-hemisphere stroke, after which the individual is completely de-
pendent on the left hemisphere to bring his body and his world into
being.

Because the concern of the left hemisphere is with the right half of the
world only, the left half of the body, and everything lying in the left
part of the visual field, fails to materialise ... So extreme can this phe-
nomenon be that the sufferer may fail to acknowledge the existence of
anyone standing to his left, the left half of the face of a clock, or the left
page of a newspaper or book, and will even neglect to wash, shave or
dress the left half of the body, sometimes going so far as to deny that it
exists at all. This is despite the fact that there is nothing at all wrong
with the primary visual system: the problem is not due to blindness as
ordinarily understood. If one temporarily disables the left hemisphere
of such an individual through transcranial magnetic stimulation, the
neglect improves, suggesting that the problem following right hemi-
sphere stroke is due to release of the unopposed action of the left hemi-
sphere. But you do not get the mirror-image of the neglect
phenomenon after a left-hemisphere stroke, because in that case the
still-functioning right hemisphere supplies a whole body, and a whole
world, to the sufferer. (McGilchrist, 2009)

Hemi-neglect is a characteristic of left side thinking, whether it is the
biological brain or the scientific mind. The left side is aware of only one half
of reality, whereas the right side must be aware of the whole.

Hemi-neglect runs right across the left side sciences. It always manifests
itself in a binary way of thinking. We have already seen this in logic, where left
side reasoning is based on the Law of the Excluded Middle. Something is either
true or false. Analytic Philosophy is full of it. Pet pre-occupations are the
Mind-Body dichotomy and the duality between the imagined and the real.
When it comes down to linguistics and semiotics there will usually be two
versions. One will be the left side version and it is always dualistic.

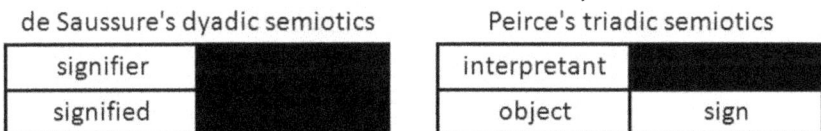

de Saussure's dyadic semiotics		Peirce's triadic semiotics	
signifier		interpretant	
signified		object	sign

Figure 13 Left side semiotics is dyadic; right side (Peirce) is
triadic.

The Tower of Babel

And the Lord said, Behold, the people is one, and they have all one language; and this they begin to do; and now nothing will be restrained from them, which they have imagined to do. Go to, let us go down, and there confound their language, that they may not understand one another's speech. (Genesis)

This is where we come to our point of wonder. The first wondrous aspect of the reality we live in is that it can be understood in terms of a single unique code formed from four letters. This was the dominant concept running through ancient civilisations, right into medieval times. The concept gains new impetus with the discovery of the genetic code. The concept will return to central stage with the development of our understanding of the *generic* code, the unique code underlying all reality, not just the animate. This four-lettered code, describes every cell in our bodies. Every cell has a copy of the same code.

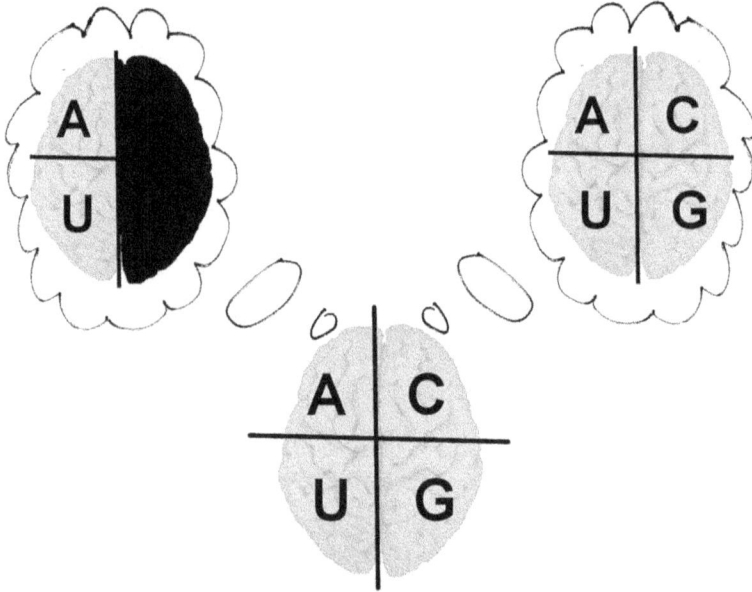

Figure 14 The generic mind: The right side is conscious of the whole. The left side has dispensed with the regulating machinery of the right side and has become open loop, relying on learnt rules. It has dispensed with the generic code and speaks the local patois.

The central theme of this book is that this generic code is the language of wholes. As such, it is the natural language of Mind, the mind conscious of itself as a whole. However, it appears only one half of mind, the right side, is based on the thinking in wholes and the corresponding 4-letter generic code. Here we come to the second theme of the book. We assert that the left side is *not* based on this four-letter alphabet. Rather than four letters, it only uses two, the two letters on the left as illustrated in Figure 14.

The right side thinks in terms of wholes and needs the full *four* letters of the generic code. However, the letters C and G relate to the Singular and the Universal. They express the requirement that the One must be One and the Multiple must be One. These are regulatory requirements. Such a mechanism can be restrictive. Like the free market economist who abhors legislation and regulation of the marketplace, left side reasoning dispenses with such constraints against individual freedoms. It becomes open loop and tries to go it alone. It doesn't need any Cosmic Reason to figure out what should be done. It just needs a notepad of rules and a belief in Providence. Totally unaware of the guiding hand of the right side, an incomprehensible entity at best, the left side thinks that it is master of the world.

7

Kant and Anti-Mathematics

Kant's Semiotic Square

Before proceeding any further, we will complete our coverage of the semiotic square. Because of the importance of Kant in Western philosophy, it is opportune to consider his implicit version of the semiotic square.

An easy way to construct a semiotic square is with two dichotomies. The hard part is choosing the pertinent fundamental dichotomies. We adapt the convention that the first and primary dichotomy provides the *left side, right side* dichotomy of the square. The secondary dichotomy provides what we will call the *front side, back side* of the square. Drawn on a piece of paper the secondary dichotomy will correspond to the two halves defined by a horizontal dividing line of the square.

In previous sections, we developed a fundamental semiotic square based on the oppositions between the One and the Multiple. This opposition was applied twice, once as a left-right side dichotomy and once for the front-back axis. In the first case, the One involves the impersonal One. The second case involves the personal One. This was repeated in another interpretation as the opposition between subject and object. The first opposition involved the impersonal subject and the objects that it subjectifies. The second opposition involved the personal subject and its corresponding kingdom.

It was argued that the most fundamental of these kinds of dichotomies was based on gender, where the masculine expressed the attribute of pure singularity free of any other particularity. The masculine was an entity in its own right. The only particularity possessed by the feminine was that it *had* this attribute. In other words, the primary opposition was between two entities of different gender. The only specificity of the feminine entity is that it *has* an attribute. The specificity of the masculine entity was that it *is* this attribute: One *has* it, one *is* it. There is nothing more elementary than this gender principle. The first gender opposition is between the impersonal feminine and masculine, the left-right dichotomy. The second gender opposition has the personal feminine and masculine for its two poles, the front back dichotomy. The gender construct underlies the fundamental typing mechanism underlying all of the unifying science presented in this work and helps explain the ancient theory of the four "letters" or the four elements. Exploiting this hyper-generic gender construct, a universal typing mechanism can be constructed, where any entity whatsoever can be described in terms of such generic types. The four binary combinations of the two gender types provide the alphabet for such a system. In a previous section, we made a preliminary interpretation of the genetic code as being such a typing mechanism. We even tentatively linked the four letters A, U, G, and C of the genetic code with their corresponding four binary gender types MF, FF, FM, and MM respectively.

How this material should be taught and at what age the various concepts should be introduced, the author has no firm opinion on such matters. The author has found that even mature people can have problems coming to terms with the concepts. Some, the highly trained academic for example, find the ideas threatening.

It's time now to look at Kant's version of the semiotic square. In the *Critique (Kant, 1738),* and particularly more clearly in *Prolegomena to any Future Metaphysic* (Kant, 1783), Kant effectively outlined two superimposed dichotomies. The primary dichotomy was between knowledge founded on *a priori* judgments and that founded on *a posteriori* judgments. *A priori* judgments are based upon reason alone, independently of all sensory experience, whilst *a posteriori* judgments require real world experience.

In addition to the primary dichotomy between *a posteriori* and *a priori* judgments, Kant superimposed a second very important dichotomy. This was the dichotomy between analytic and synthetic judgments. Analytic judgments are where the predicate is wholly contained in the subject. Synthetic judgments are where the subject is completely distinct from the predicate and so must be related to some *outside principle*.

We can visualise Kant's secondary dichotomy by complementing the left-right side primary dichotomy with a front-back side secondary dichotomy, where the synthetic is on the front side and the analytic is at the back side, as shown in Figure 15.

Superimposing the two fundamental knowledge dichotomies leads to visualising the overall architecture of knowledge as a kind of semiotic square with left and right side specialisations each with its own analytic "frontal lobes." In effect, this is Kant's version of the semiotic square. The artifice provides a way of visualising the epistemological structure of knowledge, the layout of the epistemological brain, so to speak. In our more rash moments, we claim that this also provides a sketch of the architectural and functional layout of the biological brain. In later sections, we will investigate the role that this structure plays in the science of spatio-temporality. In the broader picture, we intend to demonstrate the science behind Kant's claim that all perception and cognition takes place within a spatio-temporal framework. The first informal, intuitive acquaintance with this framework is via the semiotic square. This artifice is not presented here as a theory of the brain, but merely as a pedagogic aid to visualisation.

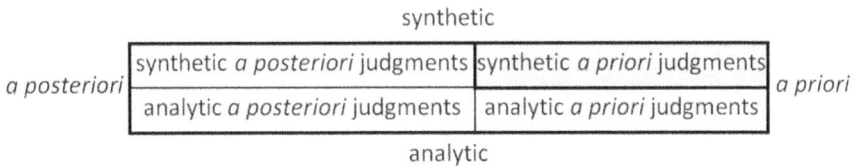

	synthetic		
a posteriori	synthetic *a posteriori* judgments	synthetic *a priori* judgments	a priori
	analytic *a posteriori* judgments	analytic *a priori* judgments	
	analytic		

Figure 15 Kant's two fundamental dichotomies can be superimposed to construct a semiotic square of knowledge. Solving the Kantian question requires knowledge of the right hand front corner kind.

As can be seen from the diagram, we end up with four different kinds of science. Kant homed in on the front right side of the diagram, that of synthetic *a priori* judgments that, in theory, should synthesise new knowledge that was necessarily true. This is the domain where the Kantian question addressed by this book, is located: How do we produce right hand, front side knowledge?

Polysynthetic Knowledge

The knowledge that we seek is based on Kant's synthetic *a priori* judgments and is, in effect, *doubly* synthetic knowledge. Using a term borrowed from linguistics, we will call it *polysynthetic* knowledge.

In linguistics, an important language classification is between analytic and synthetic languages. The difference between the two classifications is not very precise but, in general, analytic languages tend to have simpler words consisting of a smaller average number of morphemes. Also, the grammatical structure of the language is expressed more in terms of syntax based on word order, rather than inflexion and affixing and prefixing of grammatical markers on individual words. Overall, the analytic language will be synchronic in structure rather than spatio-geometric. Thus, the analytic language speaks by the intricate sequential flow of single notes of a melody. On the other hand, the synthetic language expresses its message in terms of dense, rich chords, each articulating a beautifully vivid spatio-geometric image. It does this by using words with a larger number of morphemes per word. In addition, word order is of less importance or, in some cases, of no importance.

It is interesting to note that proto Indo-European was highly synthetic; this is the reconstructed common mother language of all Indo-European languages including Latin, Sanskrit, Hindi, and most European languages. Since then most Indo-European derivatives, of which English is a good example, have become increasingly analytic with the passage of time.

At the extreme end of the synthetic scale are situated the polysynthetic languages. Examples of these hyper-synthetic languages seem to be closer to the deeper natural order of things. Examples of polysynthetic languages include languages of North America, Siberia, and Australia.

The polysynthetic nature of Australian Pama-Nyungan languages is illustrated by the example:

"...the words meaning *man* (ergative) + *see* (past tense) + *you* (accusative) + *big* (ergative) can be placed in any word order whatever; they will be understood to mean '(A/The) big man saw you.'" (Heath, 2010)

At the other end of the spectrum would be the doubly analytic forms of knowledge located on the front, left side of the Kantian square, *a posteriori* knowledge expressed in analytic judgments; the analytic *as* analytic. We will call this *polyanalytic* knowledge. This classificatory term doesn't seem to be used in linguistics and so we will refrain from rashly endeavouring to discern polyanalytic language groups. However, we can get a good grasp of what the polyanalytic entails as far as a philosophical classification is concerned. There probably would not be much objection to saying that a good example of a polyanalytic philosophy would be none other than Anglophone oriented analytic philosophy.

If we admit polyanalytic philosophy into the fray then, to be fair, we should also admit the totally opposite number, polysynthetic philosophy, yet to be born. Whilst analytic philosophy delights itself by listening to the tinkle of meanings flowing from natural language, usually English, on the other polysynthetic side, a different language is spoken where single words can be so large and the chords struck so vibrant that the music can last for a lifetime.

Epistemological Schizophrenia

So far, our approach has been informal and intuitive. Most of the material could be taught to students at a relatively youthful age in the form of semiotic analysis using the semiotic square. Students can be encouraged to look at various aspects of reality as a whole and to interpret the whole in the form of the two primary oppositions and the resulting four "categories."

This pedagogical approach is practical without being empirical. This is the form of reasoning practiced by the ancients in the times of Heraclitus as instanced, for example, in their theory of the four elements. It is a form of reasoning that seems to spontaneously crop up in the hands of the autodidact and, as we have seen, even in that hands of the businessman trying to get an overview of the business process.

As can be seen from some of the examples, the exercise can also be quite profound. An interesting topic for discussion for the more mature students is that of non-duality. The semiotic approach to wholes is applicable to any whole "out there" when applying the analysis to the concrete. However, it is also applicable to the architecture of mind itself, the whole "in here," which can also be understood in terms of the two primary oppositions and the resulting semiotic square. The basis of this identity of structure provides an intuitive explanation of the Principle of Non-Duality. (There is even a third view that is neither "out there" nor "in here," but let us not get too pedantic here)

Students can also spend time discussing the different characteristics of left side reasoning and right side reasoning. A salient point to highlight in the case of left side reasoning is abstraction. This mode of thinking is based on a simple opposition between the "real" and the abstract, between particulars and generalisations, and goes part and parcel with left side binary, dyadic reasoning. This is the Bertrand Russell type common sense kind of reasoning. How can there possibly be another way of thinking, some might ask, just as Russell did. What can be more common sense than common sense? This question should be dramatically repeated, perhaps several times to create a sense of theatre. Anything should be tried in an attempt to open up young minds to another mode of thinking than the reductionist, narrow view of the left side. Each challenge to left side thinking is a challenge to the mundane, the monotonous, repetitively every day, the world defined by customary rules, the world of the apparently obvious. Remember that the left side of the brain is completely unaware of its wide viewing sibling.

The alternative to left side reasoning is right side reasoning which reasons in terms of oppositions. Right side reasoning is expressed in terms of oppositions where the meaning of something is understood relative to something else, not relative to a label, symbol, or words as in the left side case. To understand something, one must understand *both* sides of the opposition. Moreover, both sides of the opposition are of equal value. In fact, the two sides are always the two sides of the one coin so to speak. Right side reasoning must always have the subjective present with the objective. The objectified is always accompanied by what objectifies it, the subject. It is in this way that the reasoning remains monist, always talking about the whole as a whole.

In left side reasoning, the subject and all that is subjective are continuously ripped out and discarded. This has the advantage of removing a lot of overhead and greatly simplifying the perceived world, particularly when compared to the operations at play on the right side. The subject is replaced by a surrogate subject. In the case of the left side sciences, the surrogate subject takes on the form of an abstract scientific theory. Broadly speaking the surrogate subject is a conglomerate of pragmatically determined stereotypes for understanding the world. The stereotype can even come down to something as elementary as a label or symbol. Here we can discern a most fascinating phenomenon concerning the dynamics of left side reasoning. In order that knowledge of the objectified world grows, the subjective aspect must be continuously pared away, something like the whittling away of a stick using a penknife to produce a sharper and sharper arrow point. The overall result is that, as the mass of knowledge grows, the domain of knowledge becomes smaller and smaller. Interpreted positively, one could say that the knowledge has become more focused. Speaking pejoratively, one could equally say that the process has ended up in producing tunnel vision that has become completely devoid of context.

A world uniquely viewed from the left side reasoning perspective becomes a fractured and fragmented world, lacking in any evident global coherence. It becomes a vast conglomerate of specialised, specialisations of specialisations. In the sciences of our day, no single individual can possibly master such a bewildering array of detail, or would even want to. Each science becomes a world of its own. Each specialisation within the science becomes a world of its own. This is particularly apparent in mathematics. Mathematical historian, E. T. Bell, saw the mathematician Henri Poincaré as the "last of the universalists," the last mathematician having a mastery ranging across all of mathematics. A modern Henri Poincaré is impossible in the present morass. In the absence of any unifying science, the whole apparatus seems to be heading towards epistemological schizophrenia.

Lateral thinking

As we said, our approach so far has been informal and intuitive. We have presented a key aspect of right side reasoning in the form of its two fundamental oppositions leading to the semiotic square. Any whole can be intuitively analysed in terms of the semiotic square. Even here, with this informal, intuitive, elementary artifice we start to discern a key feature of right side reasoning. It involves discerning a *common structure* in any aspect of reality when looked at as a whole. The most elementary form of this structure is the semiotic square. In previous sections, we considered many different wholes from the perspective of the semiotic square and the accompanying semiotic analysis. Repetition on constant themes seems to be our message. Looking back, it is quite remarkable when we put into the one diagram, some of the semiotic squares that we have covered so far. This is shown in Figure 16. It does not stop here. More underlying semiotic squares are forthcoming later in our development, particularly in the area of physics.

This kind of thinking can drive the classical, analytic thinker crazy. The reasoning is just the complete opposite to the narrow, focused, and focusing rationality of such thinkers. Edward de Bono called this kind of phenomenon *lateral thinking (Bono, 1967)*, although he did not offer any semiotic or other potentially scientific explanation. His main interest was to encourage creative thinking in people, thinking that he probably conceived of as more an art than a science. He encouraged right side thinking.

The author has spent a long time exploring this kind of thought while dreaming of a new kind of science, a science that will unify all others. After becoming habituated to this terrain one starts to get the impression that nothing here seems to change much, if at all. Maybe Parmenides was right: Nothing changes. Everything seems to have a *déjà vu* aura about it. It might appear to some to be a case of lateral thinking, even lateral thinking gone wild; but really, there is no lateral movement. One keeps moving around on the one spot. Moreover, this one spot has the same tried and tested square shape. Even Nature seems obsessed with this one theme. As we have seen, every genetic sequence of DNA starts by signalling the same ground, with its AUG start codon: the same ground for any biological entity whatsoever. In later sections, we see that even the apparently inanimate world of physics plays exactly the same tune.

We are just musing over our recent journeys and what it all might mean. However, the fact remains that our approach so far has been informal and intuitive. It is now time to start the road towards a more formal approach to this embryonic right side science.

If the Stoics are anything to go by, the approach will not always be pretty to explain. Many scholars have commented how the writing style of the Stoics was lacking in elegance, having an almost agricultural feel about it. As we will later explain, the author is following in the footsteps of the Stoics. The author argues that the Stoics were not left side oriented thinkers, but right side oriented. Hence, unlike so many other Western philosophers, they were not abstract thinkers. This might explain the terseness and sometime roughness of their style. In what follows, the author adopts the same kind of approach and endeavours not to slip into that dreadful sin of all, for this project at least, the sin of thinking abstractly.

The singular and its three flavours	
general	singular
particular	universal

Freud and the psyche	
Ego	self
Id	Super Ego

Godhead	
Christianity	Advaita
Buddhism	Islam

Gender Code	
MF	MM
FF	FM

Democracy	
Parliament	State
Electorate	Judicial System

Christian Godhead	
Father	Jehova
Son	Holy Ghost

Genetic Code	
A	C
U	G

Kiyosaki's cash flow quadrant	
E	B
S	I

Islamic Godhead	
Faith	Allah
Divine Koran	Divine Justice

Dialectic of the One and Many	
One is Many	One is One
Many is Many	Many is One

Ancient four "letters"	
Water	Fire
Earth	Air

Hindu Godhead	
Vishnu	Brahmana
Buddha	Shiva

Figure 16 The unifying science has the same ground for any whole, the same shape. An intuitive understanding of this common generic ground is the semiotic square in standard form. This is the true Start Codon of anything.

Formalising the Kantian Problem

There are two kinds of knowledge, Kant argues, empirical knowledge arising out of experience and the other as knowledge *a priori*, knowledge that is independent of experience. The two kinds of knowledge were both related to the experience of the subject. In one case, the subject acquires knowledge from experiencing environmental particularity. In the second case, somehow the subject arrives at knowledge without any anterior experience whatsoever. In this case, there is a complete *lack* of environmental particularity. The traditional sciences are of the first kind, and another kind of science that Kant called metaphysics is of the second kind. We have been referring to them as left side and right side sciences respectively. The Kantian problem, as stated in the *Critique*, is to provide the science of the second kind, to make metaphysics scientific.

In Kant's day, the physics of Newton provided what appeared to be an absolutely precise theory of the movement of the planets. Mathematics had not yet got to the stage of being as formalised as it is today. The axioms of Euclid were still seen as being statements of self-evident truth. These factors influenced Kant to take certain trajectories that mark his work as a product of his time, needing some renovation. In the following exploratory section, we endeavour to effect such renovation.

Our intention is to reformulate the Kantian question in more modern and formal terms. Our ultimate goal is to solve definitively, finally, once and for all, the Kantian problem. However, in the process of stating the problem in more formal terms it will seem as if we have signed our own death warrant. The author came to this hopeless realisation over fifteen years ago and fretted about it for a long time. He now wishes to share the insight and agony with his readers. Together we can all writhe in the apparent hopelessness of the situation. This truly is a hellhole. Life was not meant to be easy.

However, all will not be lost. As this book advances, glimmers of hope emerge. Finally, in the last steps we end up kicking ourselves. Everyone has been attacking the wrong castle, on the wrong side of the river. On the other side is a house full of treasure. The author finds out to his surprise that the back door of this place is not even locked. However, we have not yet arrived at this place. For now, we are still in the slippery world of pure unadulterated angst, trying to move forward.

Knowing the Unknowable

Making progress in this epistemological hellhole in which we find ourselves, is all about getting traction on a very slippery surface; how to make the seemingly intractable tractable. In order to make progress, certain missing essentials must be added and only the strictly necessary retained. It is often the author's distinct impression when visiting the terrain prepared by many traditional philosophers, that it is like trying to think while walking across a cow paddock in bare feet. It is time to clean things up somewhat.

...

An important first task is to revamp Kant's dichotomy between "empirical knowledge" and "pure knowledge." The first step is to remove any anthropomorphic cum biological bias, as it only muddies the waters. If there is any anthropomorphism or biologicalism lurking here, it will be of a much more profound kind than some semi-transcendental organism receiving and processing sense datum or some transcendental ego obsessed in speculative introspection.

Kant's basic schematic involves a knowledge-creating machine, which will only function if fed a stream of raw fodder, no fodder, no knowledge. Thus, the fodder comes first and the knowledge comes out the other end. Because it comes first, the fodder is granted the status of being *a priori* stuff relative to any knowledge created. This is quite a reasonable schematic and worthy of investigation.

However, Kant, like practically everybody else, complicates the picture with his unconscious attempt to inject anthropomorphic and biological interpretations into a domain that is inherently and brutally neutral in this respect. The word, *experience*, is probably the culprit. Experience for whom or for what? What kind of experience? Sexual experience? Real world experience? Sensorial experience? Painful experience? Rewarding, end-user experience? It is all too complicated. Age-old agricultural terms are much more precise and evocative and so we prefer the term *fodder*: fodder for knowledge creation.

However, the real problem is that Kant limits the fodder to that which is only digestible by inquisitive earthworms and empirical scientists. He ignores the desperate plight of mathematicians. They too must be fed fodder in the form of bunches of axioms in order to produce mathematical knowledge; no juicy axioms, no mathematical knowledge; no axioms, no formal mathematics.

It is for this reason that we argue for abandoning any restrictions of the dietary regime of the knowledge generation machine and admit formal mathematics into the fold. Instead of talking about empirical knowledge we should simply talk about *conditional knowledge*; knowledge that always relies on being conditioned by something *a priori* to it. This includes axioms as well as any possible sensorial information in all its guises, or even knowledge of a more primitive form than that to be produced. Remember, we are talking about *epistemology* of the everyday sciences and mathematics here, nothing more, nothing less; technicalities abound perhaps, but there is nothing much here to inspire or perturb the sleep of the common man.

With mathematics let into the fold, the term conditional knowledge can be defined with a higher degree of precision. We will define *conditional knowledge* as that kind of knowledge that is articulated or can be articulated in the language of mathematics. Such knowledge obviously includes mathematics itself. The knowledge we are talking about becomes that of mathematics, the mathematical sciences, and that which can potentially fall into the embrace of mathematical sciences.

By mathematics, we mean axiomatic mathematics and only axiomatic mathematics. It is only axiomatic mathematics that formally expresses the concept of conditional knowledge. Mathematics must be conditional mathematics - conditional on its axioms. The mathematics must be tamed and put in a bottle, the bottle of axioms.

How to Bottle Mathematics

Each of the various disciplines and problem areas in formal mathematics is characterised by its own set of axioms. The axioms of a branch of mathematics are like a bottle that contains the mathematics inside. Inside the bottle one can find the different models that satisfy the axioms, and a vast array of theorems, lemmas and other such paraphernalia.

Thus, if one looks out across the vast range of modern formal mathematics, one sees a vast array of bottles. Bottles can contain other bottles. For example, most mathematical bottles will contain inside a little bottle of Set Theory stuff so that the mathematics can deal with things and collections of things and so forth. There is also a pecking order of bottled mathematics with the more abstract mathematics like Category Theory, higher up the tree than the less abstract.

A natural question arises as to whether it is possible to take all these bottles of mathematics and put them inside one great big single bottle. Whitehead and Russell presented such a project. They claimed that all of formal mathematics could be put inside a giant bottle labelled *Principia Mathematica.* The novelty of their project was to employ a different material for making mathematical bottles. Instead of making the bottle out of mathematical axioms, as was the established custom, they claimed that they could make the containing bottle out of logic instead.

The idea of bottling knowledge in a logical framework is called *logicism*, an approach favoured by Leibniz. Whitehead and Russell claimed that all of mathematics could be tamed with logicism but they took a different road to Leibniz. How does one construct a bottle out of logic? The approach of *Principia* was to build the bottle out of a "perfectly logical language." The perfectly logical language was to be crafted out of logical axioms. The claim was that all of mathematics could be wrapped up in this perfectly logical language.

The traditional technology of making mathematical bottles is to make the bottle out of mathematical axioms. To build a bottle hardy enough for all of mathematics, the big bottle would be made, not of mathematical axioms, but from logical axioms. Thus, what is the difference between a logical axiom and a mathematical axiom? The essential difference is that a mathematical axiom has a mathematical meaning whilst a logical axiom does not. In fact, logical axioms don't have any meaning at all. In the *Principia Mathematica* logical axioms are only needed to build the syntax of the perfectly logical language. As Russell explained:

> In a logically perfect language there will be one word and no more for every simple object, and everything that is not simple will be expressed by a combination of words, by a combination derived; of course, from the words for the simple things that enter in, one word for each simple component. A language of that sort will be completely analytic and will show at a glance the logical structure of the facts asserted or denied. The language that is set forth in Principia Mathematica is intended to be a language of that sort. It is a language that has only syntax and no vocabulary whatsoever. Barring the omission of a vocabulary I maintain that it is quite a nice language. It aims at being the sort of language that, if you add a vocabulary, would be a logically perfect language. Actual languages are not logically perfect in this sense and they cannot possibly be, if they are to serve the purposes of everyday life

In order to have a vocabulary for this syntax-only language non-logical words would be added. For these words to have meaning, they must directly signify existent things or, by analytic reasoning, indirectly lead to such significations.

This is as far as we wish to venture into the logical atomism of Russel. However, what is so remarkable is the similarity of the *Principia Mathematica* project and the one reported here. The big difference is that *Principia Mathematica* reports all of its findings through the prism of the traditional left side scientific paradigm.

Basically, the philosophy behind the *Principia Mathematica* presents as an excellent formal explanation of the common paradigm underlying all the traditional sciences, including mathematics. In this book, we have been talking about the left side scientific paradigm. The tragedy is that this is practically useless in the real world, as we shall see. After the tragedy comes the farce.

What fascinates us is that despite the fact the *Principia Mathematica* is written in the left side mindset of analytic philosophy, it covers many of the important bases of right side science.

On the left side is the analytic "perfectly logical language" of the *Principia Mathematica*. On the right side is the perfectly logical language that Nature came up with, the genetic code. In the Appendices, we show how the generic code is underpinned by Stoic logic.

Working on the left side, Russell is obsessed with removing all oppositions in order to avoid paradoxes. Right side science doesn't avoid the paradox; it is founded on the apparent paradox of the Socratic Confession of Ignorance. Rather than avoid oppositions it is constructed from oppositions. These oppositions work together to enforce FC.

Driven by efforts to avoid paradox, Russell developed his Theory of Types. These types have nothing to do with typing real entities in any way. They are logical types and, in fact, nothing more than a never-ending logical hierarchy. On the right side, we find the parents of types *par excellence*, the masculine and feminine. These lead to the four bases of Nature's version of the perfectly logical language.

Once again, we make the assessment that the *Principia Mathematica* is very close to the truth, but on the wrong side of it. The true can be found on the left side, but only the right side can harbour truth.

Our basic conclusion is that the most fundamental shroud for wrapping up knowledge, like the *Principia Mathematica,* must be based on the logically perfect language. However, it is not a language devoid of meaningful semantics. It involves a language that can speak anywhere and to anything with the one Voice, the most semantically rich language possible, the generic code.

The problem now is to find a way of formalising this right side alternative to the Principia Mathematica.

...

Recapping now, the first step is to replace Kant's "empirical knowledge" with the concept of conditional knowledge. Conditional knowledge includes mathematics and is expressible in mathematical terms. Therefore, *conditional knowledge* is mathematics, or is potentially expressible in mathematics.

Now is the moment to redefine the contours of the new dichotomy, where Kant's "pure knowledge" simply becomes unconditional knowledge. We state the criterion for determining what constitutes unconditional knowledge as being equivalent to that which *cannot* be articulated in mathematical terms. Kant's pure knowledge, *unconditional knowledge*, is impervious to mathematical formalisation. This fact is so stark and so precise that it can be used not as a property of the two kinds of knowledge, but as the central distinguishing criterion.

Our new dichotomy has on one side *knowledge that is describable in terms of mathematics*. By mathematics, it is meant formal axiomatic mathematics. In this case, the kind of knowledge referred to includes empirical knowledge. However, it also includes mathematics because any mathematical formulation is itself a mathematical description. This broader definition of *a priori* based knowledge will be simply called *conditional knowledge*. Mathematics can articulate *any* conditional knowledge. Mathematics being itself conditional (on axioms) can itself be considered as articulating conditional knowledge.

On the other side of the dichotomy lies the kind of knowledge that we are really interested in, notably Kant's pure knowledge. Pure knowledge can be formally defined as *knowledge that is **not** describable in terms of mathematics*.

Anti-Mathematics

This is all quite straightforward, but can very easily lead to a popular misconception. Mathematics is the greatest, most formidable and most productive system of rational formalised knowledge known to man. Leaving aside theological constructs on the grounds of rational inadequacy, mathematics is the only known such system. The misconception is that the dichotomy between traditional *conditional* knowledge (that emendable to mathematics) and the non-mathematical *unconditional* knowledge (that impervious to mathematics) is really a dichotomy between formalisable knowledge and non-formalisable knowledge.

This leads to the notion that the home ground of Heraclitus, the crying philosopher, is that of the non-formalisable, the untameable, and some would add, the inherently unknowable. This is the land of the poet. This is the land of the logician Lewis Carol's *Alice in Wonderland*, where even scale is an uncontrolled variable; time itself in a frantic, bumbling hurry against itself, and the whole party is run by a buck-toothed madman in a top hat. This is also the land of our other favourite logician, Alexandre Zinoviev. There we find his Soviet Union laboratory of human nature. This is where the rational man comes to the end of his tether. Standing at the North Pole, he knows that the way home is to head directly south. He stands there petrified, unable to make sense of it all. Nothing has traction here. Can anyone make any sense out of all this?

Putting aside tangential prose, we come to the central proposition. We declare that even though this metaphysical pure knowledge, *unconditional knowledge*, cannot be articulated in mathematics, the most formidable formalism known, it is knowledge that *is* capable of formalisation. Such a formalisation cannot be a mathematical formalisation. However, the formalisation we require will not be referred to as non-mathematical, but rather as *anti-mathematics*.

Formal axiomatic mathematics can only articulate conditional knowledge. It does so in terms of conditional knowledge. In order to articulate scientifically the other kind of knowledge, *unconditional knowledge*, or *universal knowledge*, we need another kind of formalisation, *anti-mathematics*. In this way, we can redefine pure or unconditional knowledge as that knowledge *that can be articulated using anti-mathematics*.

This is all very exciting, if only we knew what anti-mathematics was. We will find out later, but for the moment, we simply infer that anti-mathematics requires an orthogonal way of thinking to mathematical thinking. It is not *less* precise, but rather *more* precise. we must take a phrase from the ancient alchemists and arrive at a brand of knowledge where we can rightfully exclaim:

"Truth! Certainty! That in which there is no doubt!"

...

If the reader is not keen on mathematics, maybe trying anti-mathematics might be a more rewarding experience. As such it might be worthwhile reading on further to find out. Perhaps mathematics is suited to the crystalline thinkers and anti-mathematics is right up the alley for the global fluid thinkers. Perhaps if you are left-brain dominant and not too interested in the deep foundations of things, you should stick to mathematics. If you have a propensity to be right brain dominant, fascinated with the deeper picture, and not averse to shedding a tear or two, then a career path in anti-mathematics might be the way to go.

Kant and Anti-Mathematics

For the moment, we will provide two ways of understanding the difference between a mathematical formalism and an anti-mathematical formalism. Our first approach would be familiar to students of Western philosophy. It was pioneered by Kant in the *Critique of Pure Reason*. The second approach uses the logic of arrows and provides a tractable, potentially formal, and hopefully easily understandable way of leaping over the Kantian minefield.

First of all, consider Kant's way of defining the problem. Knowledge can be articulated in terms of two different kinds of propositions: analytic judgments and synthetic judgments. Analytic judgments are tautological type propositions where the predicate is already logically implied in the subject. In synthetic judgments, the predicate is wholly distinct from the subject. An example of an analytic proposition is that "A implies A." An example of a synthetic proposition is "A implies B," where A and B are distinct.

Kant then went on to argue that his "pure knowledge," what we are calling unconditional knowledge, would be devoid of analytic propositions. In other words, propositions of the form "A implies A" are not allowed. In the pure knowledge arena, the proposition "A implies A" is not admissible. He is in fact declaring his accord with Heraclitus: "You cannot put your toe in the same river twice." It is at this point that Kant drags fundamental reasoning back into its proper roots, the realm of pure melancholy, the zone of epistemological despair. Kant is putting the angst back into metaphysics.

Kant then went on to argue that pure knowledge was uniquely articulated in terms of *synthetic* judgments. Unlike in empirical knowledge, these synthetic judgments would all necessarily have the property of being *non-negatable truths*. On so doing, Kant was opening the door towards the exciting world of anti-mathematics. From a Kantian perspective, anti-mathematics must somehow be made up of synthetic judgments that are non-negatable truths.

...

Our first exposure to anti-mathematics will not be at the fundamental level, but simply based on a comparison with the traditional way of thinking in mathematics. Practically anything that can be said about traditional mathematics becomes totally negated and usually doubly negated in anti-mathematics. However, there is no simple duality or symmetry involved, unless one can see symmetry between chalk and cheese. There are no free lunches here.

Our initial objective is to provide enough of a glimpse that the whole project starts to gain at least some semblance of credibility. To begin with, we start with a basic understanding of the nature of mathematics. Mathematics is a very diverse discipline. Mathematics is like a soup. You can add things to it or take things out, and it is still a soup. However, there are three basic ingredients in mathematics that you cannot take out and still have mathematics. If you remove these three basic ingredients, you destroy mathematics and start to raise the spectre of either rational oblivion or another way of thinking, anti-mathematics. The three essentials of mathematics consist of readymade mechanisms for establishing:

- Distinguishability,
- Identity,
- Equivalence

of entities. Without these mechanisms, mathematics cannot function.

Distinguishability

The first essential involves creating a world of *choice*. There are two questions here. Firstly, how does mathematics arrive at a world of multiplicity in which choice is possible? Secondly, if you somehow arrive at a world populated by a multiplicity of objects that you can potentially choose from, how can you *differentiate* one entity from another in order to make the choice? It is of little use if we have a world of fruit and you cannot differentiate between a banana and a lemon, or even pick a rotten banana from a good one. If you are a fruit monger, you may also want to be able to count the bananas.

Clearly, these questions are mind-boggling. They require a sophisticated cosmology, ontology, a possible theology, together with some kind of theory of cognition and consciousness in order to be answered satisfactorily. This is not the approach taken by mathematics. Mathematics can easily resolve all of this by simple decree. The decree is in the form of a single axiom, which makes the assumption that, firstly, there already exist candidates of choice and secondly, that it is always possible to distinguish one candidate of choice from another. The axiom concerned is called the *Axiom of Choice*. Without this axiom, or something equivalent to it, mathematics would be impossible.

Sidenote:

In Category Theory, a Topos is endowed with sufficient a priori structure so that the Axiom of Choice can be constructed from the Topos axioms. The Axiom of Choice is thus not required in Topos Theory. You still need Topos axioms though.

The Axiom of Choice is a bit tricky to understand in the way it is usually stated. However, what it really states is simply the assumption that any mathematical formulation starts with a big bag of mathematical objects and that these objects can all be labelled. As for the labels, there is a lemma that says that a bag full of real numbers will always be enough to do the job. Hence, labelling merely becomes a monotonous process of numbering things; not that mathematicians ever carry out such menial tasks. All that really interests the mathematician is the proven lemma that it can be done.

Sidenote:

The lemma is called Zorn's Lemma and is logically equivalent to the Axiom of Choice. The author's first contact with Zorn's Lemma was as a post-graduate many years ago in a series of lectures given by a recently appointed Professor in Mathematics at Monash University. The lectures started with the professor giving an off the cuff, innovative proof that Zorn's Lemma was equivalent to the Axiom of Choice. Presentation of the proof continued up to half way into the third lecture when somehow, the thread of the argument was lost and the proof had to be abandoned. The overall performance left an indelible impression.

Now, we can make a very interesting observation about things in the world of mathematics. Assuming that the Axiom of Choice is operative, any two entities that are different are also distinguishable. Thanks to Zorn's Lemma, they *are* distinguishable because they can be given different, hence distinguishable labels. The importance of this observation may not be apparent here, but is crucial in the overall plan. It is so important, that we will highlight it here by stating it as a fundamental principle underlying mathematics.

The Differentiation Principle

In formal mathematics based on the Axiom of Choice, two entities that are different are also objectively distinguishable.

In other words, if two mathematical entities are different, you can prove that they are different and so, as a consequence, they are objectively distinguishable. In brief, mathematics assumes that difference and distinguishability is the same thing.

The upside of the Axiom of Choice is that it addresses complex problems in a simple way, that is to say, by sweeping them under the carpet. The downside is that mathematics can never provide a theory of distinguishing, let alone provide deeper theoretical insights. The Axiom of Choice trivialises the profoundly fundamental.

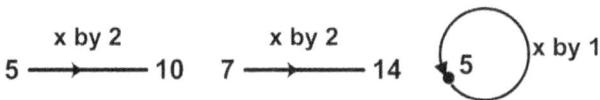

Figure 17 Arithmetic multiplication

It goes without saying that the Axiom of Choice should be anathema to anti-mathematics. No sweeping under the carpet with anti-mathematics. The upside is that the door is left open to addressing and even answering fundamental cosmological, ontological, cognitive, and even theological questions.

The downside is that this does not make life easy for the budding anti-mathematician: Providing answers to such questions doesn't look that easy.

add 2
5 ———→ 7   **add6**
9 ———→ 15   (5) **add0**

Figure 18 Addition also needs an identity arrow. The number zero is required.

Identity

Mathematics deals with manipulating mathematical objects by mathematical operations of some kind. Consider arithmetic, for example, and the mathematical operation of multiplication. Multiplying two numbers together is a mathematical operation. Consider the case of multiplying a number by the number 2. Multiplying by a number can be represented by an arrow, as shown in Figure 18, which shows the cases of the "multiply by 2" arrow being applied to the number 5 and to the number 7.

Clearly, these arrows can be generalised to handle multiplication by any number. Of particular interest is the "multiply by 1" arrow. This is an example of an identity operation. Mathematicians argue that in order for arithmetic multiplication to be acceptable as a complete mathematical system, it must possess an identity.

The requirement that any mathematics must possess an identity operation strongly suggests that if we are to formalise our anti-mathematics, then it should be based on the *negation* of the identity. Anti-mathematics must reject the identity operation. There is no identity operation in anti-mathematics. We note in passing that the identity notion comes into conflict with the Heraclitus dictum that everything changes. You cannot put your foot in the same river twice. For Heraclitus, if you have already been there once, there is no coming back. You can have at most only one twenty first birthday. You only have one shot. The next time is always different, never the same and never, ever, ever identical. Was Heraclitus one of the first anti-mathematicians?

Another common mathematical operation is that of arithmetic addition, adding two numbers together. Most people are familiar with this operation. The operation can also be represented by arrows as shown in Figure (b). In this case, the identity becomes the "add by zero" arrow.

Anti-mathematics must take the opposite position to that of mathematics. Whereas mathematics declares that it needs the number zero to exist in order for its arithmetical system to be complete, anti-mathematics cannot enjoy such a luxury. For the zero to be admissible in mathematics and hence be free of any taints of the *a priori*, it must be capable of being constructed. Constructing zero would appear not to be a simple matter. Perhaps we can do without zeros. This might seem worrying, but it did not seem to bother the ancients. In

Greek and Roman times, the number zero had not even been yet invented, or if it had, people in those times seemed to be quite wary of it. After all, is zero an allowable measure in reality? The lack of zeros also would have pleased Heraclitus.

Equivalence

Mathematicians of our time have the same mental outlook as Democritus, the laughing philosopher. One reason for their glee is that they know how to take plane trips. They even know how to make life easier by taking short cuts. For example, consider taking two plane trips. The first trip is from airport A to airport B. Then the second trip is from airport B to airport to airport C. Now if a single flight existed directly from A to C, then undoubtedly you would have taken it, as this would have made life easier. In mathematics going from A to C is equivalent to the combined AB and BC trips and can be represented by the triangle of arrows in Figure 19(a). This diagram expresses the most elementary structure in mathematics. It is called associativity. In order to get non-trivial algebraic structure mathematicians need associativity, somewhere, somehow, no matter what.

Associativity

A simple example of associativity is arithmetic addition. Figure 19(b) shows an associativity diagram with an "add 3" arrow applied to a number object with the value 4. Not surprisingly, this gives the number 7. Applying an "add 2" arrow to that number gives the final number 9. Now if arithmetic addition is associative then there should exists an "add something" arrow going directly from the number 4 object to the number 9 object. Happily, an "add 5" arrow exists in the addition arithmetic and so neatly accomplishes the task. We are all laughing.

At this point, we should spare a thought for the anti-mathematician. The anti-mathematician has the same mental outlook as Heraclitus, the weeping philosopher. It is not hard to see where the weeping comes from. Anti-mathematics must reject *all* constructs that depend on any entity that is *a priori* to it. Consider the difficulty facing the anti-mathematician that wants to fly from A to C with a stopover at B. In order to fly from B to C he must have already flown from A to B. The BC flight is impossible unless preceded by the AB flight. However, this violates the anti-mathematics edict that no entity should have its existential status dictated by something that is *a priori* to it. Flight BC cannot exist without the preceding, hence *a priori,* flight AB. Sorry; the anti-mathematician can never take flight BC. He suffers the typical dilemma of all anti-mathematicians and for that matter, any Kantian metaphysician worth his salt: He is stuck in the one place.

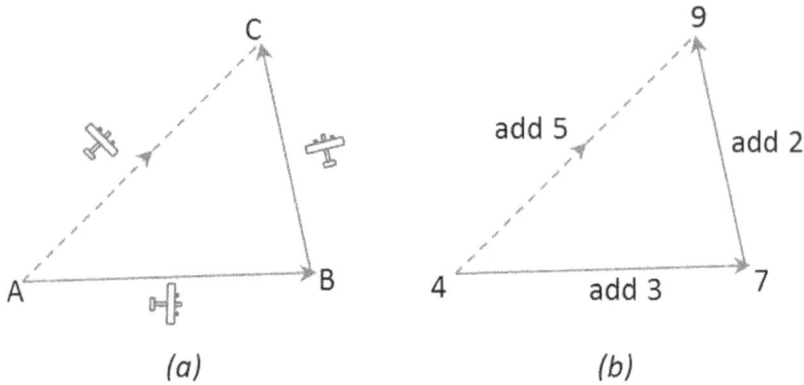

Figure 19 The most elementary and essential structure in mathematics illustrated by the composition of two arrows to give a third. Mathematics is impossible without this kind of structure. Anti-mathematics does not allow any of this.

This concatenation of the AB and BC flights to define the equivalent shortcut flight AC is an example of what the mathematicians call *composition*. Armed with compositions and the associativity assumption, mathematicians can add up, multiply, and engage in all sorts of symbol manipulation feats. It is enough to calculate your way to the moon and back.

Sidenote:

> It is easy for the reader to loose traction at this point, as we delve into the apparently absurd. Proposing a reality where even a translation in space from one point to another seems impossible appears ridiculous. However, truth is stranger than fiction. As we demonstrate in the Appendix, the superb graphics on your fast moving computer game was probably developed using a geometry where nothing ever moves from the origin and there is no absolute scale. This (anti-)mathematics is based on Conformal Geometric Algebra where what appears as something moving from A to B in two dimensional space is actually a rotation about a generic origin in a special 5 dimensional space. Things may look bleak at this stage, but there is light at the end of the tunnel.

This is all happy days for mathematics. The anti-mathematician has none of this free baggage. All concatenations, all compositions, are all out of reach. He cannot even add up. There are no shortcuts either; it is impossible to have a shortcut if there is nothing to shortcut to begin with. This means that any thought of associativity property is also well and truly out the window. He is certainly stuck in the one place. This might start one thinking that the identity construct, like that mentioned above might be permissible: at least the anti-mathematician could amuse himself by multiplying things by one. However, even that is not allowed. The identity assumption that A=A is something that no anti-mathematician worth his salt could ever contemplate, and that goes right back to Heraclitus. All of this is enough to make a grown man cry.

In brief, things look decidedly bleak for anti-mathematics. Then comes the final straw that could really break the camel's back. Can anti-mathematics even have *arrows*? Are arrows also out the window? The author can state with absolute conviction that if arrows are also out the window then we can kiss anti-mathematics goodbye, together with any chance of answering the Kantian question, and so construct the new science that we have been talking about. Nevertheless, the arrow as an admissible component of anti-mathematics looks an extremely wobbly proposition. Just look at it. An arrow has two ends: one end at the beginning of the arrow and one end at the other end. The beginning end comes first, *before* the pointed end. How can we possibly say that one end definitely comes first and the other second? Remember, we are aspiring anti-mathematicians and must have a science devoid of any *a priori* assumptions whatsoever. The simple arrow is a formal representation of everything that we wish to avoid. There it is, taunting and mocking us with one end *before* the other: Eat your heart out! This is exactly what we do *not* want. This little arrow is our only hope!

Anti-Formal Mathematics and Intuitionism

Formal mathematics is situated in the left front side of the Kantian square of knowledge. It is the quiescence of poly-analytic science. As such, formal mathematics is absolutely non-synthetic and hence non-constructionist. It wasn't until early in the twentieth century, that the mathematician L. E. J. Brouwer and his student Arend Heyting developed an alternative approach to formal mathematics. It was called *intuitionist* mathematics and was character-ised by its rejection of non-constructionist proof. The starting point for any constructive demonstration was from simple intuitive concepts, which were then constructed into concepts that are more complex. With its attendant psychologism, the approach could never attain the rigorous status of axiomatic mathematics. In fact, intuitionism never entertained this ambition; instead, it presented itself as an *alternative* to formalisation, as an *anti-formal* methodology.

In this work, we take a different approach. Instead of constructionist mathematics being conceived as intuitive, informal mathematics, we must develop a constructionist, formal system that we provisionally call *anti-mathematics*.

Life on the Other Side

Anti-mathematics takes us right to the brink of epistemological oblivion. It is now time to claw our way back. It looks as if we are stuck with the arrow as a representational tool for our new science; otherwise, the situation is just too hopeless. On the face of it, the very fact that the arrow has a start and an end point seems to violate the very principle which we want to construct, a principle free of any such ordering of one thing at the expense of another. We have to find a way around this impasse. A good way of finding the right approach to take is to look at how traditional scientific and mathematical thinking tackles things. In other words, we should look at the left side approach to what we should do, and do exactly the opposite. In this respect, we should look at how left side, traditional scientific thinking treats the humble arrow.

When confronted with a thing it wishes to describe, left side thinking will start by putting labels all over it. In the sciences, a similar approach is used by putting probes all over it and even into it. Mathematics uses labels; the empirical sciences use measured attributes. According to left side reasoning, the labels are arbitrary and devoid of any intrinsic meaning. Thus, choosing which label to use is a piece of cake. Presented with a humble arrow, the start of the arrow can be labelled A and the pointed end B. This leaves the shaft of the arrow, which can be labelled f. Why not? This labelled arrow, believe it or not, just about characterises most of the dominant aspects of left side thinking. To begin with there is the arbitrary labelling, a sure sign of left side methodology. Then we can ask what actually lies at the end of arrows. The answer is simple: there are two objects at the end of an arrow. These are the two things labelled A and B. It is important to note that the thing labelled A and the thing labelled B are different and distinguishable, not only from each other but from everything else. "How can you distinguish between these two objects?" you might ask. Once again, there is a simple answer. You can tell the difference between them by their different labels. Remember, one is labelled A and the other B. Alternatively, *using* the arrow could possibly tell the difference. This would work as long as there is only one arrow labelled f.

Another characteristic of the left side reasoned world is that it adheres to the atomism paradigm. Atomism goes hand in hand with its sister paradigm, dualism. Atomism and dualism go together. The dualism of traditional mathematics is evident in mathematical Category Theory. Category Theory is the most abstract form of mathematics and, in fact, can be considered as an abstract theory of mathematics itself. Central to Category Theory is the arrow theoretic method that we have tried to informally exploit here. The fundamental dualism of mathematics can clearly and formally be seen in Category Theory in the way its treats its arrows. Category Theory, this abstraction of mathematics itself, expresses the fundamental dualism of mathematics right at its very foundation. Category Theory is defined from two distinct collections of entities: a collection of atomic objects and a collection of arrows between these objects. This is the abstract way that Category Theory formalises the fundamental dualism pervading all of mathematics, the dualism between things and relations between things.

It all sounds excruciatingly boring as we go about dealing with things everyone takes for granted. Mathematics says that its world is made up of atomic things on one side and a heap of atom manipulating bits and pieces, arrows and so forth, on the other side. Doesn't that seem perfectly reasonable? Also, is it not reasonable to be able to plan airline trips with and without stopovers? Is it not reasonable to be able to walk from A to B? It all sounds reasonable. However, imagine the situation where all of the things we take for granted are no longer valid when we voyage to another land. Perhaps we will be like the tourist. The tourist does not understand how these natives can possibly get by without even having a dishwashing machine. The tour guide then points out that this is not a problem because in this neck of the woods they do not have dishes.

Some people travel well, others with great trepidation. Just imagine for a moment having to travel to a world that did not even have labels. Imagine communicating when the inhabitants of this world speak a language where there is no distinction between noun and verb. Try mulling over the fact that, in this land, you actually cannot walk from A to B. You can't walk anywhere. You can't walk across the land as we do in the world of Common Sense. Instead, anyone that wants to progress here has no choice but to take their world *with* them. Then it becomes a question of who is leading whom. Yes, this is the non-dualist world of *Monism*, the most mysterious and least understood of all the lands. It is the land on the right side. The poets love it.

Making Anti-Mathematics Possible

Mathematics articulates a comforting world. It is a world where the home you left in the morning is identical to the place you will come back to in the evening. It is a world where you can even walk to work, one step after the other. You can be confident that you will be able to distinguish your left foot from your right and even each step from the next. All of this activity can be conveniently described in the parlance of mathematics. This is the realm of conditional knowledge, empirical reality, the shallow reality. Our aim is to come to terms with Deep Reality.

However, enough of the musing, back to business. We are not poets, but aspiring anti-mathematicians. Crack the anti-mathematicians problem and you crack the Kantian problem: the floodgates open to a new unifying science and a very exciting way of understanding reality, the only way of *understanding* reality. Crack this science, crack its algebra and you crack the God Code. This is serious stuff.

The elementary building block for anti-mathematics is the arrow. However, we have immediately run up against an obstacle; the arrow has two ends to it, one that comes before the other. We have seen that the arrow in traditional mathematics has an object at each end of the arrow. This does not worry mathematics, as it has no ambition to solve the Kantian problem. There, it does not matter if one object is impaled on an arrow held aloft by another. Mathematics has no fear of the *a priori*. In mathematics, everything is even predefined *a priori* in the axioms. However, for anti-mathematics there can be no *a priori*.

Once the problem is defined thus, the next step is quite simple, even though the final solution is far from simple. If anterior and posterior objects are a problem, *get rid of the objects*. This means that all we are left with a bunch of arrow having a very peculiar characteristic: the two ends of the arrow are part and parcel of the one thing. Rather than two things at the ends of an arrow, the arrow is one thing with two sides to it. This is a *monistic* arrow. The ancient Pythagoreans called it the *indeterminate dyad*.

The indeterminate nature of this arrow becomes apparent once one starts to try to nail it down. One way of looking at the monistic arrow is to attempt to type the two ends. This requires two different types, one for each end. Now the only typing system that does not rely on experience and measured attributes is the one we have already mentioned in previous sections. This involves typing by *gender*, where gender is defined in a very precise, ontological manner. The entity that is totally *devoid* of specificity is an entity of feminine gender. Such an entity has no determined attribute. Nevertheless, it does have *an* attribute, albeit an undetermined one, the attribute of being totally devoid of specificity – an extremely rare condition. In this non-dualist, monistic world, this attribute of total non-specificity must be considered to be an entity in its own right on an equal footing with the feminine entity. This attribute entity is said to be of masculine gender. Thus, the feminine *has* an attribute and the masculine *is* the attribute. Understanding this remarkable formula is the first real step along the path to understanding right side science and its anti-mathematics foundation.

These two entities that only differ by simple gender are different because they differ in gender. However, they are totally indistinguishable. Distinguishability requires a difference of attributes. In our case, we have two entities and only one attribute between them. They are thus quite irrefutably indistinguishable.

It is with the two types that we can ontologically type the ends of our arrow. Being rash, and possibly secretly wishing to favour the author's own biological gender disposition, we adopt the convention that the beginning end of the arrow is of masculine gender, with the orientation placing the feminine in second position. We can now sit back and say that we have constructed a thing-in-itself that is totally free of any entity being ontologically, or epistemologically or in any other way *a priori* to any other. Of course, there will be those people that are never satisfied who will argue that we have violated the Kantian *a priori* condition by placing the masculine before the feminine on the Monistic Arrow. We could become very smug. We could grant this point, but then ask the question, "would anyone ever notice?" No one is likely to come along and ever experimentally refute our choice of gender supremacy. No one can ever show the masculine does not in fact predicate the feminine and that it is really the other way around. If the system is flawed, but no one can ever detect the flaw, then maybe that is good enough.

Despite our smugness and secret wish for ontological gender dominance, the final arbiter of this question must be Pythagoras himself. How would Pythagoras react to the idea that his indeterminate dyad has been replaced with a shiny, uni-directional arrow with some clever relativistic typing based on gender? Through the shroud of history, you can see him tapping his fingers on his desk. He does not seem too impressed. Thinking about the matter for a while, one can see why. The system cannot be flawed. The only way out requires an approach much like that in Quantum Mechanics, an approach based on *superposition*. This quickly leads in to a can of worms that we discuss elsewhere, but the final result is that the gender of each end of the arrow must be indeterminate. In the final analysis, we end up once again, not with *one* solution but *four*. Instead of our single shiny MF typed arrow we end up with MF, FM, FF and MM typed arrows, the four letters of the Generic code, the Genetic code, or the God code, whatever you want to call it.

The indeterminate dyad of Pythagoras is the superposition of these four arrow types. We think Pythagoras might be happy with that. As for the dictum that "everything is number," that will need some clarification and a little more anti-mathematics than we have developed so far.

The Anti-Mathematic Triad

Anti-mathematics is the right side alternative to traditional mathematics, which is a left side science. Anti-mathematics cannot enjoy all of the things that left side science takes for granted. As illustrated in Figure 20(a), identity is not allowed, the composition of two things to produce a third is not allowed and even concatenation is not allowed. However, the arrow structure itself remains a possibility for this right side science, provided the two ends of the arrow make up the same inseparable entity. This calls for a radically different kind of typing mechanism from that found in the left side sciences. All difference must be articulated in terms of ontological gender. Gender is the only typing mechanism that is compatible with Kant's condition for non-conditional knowledge.

For the moment, we dispense with gender and just concentrate on what kinds of arrow structures are compatible with the draconian requirements of right side science. Since any concatenation of arrows violates the Kantian condition, the only arrow structure permissible is one where the arrows are either head-to-head or tail-to-tail. In this way, no arrow can be said to be *a priori* or *posterior* to any other, thus satisfying the Kantian condition.

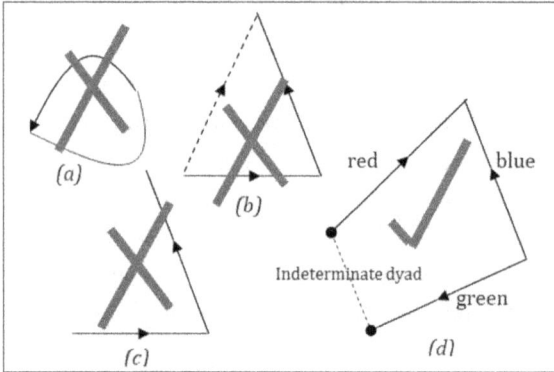

Figure 20 Mathematical structures violate the unconditional Kantian condition. (a) identity is not allowed (b) the composition of two things to produce a third is not allowed (c) even concatenation is not allowed. (d) Only head to head, tail to tail structure is allowed. The triad is the only such structure that uniquely identifies the individual arrows without using labels.

The System Labels Itself

Another implication of the Kantian condition is that, unlike all traditional science and mathematics, there are no readymade labels that can arbitrarily be stuck onto things. Thus, without labels, how is it possible to distinguish one arrow from another? Fortunately, we have avoided going down Bertrand Russell's atomism road. If you have a Bertrand Russell atomism, you have to a have a strictly nominalist Bertrand Russell labelling system. Right side science does not work with atoms, but only with things that have *extent*. In right side mathematics, these elementary bodies take form in the guise of arrows. Thus, the temptation to label things can be averted. With arrows, we can *construct* our own labelling system. The system is capable of labelling itself.

Remarkably, the self-labelling problem has a unique and very simple solution. Labels are used to distinguish one entity from another. Thus, we cannot build a self-labelling system using only one arrow, as there is no point. Next, try a system of two arrows where the arrows are head to head. Self-labelling will not work here, as there is no way of objectively distinguishing one arrow from the other. The same applies to the other possibility of two tail-to-tail arrows.

We then come to a *triad* of arrows and this is where the answer lies. As can be seen in Figure 20(d), each one of the three arrows is structurally distinct from the other two. There is no ambiguity whatsoever. Even though any explicit ordering of the arrows is not allowed, an *implicit* ordering raises its head. There is a natural first, middle, and last arrow. We have named these positions as red, blue, and green respectively. Note that the structure determines the "colour." These "colours" are not prefabricated labels. They are *fabricated*.

Continuing on to the case of four arrows, there we find ambiguity creeping in. There are two possible ways of projecting our constructed triadic labelling scheme onto such a structure. There would be three coloured arrows plus another. It is impossible to eliminate the ambiguity concerning what makes up the coloured triad and which arrow is left over as the other.

A similar argument applies for a structure of five arrows. It is only when we come to six arrows and any other multiple of the basic triad, that ambiguity is eliminated. From this discussion, we see that the generic form of arrow structures is in multiples of RGB triads. This provides the generic system of placeholders. Such a structure is composed of untyped arrows. The only typing system compatible with the Kantian condition is that based on ontological gender, already discussed.

At this point, it should be noted that we seem to be talking about two things in the one breath. We are talking about *generic structure* and we are also talking about a *generic code* for describing this structure. We are also talking about non-dualist, monism. The structure and its coding go hand in hand as two different sides of the one thing.

So far, the coding appears as a sequence of RGB untyped triads. The first step is to replace the RGB terminology with generic gender based terminology. The first triad in any biological genetic coding sequence is always the start codon defined by the AUG triad. One could interpret this operation as setting the reference frame, the common ground, for the triads that follow. The start codon types the common ground for all the arrow triadic codons that follow. Each arrow triad following the start codon will be gender typed. In the biological genetic code, each codon can be made up of any combination of three bases from the four letters A, U, G, and C.

According to our interpretation of right side science, these four "arbitrary" letters hide an underlying gender typing mechanism. The correspondence is that A, U, G and C correspond to the binary gender typing MF, FF,

FM and MM respectively. The generic code hides an underlying gender code, the God code.

8

Stoic Structures

Why then do you strut before us as if you had swallowed a spit?
My wish has always been that those who meet me should admire me,
and those who follow me should exclaim, 'Oh, the great philosopher.'
(Epictetus, 55-135 AD)

The most fundamental epistemological opposition is that between the impersonal subject and the kingdom determined relative to it. This primary opposition constitutes the left-right epistemological divide of science can be understood in many different ways. Consider the following story of a band of misfits on a sheep station in outback Australia. It seems that they too are engaged in the search for higher knowledge.

First Classness and the Shearing Shed

The shearing shed was alive with curiosity and excitement. Times had been tough and so the crew had given up the wool trade and had gone off trying their hand in markets more in tune with the twenty-first century. Everyone was there, the Ringer, the Jackeroo, the Tracker, the Roustabout, the Sheila who did the books, and of course the CEO.

The CEO gave a short speech introducing the speaker and the overall direction of the company. The company was to go into the business of explaining reality. Apparently, the question of what's real and what is not was of growing concern among the general populace. The seminar speaker was to be Jackeroo, who had been doing some background research on the problem.

Jackeroo started off awkwardly, "The problem is to find the underlying principle governing the universe."

"What if there isn't any underlying principle?" asked Ringer dryly. "If there were no principle then the world would be in total chaos" answered

Jackeroo. "To me, the world always looks to be in total chaos anyway," piped up Roustabout.

Jackeroo was beginning to get mired in technicalities before he had even started. In desperation to cover all angles, Jackeroo blurted out, "The universe is organised on a principle which is not a principle," and then added as a consolatory but desultory explanation, "It's a principle, which negates itself." A baffled look of glazed eyes flashed across faces in the audience.

"Well that quietened every one down," commented Ringer. "A principle that isn't a principle but negates itself, eh? Well, good luck."

Jackeroo took a deep breath and, reading from rough notes, launched into his monologue. "We cannot make much headway without starting to come to grips with a very fundamental concept. Borrowing from Computer Science terminology, we call it *First Classness*. Glimpses of this concept can also be discerned in String Theory in the guise of some kind of 'democratic principle.' Of particular interest will be the notion of First Class systems and those systems, which are not First Class, that is to say, systems based on Second Classness. There is no notion of Third or Fourth Classness. First Classness introduces a strictly binary notion; you either make it through the pearly gates of heaven or you don't.

One system that doesn't make it through the pearly gates is formal mathematics. Mathematics is fundamentally wallowing in hardwired, incurable Second Classness. This is due to the fact that the only candidates for being First Class entities in mathematics are the axioms. Everything else in any mathematical formalism is qualified and predicated by axioms and hence is Second Class. This includes even the entities defined in the axioms, as also any theorem, which can be deduced from them. This absolutist, undemocratic structure banishes all those entities that are dominated by the emperor axioms to stagnate in a static, dead world of Second Classness. Mathematics is not based on First Classness. Mathematics is a Second Class system."

At this point the CEO interrupted, "Well that sounds all fine and noble but where's the business opportunity?" Jackeroo, starting to get excited, and exclaimed, "There it is! Clearly, mathematicians have been flooding the formalisation market with Second Class systems for years. Surely then, there must be some people out there that would snap at the chance to take possession of a totally pure First Class Formalisation System. When offered the choice between a Second Class banana and a First Class banana, which one would you take? Cursory market research will show that most people will choose the First Class banana over the Second Class, even if only because it just sounds better."

This small team of former farmhands was fast transforming themselves into a team of entrepreneurial metaphysicians. They decided that there was a market for this First Class product. But, it is here that they met a snag. There was a market, but they didn't have a product. Presently the market was being flooded by products based on Second Classness, notably mathematics and the mathematical sciences. Mathematics is fundamentally riddled through and through with Second Classness. What they needed to put on the market was a formalisation system based uniquely on First Classness. They needed something that was entirely the opposite to mathematics, something that didn't rely on *a priori* assumptions like axioms and data and so forth. The needed something that could be built from reason alone, something like what Kant was talking about.

The CEO suddenly rose to his feet, almost delirious with excitement. "And so what we need is…" he yells, but doesn't have time to finish the sentence. His voice is drowned out by an immense shout from the floor. "We need anti-mathematics!" everyone shouts in unison. And so it came to pass that the case for anti-mathematics was proved: by general acclaim. The shearing shed would never be the same.

The CEO was all fired up by the idea of launching his First Classness supercharged anti-mathematics onto the world stage. As the excitement died down, the CEO turned to his Ringer, who was his acting CTO. He asked in a whisper, "What exactly is First Classness?" The Ringer shrugged his shoulders, admitting that he had no idea, but maybe the Rouseabout might know, as he seemed to know a bit about everything.

In the weeks and months that followed, the CEO asked many wise and learned people the same question. Each time he got the same negative response. The only remotely promising response was from an aging computer scientist who said that First Classness was Good. His eyes glazed over and he then entered into an explanation that was totally incomprehensible to the CEO.

Finally, in desperation, the CEO decided to pose the question to a mysterious Oracle who happened to be passing through town at that time. The Oracle replied enigmatically, "You will find your answer by taking on the complexion of the dead."

The CEO was rather shaken by this, but after some reflection, he decided that this meant he had to read about the ideas of dead people. He started off by reading about the ideas of *very* dead people. In fact, he started reading about the ideas of Zeno of Citium, born in 334 BC. Coincidentally, it appears that Zeno also had a similar experience with his Oracle. Zeno, of course, was the founder of Stoicism.

Hellenistic Tennis

After the life of Socrates, Hellenic philosophy started a process of splitting into two poles. The early signs of the process were already becoming apparent with the differences between Aristotle, and Plato his teacher. By the time it came down to the philosophies founded by Zeno of Citium and by Epicurus, the separation was complete. The aim of philosophy in those days was to resolve the central problem of man, notably how to achieve happiness. Unlike any of the world religions that came later, both philosophies addressed how to achieve happiness, not in the afterlife, but now in the present. Philosophy became the art of living happily. Both philosophies agreed on the aim, but believed that the means to achieving this aim was located on different sides of the tennis court. Let the game begin.

On the left side of the tennis court are the Epicureans, inspired by the ancient philosopher Democritus. On the right side of the court are their arch enemies the Stoics, inspired by the ancient philosopher Heraclitus. It's a familiar sight then, with the merry making atomists on one side and the brooding holistic thinkers on the other. In the middle, sitting in the umpire's seat, are the Sceptics. The Sceptics, in their attempt to be absolutely objective, have suspended judgment and sit with their backs to the game. Despite a verbal hand grenade being tossed over the net from time to time, play is slow. The object of the game is the pursuit of happiness.

Epicurean Tennis

The Epicureans have set up a dinner table on their side of the court and are enjoying themselves with pleasant chit chat, pleasant drink, and pleasant food. Epicureans love bathing in pleasantness. All their friends are pleasant people who all behave pleasantly at all times. For them happiness is to enjoy oneself pleasantly. Happiness is synonymous with pleasure. Pleasure, however, does not mean unrestrained hedonism, as the excesses involved inevitably leads to unhappiness, which is contradictory to the basic intent. As Epicurus himself remarks, "It is not an unbroken succession of drinking-bouts and of revelry, not sexual lust, not the enjoyment of the fish and other delicacies of a luxurious table, which produce a pleasant life."

The tension between the Epicureans and the Stoics on the other side is intense. However, despite the deep rivalry, the two schools share a lot in common. Both sides are dogmatic materialists in belief. Both sides are also in agreement that the fundamental aim in life is to live happily. Furthermore, they both unify and justify their doctrines by turning to the science and structure of nature and reality itself. However, it is at this point that they part company. The Epicureans are atomists, whilst the Stoics are monists.

Epicurus was a great cosmological pastry cook. His strictly materialist creation was a recipe for responding to any question under the sun. The adherent, armed with such a world view, is thus free to lead a life unencumbered by doubt or fear arising from the metaphysical. The task was to be accomplished without recourse to the heavy hand of necessity, so popular in other brands of philosophy. His was to be a world of the *laissez faire*, where even the gods went about their daily business without interfering with human affairs.

A perennial problem for materialists is how to allow a world that admits of beings, which somehow behave in a way contrary to the absolute mechanical determinism of matter in motion. How can you have free will in such a world? Epicurus came up with an innovative response, something that could be very useful on a tennis court moreover. He invented the *Swerve*. All bodies consisted of matter made up of atoms. The space in between atoms was filled with the void. Atoms moved about and interacted with each other in a very deterministic manner, except now and then there was an exception to the rule. An atom would spontaneously make a tiny imperceptible swerve from its deterministic trajectory. This explains how the universe gradually microswerved to its present state and also the spontaneity of movement in animals and man.

It is interesting to note that Darwin's theory of evolution introduces the Swerve into the reproductive process of living organisms. Each child organism may differ slightly from its parent or parents, explained by a swerve arising from the latent indeterminacy involved in genetic coding, which in its turn arises from combinatorial variation and accidental mutation. Some swerves are successful and the organism lives on to reproduce. The unsuccessful swerves lead to failure of the organism to propagate. Evolution thus becomes the sum total of the successful swerves.

Some writers of recent times working under the banner of Atheism want to push this process further back to a time when the only matter that was, was dead matter. They postulate that somehow dead matter experienced swerves that lead it to leap the bridge from the dead to the living, from the inanimate to the animate. This is all part of the declared war with the stalwarts of Creationist Theory. The Creationists need God to create the world. Like Epicurus, the new Atheism only needs the Swerve.

Swerve theories have taken different forms across the ages to express that allusive difference between strictly mechanical deterministic behaviour and the observed spontaneity of the animate. One non-materialist approach proposed by Bergson postulates an *elan vital*, an underlying "current of creative energy operating on matter directed to the production of free acts." And so, the

Epicurean Swerve becomes powered by an *elan vital*. However, as Julian Huxley dryly remarks, the *elan vital* is about as illuminating as describing a locomotive as being powered by an *elan locomotif*.

Epicurus' cosmology starts off with a universe of atoms all moving vertically downwards in straight lines. The idea of the predominance of an absolute vertical up and down axis in the Cosmos might seem curious, but is easier to grapple with if one considers that the world may have been flatter in Epicurus' neighbourhood. His Swerve was necessary to explain how the predominately vertical state of affairs could possibly end up in the complex structured world around us. The world became the way it is by trillions upon trillions of micro-swerving atoms. In addition, the Swerve was to be the genesis for explaining non-mechanistic animal and human behaviour. Nowadays modern science has replaced the indeterminacy immanent in the Epicurean Swerve with the fundamental uncertainty that reigns in Quantum Physics. This is summed up in Heisenberg's Uncertainty Principle, a fundamental tenet of Quantum Theory. In the Uncertainty Principle, we find the most fundamental expression of the Epicurean theory of the Swerve.

Stoic Tennis

While the Epicureans quietly party on the left side of the court, the Sceptics find their particular brand of happiness in their customary fashion by always sitting on the fence. In that way, they experience the comforting satisfactory glow of never going down the wrong path, which is their way to a particular kind of happiness,

On the right side of the court, we find an entirely different ambiance. Over on the far corner, a Stoic called Leon has been captured by the enemy and is being tortured on the rack. The Torturer, a tattooed, seedy looking creature, leers down at Leon and taunts, "I bet you're not feeling so good now." "Perfectly good thank you," replies Leon, "Quite happy." "Happy?" exclaims the Torturer. "How can you be happy being tortured on the rack?" "I am always happy as this must have been meant to be. Things might appear to be going badly for me but that is only how it appears when, in reality, things are going perfectly well. Things couldn't be better, in fact." The Torturer was a bit taken aback and countered by boasting, "You know, I can take your life on this rack." "Yes, you can take my life," declared Leon, "but you can't take my soul. If taking my life profits you then take it." This was too much for the Torturer. He gave an almighty twist to the rack and watched to see how the Stoic reacted to real pain. Sweat broke out on the Leon's brow as he quietly muttered between his teeth, "My friend…" "I'm not your friend, I'm your torturer!" came back the snarl. "I know," said the Stoic "but Dion, that person standing right behind you is my friend." The Torturer spun around to come face to face with an Athenian soldier in the process of pulling out his sword. He gave a blood curdling scream and ran off.

Dion cut Leon loose from the rack, rubbed down his poor twisted limbs, and the two of them rambled off. "It was lucky that I just chanced to be passing by," commented Dion. "That was not chance," replied Leon, "it was fated." Leon was starting to clear his head and muttered, "What appears as chance is caused but beyond our comprehension."

They kept walking until they came to the home of Chrysippus where they stood, hesitating at the open front door. They could see Chrysippus in the kitchen inside, warming himself in front of the stove. Chrysippus beckoned to them: "Come in; don't be afraid: there are gods even here." As they walked inside Chrysippus laughed out loud, "I've always wanted the chance to say that. Those were not my words but those of the ancient Heraclitus."

They sat down at the table and Chrysippus served up a plate of dried figs, his favourite. They started talking and Dion was curious to know how the Stoics related to the gods. Chrysippus explained that men were on the same levels as the gods. There was no friend behind the scenes. Zeus was a friend to men as men were a friend to Zeus. Chrysippus went on to explain the universe and how it was governed.

Dion, who had always been curious about Stoicism, asked Chrysippus a question that had been bothering him for ages. "Chrysippus, my dear friend," asked Dion, "what is virtue?"

Chrysippus paused and said that virtue was the cornerstone of Stoic philosophy and demanded careful explanation. He drew in a breath and started his small lecture on the subject.

"From Parmenides we learn the only real truth is founded in that which exists in the eternal present. Nothing else exists, neither in the past nor in the future. Existence is limited to the pure Oneness of the present. Everything might appear to change, but that is only in appearance. In reality, nothing changes. That is the truth. The ultimate knowledge is knowledge of this truth, according to Parmenides. Heraclitus taught that such knowledge could only be understood in terms of pairs of opposites. He described the oneness of the world as ever-living fire saying:

This world, which is the same for all, no one of gods or men has made.
But it always was and will be: an ever-living fire, with measures of it
kindling, and measures going out.

The ever-living side of the world is Nature; the fire side of the world is Zeus, the only immortal of the gods. Zeus and Nature are two sides of the one reality. They are both expressions of the masculine and the feminine encompassed in the Gender Principle. The Gender Principle expresses the opposition between the singularity of subject – the masculine – and the expansiveness of what accompanies subject – Nature, the feminine.

Oppositions even have oppositions. In opposition to the Gender Principle is the opposition of the Active Principle and the Passive Principle. The Active Principle and the Passive Principle are the *personal* expressions of the masculine and the feminine principles. The masculine and the feminine principles are the *impersonal* expressions of the Active and Passive Principles. This is how everything can be expressed in terms of oppositions. Gender is impersonal Active-Passive and the Active-Passive is personal Gender. As every Stoic knows, these two oppositions explain the four letters."

Dion interrupted, "Chrysippus, are these the four virtues?"

"Not directly. The four virtues are the singular form of the four letters. The non-singular form of the letters leads to the four elements. These are Fire as masculine-active, Air as masculine-passive, Earth as feminine-passive and Water as feminine-active. To understand virtue, you have to understand the elements. The four elements, starting with Fire, provide Nature with the power and propensity to diversify from the fiery One to the Many. The Elements provide the possibility of body having parts. Harmony requires a balancing influence that acts in the opposite direction to the tendency to multiply and diversify. To balance the One dividing into the Many, there must be harmonisation so that any of the many Ones remain at one with the original One. The particular separates from the singular because of its particularity. The particularities must reunite with the pure singularity, to maintain a common world.

The pure singularity, the pure One has no particularity except its singularity. The One simply *is*. The One *is* singularity, nothing more specific than that. However, the particularity of the particular One, as distinct from the pure One, is due to its own peculiar configuration and composition of the elements.

This movement from the One to the many Ones through diversity of the Elements must harmonise with its opposite, an attractive influence of the One and the many Ones to remain at one with each other. This attractive influence is called *virtue*. There are four virtues and, unlike the elements, they are inseparable. This forms the foundation of our physics and our ethics"

Dion was concentrating intensely. The One having no specificity but that of being, intrigued him. He still wasn't clear on what virtue really was, however. "Chrysippus, from what you are saying the virtues are the four forces of Physics that maintain everything in order."

Chrysippus thought for a moment and made a doctrinal correction. "The virtues are material bodies like everything else. Forces act through bodies interacting with bodies. One cannot talk about forces without talking about bodies."

After another pause, Chrysippus continued, "A lot of my students have problems understanding the difference between bodies composed of the elements and bodies composed of the four virtues. Compound bodies made up of the elements are easier to understand than the body made up of virtue. Of course the virtues can be thought of as the four elements in a singular configuration. As such, the virtues are the four elements as the common generic ground of any being whatever, including the unique One, which is Zeus. They are the elements as *generic ground*. This is why a Stoic sees Virtue in whatever he contemplates. Virtue forms the generic ground. The elements, considered as distinct from the virtue ground, are not restricted to be in the same singular configuration. They play the role of *figure on ground*, and are unlimited in diversity of combination. The Stoic sees the diversity harmonised by common virtue, a commonality that extends to all scales, including Zeus."

Chrysippus then went on to explain that living in accordance with nature meant living in accordance with virtue. The individual shares the present with Nature. In this way, one could lead a happy life. There was no reason to fear the past as the past no longer exists. One can also be fearless of the future, for the same reason that the future does not exist. There is no one hiding in the past, hiding in the future, or anyone in the shadows pulling the strings, and so there is no fear coming from any of these sources. In fact, there is no reason to fear anything.

Nature exists as a single body and is made up of individual bodies, free of voids. Nature is not only regulated by the virtues in the sense that it *has* virtues, but it is also the case that Nature is nothing other than pure virtue itself. It *is* virtue. In this way Nature can demonstrate what virtue is by its own example. Nature can thus be said to "live virtuously," and it does so without compromise.

Zeno took these intuitively understood virtues, these pivotal, fundamental processes of physics, and applied them to individual man as ethics for human behaviour. He thus founded Stoicism on the mantra that man should live in accordance with Nature. Man should aspire to live like Nature. Man should aspire to live virtuously.

Chrysippus then went on to explain each of the virtues in an analogous manner to each of the Elements. The virtues are interdependent and can't exist independently. The violation of one virtue is tantamount to a violation of all virtue. Where one virtue is, all must be. "Many people have problems in understanding this principle," he remarked.

Chrysippus continued, explaining that Zeno also pointed out the existence of the opposites of the virtues. These are vices, which are themselves, bodies and exist internal to a being. There is no vice or evil exterior to a being, only interior. Vice is relative to an individual being. Vice is only in the private possession of that individual. The vices are created affects due to the individual attempting to run counter to virtue. There is no evil or vice *out there*. To react or succumb to some perceived exterior evil or vice, that is the *real* vice. Moreover, that real vice is well and truly *interior*."

Finally, Dion asked Chrysippus to explain the core concept that distinguished the Stoics from the Epicureans. Chrysippus replied that obviously the Stoics taught the system as a whole from a monistic perspective whereas the Epicurean view was atomistic, piecemeal, lacking cohesion and any unifying principles. But in the final analysis it came down to how to handle concepts. The Epicureans were content to label things; once labelled, the thing could be put away as known. The Epicureans were extreme nominalists. The Stoic way of reasoning goes back to Heraclitus, and involves expressing all concepts in terms of oppositions. To know something, one must know its opposite.

For the Epicureans everything has a label; for the Stoics, everything has an opposite. What's more, the opposite is on equal terms, on the existential level at least. The opposition between the individual and God is one of difference, not of hierarchy. Both sides of the oppositions have equal value. They need each other.

In Physics, an entity has a property. This involves an opposition between entity and attribute. The attribute must be admitted as an entity *in its own right*. The Epicureans wouldn't admit this as they only recognise atoms and their attributes. For the Epicureans, attributes are not atoms.

For the Stoics, there is no second class tier of existence. All entities enjoy first class status. This goes right across the board. That's why slave and the emperor, man and woman, woman and child are all seen by the Stoics as having souls and having rights as human beings.

This does not necessarily mean that free men, slaves, women, and children should all have equal *civil* rights. That is a political matter, not one that can be settled by objective moral reasoning. However, contemplate the situation where the slaves, the women, the children are all considered as mere attributes of some person. You are not a being, you are merely owned by a being. Imagine a slave or woman considered merely as a contribution to the owner's grandeur, nothing more. It's an idea, which continually raises its head, no matter what the époque. Imagine, you are not even a being but a property of a being. The Stoic position is that the slaves, the women, the child, are not merely the attributes of someone; they are beings in their own right.

First Classness, virtue, it is this simple but profound concept that holds the whole system together. In the final analysis, that is how nature expresses virtue, a system based on justice where there is equality in front of the law. However, justice, even First Class justice, does not infer a fair and level playing field. Justice does not change the lot of an individual.

Suddenly Chrysippus cut short the discussion. He jumped up, remarking that he had to go as he was expected to attend a social function. Chrysippus thanked them for coming and picked up what was left of the figs. "You could feed them to the donkey," Dion joked, "And maybe you could give the donkey a glass of wine to boot," added Leon. Chrysippus chuckled, "I would die laughing if I did that!" and then left, heading for the party, chuckling all the way.

. . .

The CEO woke up with a start. He had been reading and dreaming Hellenic philosophy. He was beginning to understand First Classness.

The Epicurean and Stoic Paradigms

The history of philosophy is constantly punctuated with battles between two practically orthogonal ways of thinking. A case of this philosophical dichotomy that was particularly well thought and well fought was that between the Epicureans and the Stoics. This ancient joust of ideas is quite pertinent today. Epicureanism, with its atomism, dualism, and extreme nominalism, can be taken as a roughhewn template of the thinking of the modern sciences. Charles Sanders Peirce remarked on this opposition between the Epicureans and the Stoics and noted "Epicureanism was a doctrine extremely like that of John Stuart Mill." In the twentieth century, English philosophers like Bertrand Russell and Peter Frederick Strawson took up the relay.

As for the Stoics, Peirce was scornfully dismissal. However, the author takes a different view and claims that the Stoic paradigm provides a template for a new science capable of unifying all of the disparate present day sciences. This is the way the Stoics themselves saw their doctrine, as a unifying holistic view unifying all fields of thought from ethics, linguistics, and physics through to logic.

Now it's hard to find a philosophy more diametrically opposed to Epicureanism than Stoicism. Both philosophies might claim to be materialist and to admit a fundamental role for determinism. Both share the view that the ultimate goal of philosophy is to explain how to live happily. However, in the details and every other aspect, they are diametrically opposed. To one commentator (Kleve, 1983) the differences run so deep that the Epicureans, compared to the Socratic-oriented side of humanity such as the Stoics, appear to be "different human types."

What fascinates the author is that it appears possible that these "two different human types" can even be medically induced. This may sound grotesque, but by disabling one hemisphere of the brain, the resulting personality and competences of the viable hemisphere will either resemble a cocky tunnel vision, Epicurean for the left side, albeit rather intellectually diminished. Alternatively, disabling the left hemisphere can lead to the globally aware but rather awkward and compromised Stoic. Our schematisations here are rather extreme as no left dominant or right dominant personality type can render the other hemisphere inoperative, but merely subservient. Nevertheless, there remains a grain of truth in this extreme point of view.

The disabling of a hemisphere may be the result of a stroke, but can nowadays be medically induced at will. Psychiatrist Iain McGilchrist, with twenty years of experience ranging from stroke victims to Schizophrenic patients, examines in depth this paradigmatic specialisation of the hemispheres in his book. He does not make the link to the Epicurean Stoic divide, but does provide an in-depth picture of the two radically different paradigms at play in the divided brain. He makes the claim that a similar paradigmatic divide has coursed through history of Western civilisation. Modern times have seen the increased dominance in human affairs of the left side hemisphere, the side that we ascribe to the Epicurean style paradigm.

McGilchrist suggests that knowledge of the specialisations of the divided brain can throw new insights into the philosophical arena and *vice versa*.

The different ontological status of the two hemispheres impinges on the meaning of all the philosophical terms that are used by us to understand the world, since both hemispheres think they understand them, but do so in different ways. If there *is* this correspondence between the biological divided brain and the historic philosophical divide then it brings in a new kind of critique: these two philosophical paradigms are opened up to observational and experimental study in the clinic. McGilchrist describes many of what have become rather stereotypical differences between the hemispheres. However, he also emphasises one of the most dramatic difference between the hemisphere paradigms. Unlike the right side, the left side literally is conscious of only one half of its world, a phenomenon that the psychiatrists call hemi-neglect. It seems that the left hemisphere, the "Epicurean" side, is making do with a cut down view of the world, a half world even.

The Epicurean paradigm is not only observable in the biological brain, but also in the political brain where the left hemisphere dominant political stalwart tends to advocate right wing politics and *vice versa*. In the political arena, the Epicureans are often thought of as apolitical. Perhaps a more succinct characterisation is that they are implacably against government interference and regulation in all its forms. A modern Epicurean would prefer market forces to the political. There would also be a visceral opposition to the Supreme Court taking an innovative and creative interpretation of the Constitution. The Constitution is a set of rules that should not be changed except under duress. The Epicurean paradigm is rule based, not system based. This paradigm is implacably for small government. R. W. Sharples remarks, "Epicureanism does seem to have about it something of the closed world of the religious sect." In the limit, they can become like the current Tea Party in the US and its look-alike phenomenon in Australia. The paradigm has broken lose and is out of control. It challenges the very legitimacy of the other hemisphere. It wants to shut down the right hemisphere completely. Who needs the right hemisphere?

One key feature of the Epicurean type paradigm is its denial of the existence of any dynamic regulatory system in its sphere of consciousness. It replaces the regulatory with a surrogate system in the form of a set of rules. For the Epicurean type paradigm, to every systemic action there are no widespread systemic repercussions: there is only the rule that if you kick a rock it will hurt your toe.

On the biological brain front, McGilchrist remarks on the left hemisphere's tendency to confabulate when confronted with its own ignorance and then add the fable as a valid rule although it may not be true. The off the cuff confabulation quickly becomes an integral part of the left side paradigm's theory of its world.

The Epicurean paradigm dominates present day sciences. This is particularly the case for mathematics. Every branch of formal mathematics is totally determined by its own set of axioms. Modern mathematics is totally rule driven. Epistemologically speaking, every branch of mathematics forms its own little Epicurean rule-doting sect. This is a serious matter.

By understanding the Epicurean paradigm, its strong and weak points, and particularly what it lacks, one can arrive at some understanding of the other paradigm that stands so diametrically opposed to it. However, this is not a war zone, no matter what some on the left hemisphere side might think. An essential component of the right side paradigm is a deep understanding of the role and importance of the left side paradigm.

Unlike the Epicurean like left side paradigm, the right side does not rely on static rules for the basis of its reasoning. The surrogate rule based system is replaced by the real thing, a dynamic reasoning system interacting and aware of a changing world. This comes at considerable cost. The overheads are high. In a slowly changing world, a snapshot rule system may be sufficient and free of the onerous overheads. This is the strategy of the left side paradigm. The problem the left side encounters is how to obtain updated rule systems. Healthy practice is to borrow from the right side. But there can be a tendency to go it alone and not even change the rule set. The other possibility is empirically to generate the rules set.

However, there may be more than pragmatic, performance considerations at play. The holistic right side view of reality might provide awareness of the whole situation, but this is ground, not figure. For many situations, figure may be more pressing than ground, provided it operates within the guidelines of the Master hemisphere, the right hemisphere.

A brief summary of the Stoicism paradigm must take into account the central position occupied by their physics. Now, many people sympathetic to Stoicism often add in the disclaimer that, of course Stoic physics was all wrong and nothing to do with the science of our day. Such a comment is curious, considering that the central tenet of Stoicism is that the only rational basis for ethics is that supplied by physics, ethics being a key aspect of the Stoic doctrine. If their physics is hopelessly awry, then their ethics and even their very reasoning must be in the same boat.

Naïve Modern Day Stoics

One recent writer, Lawrence C. Becker, makes this very apology. His literary strategy is to pose as the descendant of the Stoics. Faced with the unchallenged ascendancy of modern science, this latter day Stoic is forced to make doctrinal accommodations in order to fit in with the prevailing scientific wisdom. He writes:

> Modern science presented significant challenges to our metaphysical views, and during the seventeenth and eighteenth centuries we gradually abandoned our doctrine that the universe should be understood as a purposive, rational being...Natural science no longer gives grand teleological explanations. Thus, we cannot plausibly propose to "follow" nature, as the ancient motto had it (Becker, 1998).

In this view, the science of the ancient Stoics was naïve and erroneous. Modern science has come up with all the answers. The universe is not based on any life principle but, as "natural science" will have it, is just the ashy aftermath of a great dirty explosion. The end result is that, despite all of Becker's creative attempts to revamp a new Stoicism compatible with current science, he necessarily ends sliding into the Epicurean paradigm, which is diametrically opposed to his original intention. Current science is atomistic, dualist and a philosophy espoused by Bertrand Russell, Peter F. Strawson, and so many others of modern times; the same position espoused by the ancient Epicureans.

As Becker well knows, the *"doctrine that the universe should be understood as a purposive, rational being"* is central to Stoicism. Throw it into the dustbin of history and Stoicism goes with it. Becker, or any of the moderns for that matter, has not even attempted a case against the ancient scientific doctrine. The only argument seems to be is that if it is ancient then it must be inferior to the science of us moderns. Why, in those days, they did not even have internet shopping and motor cars! An alternative approach, and the one promoted here, is to resurrect the ancient science of antiquity and in particular the Stoic version.

The Stoics conceived a world free of anything or anybody behind the scene pulling the strings. There can be no outside organiser. This world is totally unconstrained. There are no pre-ordained rules. The only principle constraining the world is that it *be* totally unconstrained. Thus, the central principle is of a rational kind. The system is constrained by its own Logos. This principle applies to the Cosmos, and equally to any being of the Cosmos.

It is important to note a key subtlety here that distinguishes the Stoic mode of thinking from that of Aristotle. The principle doesn't apply to Everything, as Everything is not a thing. It applies to *Anything*. The Stoics never proposed a TOE (Theory of Everything). Instead, we argue that they proposed a TOA (Theory of Anything). This subtlety rests on the difference between thinking abstractly and thinking generically. The Stoics were not abstract thinkers like most modern Western thinkers. They were not generalists; they were generic thinkers. As for Aristotle, he always seems to have had a bet each way.

Stoicism and First Classness

There is a modern way of understanding the central tenet of Stoicism. The concept is called First Classness (FC), a term borrowed from Computer Science. FC is a fiendishly difficult concept to formalise and even more difficult to make iron-like. Many examples of FC abound in Computer Science but won't be mentioned here. Remember though, that asking your local guru on the matter will only result in a partial answer. The only thing we will say about FC is that it leads to the common perception that developing systems that do not violate FC is considered Good. Whenever a software developer violates the principle in his work, he knows deep down that he is committing a sin which one day will come back to haunt him. Thus, in the field of software development, the FC concept provides a moral compass to software engineers. It is perhaps, the only moral compass.

Returning to the Stoics, a basic expression of FC in Stoic physics is that the *property* of an entity must be considered *an entity in its own right*. If this were not the case, an entity would be ontologically more important than its property, or perhaps *vice versa*. FC demands a dictator-free zone: no one definitively calls the tune.

Now Becker argues that, in the centuries following the Renaissance, the advancement of science put Stoicism on the defensive. This is certainly the case in the emergence of classical physics. Classical physics violates the Stoic FC requirement concerning entities and their properties. Classical physics opted for Second Classness constituted of two modes of being in the form of particles and force fields between particles. The force fields were not particles in their own right but a kind of something else. Stoic physics was dead in the water. Physics had become an Epicurean atomistic dualism.

Then, in the twentieth century came the rise of Quantum Mechanics. The tide was turned somewhat. Everything became particles. An individual particle appears to behave in a not completely deterministic manner. However, from an alternative point of view the particle appears to be a wave as explained in quantum field theory. There is no duality present, except the two points of view, the wave, and the particle view of the same thing.

As an example of First Classness, the forces between particles can be explained by an exchange of other particles called gauge bosons, particles in their own right. The movement back towards the original Stoic FC doctrine had started. However, Quantum Mechanics, although incredibly successful, is still only an empirical "suck it and see" science. Its findings cannot be derived from first principles. The source of any such principles of course can only be from the dictates of FC, a slippery beast but not an impossible one.

FC is a principle and has no explicit structure. However, a certain kind of structure is necessary in order not to violate FC. Some of the ancients knew this going right back to Empedocles. Empedocles called them the four Roots. Others have called them the Four Letters. The Stoics saw it as the theory of the Four Elements. In this book, we argue that the Four Letters can be interpreted as a unifying science that includes the four-lettered genetic code as a special case.

That any entity in reality can be coded by a four letter, generic code has stupendous implications. That every cell of an organism must contain a copy of the same code for the whole organism is incredible enough. What is even more stupendous is that this genetic code should be more generic than just for the animate, applying right across the board, applying to *anything*. None of this would have surprised the Stoics. This was also the way that they saw the world. Present day traditional sciences might be devoid of any "grand teleological explanations" as Becker states; however, things will be different for the resurgent new unifying science based on the ancient Stoic doctrine.

The naivety of the theory of the four Elements lies more in the naivety of the eyes of the modern beholder than in the ancient theory. The naivety of modern science lies in its belief that the only source of scientific knowledge is empirical measurement. This naturally leads to a dualistic way of seeing the world: the world is populated by material entities, which possess sensual properties; the sensual properties are not material and are not even entities. This is a valid and useful way of conceiving the world. It is particularly useful as it is free of any costly overheads that may come with a more elaborate conceptual schema. All that really matters are the data. However, it is *not* the only way, and certainly not the most fundamental way of conceiving the world.

The empirical view of the world violates FC because of its dualism between entity and property. Ignoring attempts by Quantum Mechanics to buck the trend, modern science is based on Second Classness.

This acceptance of a Second Classness view of the world can be traced back to Aristotle. Aristotle interpreted the four Elements of antiquity in a way that would be somewhat similar to the moderns. The four elements were the elementary constituents of matter. Matter was a simple mixture of these four elements.

Aristotle retained some of the ancient form of reasoning in terms of oppositions, but otherwise his approach resembled the moderns. For him each of the Elements possessed particular, plain, and ordinary properties. Thus, according to Aristotle:

- Water is primarily cold and secondarily wet.
- Earth is primarily dry and secondarily cold.
- Air is primarily wet and secondarily hot.
- Fire is primarily hot and secondarily dry.

The Elements were material whilst their properties were not. Aristotle's system was definitely incompatible with FC. The science of the moderns replaces the four Elements with a much more extensive list with more elaborate properties. However, the underling acceptance of the Second Classness remains the same as for Aristotle. In this sense, modern science can trace its lineage right back to Aristotle.

Stoics Reject Aristotle's Version of the Four Elements

This is *not* the case with the Stoics. They rejected Aristotle's perspective and returned to the pre-Socratic paradigm. They refined it and made it their own. Unlike Aristotle, the Stoic embraced a worldview, which did not violate FC.

First Classness cannot be built out of empirical attributes. Attributes in a system satisfying FC must be arrived at, not by measurement or by the senses, but via reason. The age-old solution to this problem, across many cultures and civilisations has been in the form of gender. Entities in a First Class reality are not typed by perceived qualities: they are gendered. Gender is not a perceivable quality. To understand FC, a deep, fundamental understanding of ontological gender is required. Ontologically gendered entities provide the building blocks for any autonomously existing being, any living being.

A detailed, advanced explanation of ontological gender can be found in a later section of this work. It is a prodigiously deep and fecund construct: there is nothing more fecund than the interplay of the genders. For the moment, it suffices to know, as mentioned previously, that a purely feminine entity would be one that *has* the attribute of being absolutely devoid of specificity; the purely masculine entity *is* that attribute. Two entities and only one attribute in play: one *has* it, the other *is* it.

Which comes first, the masculine or the feminine? The four Elements provide the answer in terms of FC. Nothing and nobody comes first or can come first in a system respecting FC. There are no winners. For that matter, there are also no losers. The first step of solving this riddle is to replace Aristotle's definitions of the four Elements by a system based on gender, as follows:

- Air is primarily M and secondarily F.
- Earth is primarily F and secondarily F.
- Water is primarily F and secondarily M.
- Fire is primarily M and secondarily M.

This might sound obtuse; however, one should remember that this is not *abstract* thinking. In fact, those with even an elementary knowledge of computer languages should find the concept familiar. The following sidenote explains one link and is well worth reading for anyone wanting to understand gender and delve into the unconscious misogynous behaviour of early computer scientists.

Sidenote:

The first scientific programming language, developed in the 1950's was called FORTRAN. A fundamental aspect of all programming languages is the typing of the objects of programmation. What is of interest to us is typing by gender. Is there any sign of a gender typing in the original old FORTRAN? The answer is in the affirmative, but only one gender is apparent, a rather de-sexed and emasculated masculine gender. For example, the programmer wants to write a formula that uses the variables speed, time and distance. These variables will have human readable names that ultimately refer to places in the computer memory from where the variable values can be stored and read. The attributes of these memory locations can be thought of as the values of the variables. These are the only entities that FORTRAN allows direct program access to: these are the entities of "masculine" gender. The language does not allow programs directly to manage the memory of the computer. The language does not allow program control of where and how the "feminine" gendered placeholders are managed. This meant that the FORTRAN language was infertile: It is impossible to write a native FORTRAN compiler in the language FORTRAN. For a long time the only way to write language compilers was to use a lower level programming language close to machine language. Eventually came programming languages like C and C++. These languages combined high level FORTRAN type computational competence with high level constructs capable of directly managing the computer resource itself.

The typing technique was astoundingly simple. Firstly, we ignore the usual traditional typing for numbers and characters, these are the hard coded types and, from a gender perspective, not fundamental. They essentially reek of Second Classness. What is interesting from a gender perspective is a radically different type called **void**. In principle, an entity with the type **void** can be a value of any kind whatsoever. This is a non-trivial case of the implementation of a "masculine" value type. In addition to the **void** type, there is a second type which has no name and no direct implementation but is written, in combination with other types, as simply a ★ wildcard symbol. An entity with this type is not a value, but a pointer. An entity of this type is not a value but can potentially have a value by pointing to it. This provides a concrete illustrative example of a feminine entity. The C and C++ languages implement a limited form of compound gender typing. For example the mixed gender MF type would be written as **void★** and is commonly used. It corresponds to a pointer to a value of unknown type. The type **void★★★** corresponds to a pointer to a pointer to a pointer to a value of unknown type. Armed with this apparatus, it is possible for a C or C++ compiler to be written in itself, and thus reproduce native versions of itself, something that ancient FORTRAN could never do.

Typing in present day computer languages is a long way away from a true gender typing system, one that the Stoics would claim could implement a living organism, big or small. The Stoics were very sketchy on the details of such a scheme, but they had a firm understanding of many of the key concepts. Thus, what are some of the broad characteristics of this gender based cosmic algebra based on the four Elements. All of these characteristics can be explained in terms of the overarching FC requirement. This has led the Stoics to make some claims that appear to be quite outlandish.

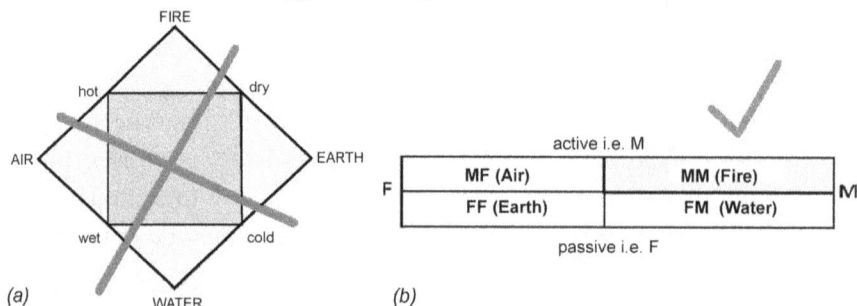

Figure 21 (a) Aristotle's four material Elements had properties which were immaterial and hence violate First Classness. (b) The Stoic system must be based on First Classness where anything must be body, including attributes.

Outlandish Claims not so Outlandish

The first outlandish claim was that not only are all entities bodies, but their *properties* are also corporeal bodies. Our discussion above indicates that such a claim is quite reasonable and even necessary if FC is not to be violated. This notion start to become much more formalised when we realise that the very algebra of the system is built from entities, where the entity and attribute are indistinguishably intertwined, as spelt out in the algebra of gender.

A second seemingly outlandish claim of the Stoics is that *actions* are also corporeal. Some commentators, and even some Stoics, have taken this too literally mean that the very act of sitting on a chair is a corporeal material body. The mind boggles. It is becoming very clear that a science based on FC, as demanded by the Stoics, will be radically different from present day sciences that bathe in Second Classness. It demands a different way of conceiving reality and an equally different way of describing it.

In order not to violate FC, we saw that the *property* of a material body had to be a material body in its own right. The fundamental mechanism for achieving this FC requirement was by gender typing. This leads to the four Roots of Empedocles, each Root being typed by one of the four binary combinations of mixed gender. From a linguistic perspective, each Letter can be considered as combining the entity and property, noun and adjective, in the same body. Each Root articulates one way to resolve the dichotomy between entity and its attribute, noun and its adjective. There are four ways of resolving the dichotomy. This elementary structure in itself is not sufficient to achieve FC; however, the structure is absolutely necessary. There is no way of bypassing the Four. The only root to it all is the Four Root system of antiquity, the necessary but not sufficient structure that respects FC.

This may seem all very strange to the reader, perhaps even incomprehensible. One way of understanding it, rather than trying to construct the universe parting from the four Elements, is to think in terms of constructing a language to describe and explain the universe. To be more precise, we want a language that can describe a whole living entity, *any* living entity, any entity that must rely on itself to maintain the integrity of self. No one is behind the scenes pulling the strings. The responsibility of being in existence is the ultimate responsibility of the being itself. According to the Stoics, the universe is one example of such an entity. Us humans and all other life forms are also a part of the mix. The underlying organisational principle for such entities is FC. We claim that this is also the underlying unifying paradigm of Stoic thought. The Stoics were FC thinkers.

From modern biochemical research, we now know that there actually exists a common, universal organisational language applicable to all life forms. Moreover the language is based on an alphabet of four letters. This is the genetic code. We have already staked the claim that the genetic code is a *generic* code and hides an underlying gender based structure. We made the association between the four letters A, U, G, and C of the genetic code and the gender coding MF, FF, FM, and MM respectively.

From a left side science perspective, the genetic code seems to be little more than a transcription language mapping sequences of letters in the DNA to amino acid building blocks of protein. Not all sequences code protein; coding for some other functionalities is little understood. This is the typical left side science view of language, any language. Left side linguistics is intrinsically rule based. It models a language in terms of rules of syntax with an additional set of rules for modelling semantics, usually not much more than mapping rules for a lexicon. The approach reveals non-trivial syntactic structure for highly analytic natural languages like English. However, for the more synthetic natural languages, like the Dravidian languages of Southern India and the Basque language in Europe, the pickings are more meagre. When it comes to the most synthetic language of them all, the genetic code, the biological instance of the generic code, the approach produces only the most trivial transcription rules.

A basic tenet of the work reported here is that right side science, that missing universal unifying science, is based on a unique equally universal generic language. The genetic code is a biological instance of this language.

This language is extremely widespread. The generic-cum-genetic description of all living creatures is expressed in this language, without exception. For multi-celled animals, the code is replicated in every cell of the animal. The principle underlying this code is the requirement of FC. According to the Stoics, the living world does not stop with biological. It continues to all aspects and all scales including the universe itself, which is considered a living entity in its own right. A world split down the middle between living stuff and dead stuff creates a dualism. The principle of FC rejects *any* such dualism. The existential preoccupations of the universe are not different, in principle, from any other autonomous, self-sustaining entity. We all share the same preoccupation.

Thus, one code codes all biological life forms: that is well demonstrated by empirical science. Ancient thinkers like the Stoics declared simply that there is no restriction to the scope of language of the four Letters. This is not a problem-oriented code. The code is generic. The code is capable of coding *anything*. As we have been saying, it speaks in one Voice. According to the Stoics, only bodies exist. Any particular body is made up of the four elements. Any such body can be described, generically specified, by the four letter generic code.

Returning now to the Stoics and their mind-boggling claim that actions are bodies. In making this declaration, they had no choice: FC obliges. From the perspective of traditional left side science, this claim appears ridiculous. Left side science simply sees a world of entities, which can and do act on one another. The entities are material, but the acts are not. From a *right* side science perspective, that of the Stoics, FC demands that the acts are bodies in their own right.

A world of bodies, where the properties of bodies are bodies and the acts of bodies are themselves bodies, demands a language that speaks uniquely in terms of bodies. In such a language, the nouns can also serve as adjectives and also as verbs. An example of such a language is the genetic code. Such a language will be truly polysynthetic. Detailed discussion of the polysynthetic nature of the genetic-cum-generic code will be covered in a later section. It is by understanding this language that one starts to understand the truly revolutionary nature of right side science. The Stoics entertained an embryonic form of the science. After two millennia, it is now time to deliver.

Physics and the Moral Compass

Zeno began by asserting the existence of the real world. "What do you mean by real?" asked the Sceptic. "I mean solid and material. I mean that this table is solid matter." "And God," said the Sceptic, "and the soul? Are they solid matter?" "Perfectly solid," says Zeno; "more solid, if anything, than the table." "And virtue or justice or the Rule of Three; also solid matter?" "Of course," said Zeno; "quite solid." (Murray, 2007)

The Stoics certainly were not adverse to making apparently outlandish claims. Finally, most outlandish of all, was their claim that Virtue was a material body. Virtue is the keystone of Stoic philosophy and the basis of their ethics. To the Stoic, the purpose of life is to live according to Nature. The key qualification of nature is Virtue. Thus, to live according to nature is to live virtuously, just as Nature does.

Letter as Material Body	Material Body as Letter
Body as material entity	Letter as Noun
Body as property	Letter as Adjective
Body as Action	Letter as Verb

Figure 22 The Stoics claim that anything that exists is a material body. Properties and actions exist and so must be considered as material bodies, the same for actions, they also are bodies. Elementary bodies can be considered also as letters of a four letter alphabet. The same letter can be noun, adjective, and/or verb. Such a language would be totally poly-synthetic.

The Stoics claimed that Virtue is the key organisational fulcrum of the universe. It is the ultimate expression and arbitrage of rationality. They claimed that such an entity exists. Since only bodies exist, this Virtue-thing must be a material body. This body consists of four parts, notably the four cardinal virtues. An important dictum of the Stoics was that the four virtues formed a unified whole. It was thus impossible to waver on this principle of the unity of the virtues but the early Stoics, particularly Chrysippus, were adamant about it.

Stoic Extravagance and Virtue

Our approach in this work has been rather Stoic in more ways than one. Plutarch claimed in one of his essays that the Stoics talked more paradoxically than the poets. The Stoics certainly were involved with the paradoxical and were often criticised for their extravagance. The height of extravagance is the Stoic claim that the four cardinal Virtues are material bodies and are inseparable. The burden now falls upon the author to endeavour to throw some modern illumination on the claim. It certainly does not look like an easy task.

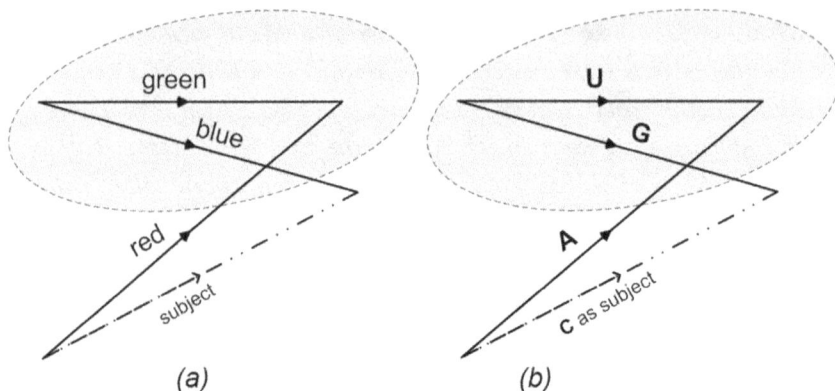

Figure 23 The generic RGB ground structure for all triads. The subject is determined by the three distinct and inseparable attributes. (b) The AUG start codon proposed as the generic ground for all codons. The start codon determines the ground for the subject. AUG also codes the amino acid methionine.

We must resist any attempt to lapse into abstraction, a decidedly non-Stoic occupation. Abstraction involves shallow, linear thought forms, left side thinking. The Stoics were right side thinkers, they thought generically, not abstractly. One way of avoiding abstraction is to employ concrete examples that illustrate the generic. When it comes to the generic, a good place to start is with the genetic code. We interpret it as being the generic code, a language that can code not only the animate but also any material body whatsoever. As for the genetic code, it has been convincingly shown to code any living biological body whatsoever. By relaxing the constraints on implementation technologies, and by adopting a broader notion of what constitutes life, it is not too extravagant to contemplate that the genetic code may be open to a much more generic interpretation than is currently the case.

The genetic code, as a language, differs from all other known languages in that its words, its morphemes, are all made up of material bodies. In fact, all the words and morphemes are made out of deoxyribonucleic acid (DNA). An intermediate material support for the language is messenger strings of ribonucleic acid (RNA). In this work, we use the RNA version of genetic coding with the letters A, U, G, and C instead of A, T, G and C. In brief, the biological world of organic material is coded in terms of organic material. Body and its language are all made of the same stuff, material stuff; this is a good example of FC in Nature.

Virtue as an Organisational Principle

According to the ancients, the four elements can produce the necessary variety of combinations to articulate the structure of the world around us. They articulate the figures of a multiplicity that is inherently unified. The theory of the four elements does not guarantee unity in a world of multiplicity. This is explained in terms of a less qualified version of the four elements, something in common, generic, no matter what element combinations there might be. Unification of the system required a common, generic ground. This generic ground, according to Zeno, is Virtue: there are four virtues and their specificity is determined by their relative position to each other. This results in the formal unity of the virtues. The semantic coherence of the whole demands the rigidity of "virtues" relative to each other.

For the Stoics, Virtue is the central organising principle of Nature. Virtue is a material body. There are four distinct Virtues and they are inseparable from each other. If we look at the structure of the genetic code, we see one structure that may be associated with the Stoic's organisational principle: it is the start codon AUG. This structure has permeated all of our previous discussions. Codons are triadic structures made up of three letters from the four-letter genetic code alphabet. In our discussion on anti-mathematics, we examined the generic, arrow structure shown in Figure 23. The demands of FC forbid the simple sequencing of elementary arrow elements. However, the RGB arrangement provides a synchronic ordering of elements whilst avoiding any diachronic ordering of arrows. Traditional left side mathematics is based on diachronic ordering and expresses the concept in not only allowing the composition of arrow, but demanding it.

The RGB triadic structure of non-sequential arrows determines the relative sequence RGB. R comes first, then G, followed by B. This is interpreted as the common generic ground for the generic code. The RGB entities are undetermined placeholders for determined types based on gender. The typing, as we have seen, is in terms of the four binary combinations of gender: notably MF, FF, FM, and MM. These binary structures provide the four "letters" for the generic code. We have associated these four structures with the letters A, U, G, and C of the genetic code.

The start codon has some unique characteristics. With the exception of the UGG codon, AUG is the only other codon in the genetic code with a one-to-one relationship between codon and amino acid. According to our interpretation, the AUG triadic structure is made up of three *synthetic* attributes. This is to say that the attributes are not empirically derived; they are derived from pure reason, as are all attributes founded on the ontological gender typing system. Any triadic combination of the A, U, G, and C "attributes" organically determines the body to which they belong. In the case of the AUG codon, the three synthetic attributes MF, FF, and FM determine MM, the subject *as* subject, the masculine *as* masculine, and the singularity as singularity. In the genetic code, they determine the base C *as subject*.

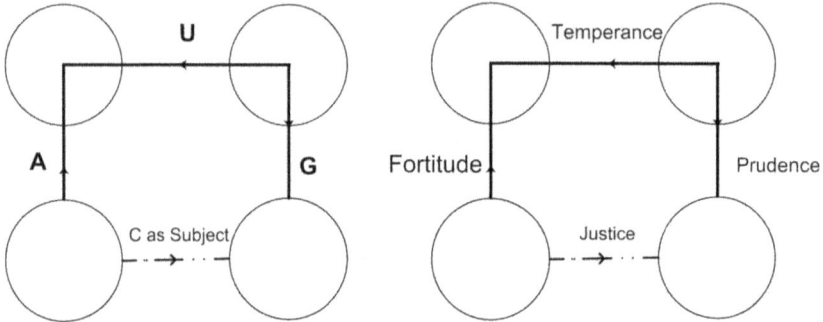

Figure 24 The Start Codon interpreted as an integral part of an organisational principle of coherence of Self, interpreted in the language of the Stoic Virtues.

What is at play here is that the generic ground RGB triad, as ground, is typed with the AUG, the MF, FF and FM bases. The Start Codon determines the generic ground for a body; it provides it with its generic frame of reference. Perhaps the Stoics would have gone one-step further to say that the AUG codon forms an integral instance of the organism's *moral* compass. As ground, the triad becomes one of fortitude, temperance, and prudence that determine an overall organisational principle based on justice.

9

Abstraction versus the Generic

A few years back the author was listening to a radio interview of a female intellectual from the Middle East. She was asked the question "What do you find is the most seductive thing about Western culture?" Her response was direct and succinct. "Abstraction," she replied, without any hesitation. Her reply stuck with the author and added to his torment on this question. Is abstraction a fundamentally Western construct? Is not abstraction the highest form of thought? If not, what is the alternative? Does abstraction have a sibling?

The author has been concerned about abstraction for many years. His approach has been to attempt to find an alternative paradigm. Back in the early eighties, the author was attending a philosophy course given by François Châtelet at Vincennes in Paris, who sadly passed away a few years later. The author volunteered to give a seminar on what he conceived at the time, to be an alternative mode of thinking to abstract thought. Instead of abstract theory, he rather naively proposed *concrete* theory, theory you could construct. The author was greatly influenced by his work at the time in Computer Science. He had started thinking that an alternative to abstract theories was a theory that one could construct with computer programs. Rather than the theory being abstractly described in static pages of a book in the library, surely we can construct theories with computer programs. Instead of reading the theory, one can actually executes it. The author gave the seminar. He though it went off rather well but to his horror, and to the amazement of the class, Professor Châtelet became extremely agitated and attempted to destroy the author's argument in the most emotional terms. After the seminar, a group of students of Middle East origin, mainly Algerian, came to the author's defence saying that they understood what he was saying and were in complete agreement. Like him, they couldn't understand why it sent Professor Châtelet off the rails.

Since then the author came across Hegel's public lecture *Who Thinks Abstractly?* (Hegel, 1966 (1808)), His speech was a real gem. Hegel remarks that it is often thought that abstraction is the affair of the educated and cultured man and that it is "presupposed in good society." Hegel observes that the community:

> *at least deep down, it has a certain respect for abstract thinking as something exalted, and it looks the other way not because it seems too lowly but because it appears too exalted, not because it seems too mean but rather too noble,*

Having played his audience one way, the rueful Hegel cuts to the chase:

> *Who thinks abstractly? The uneducated, not the educated. Good society does not think abstractly because it is too easy, because it is too lowly*
> *...*

Skirting the outrageous, Hegel must come up with proof; it is not far away:

> *The prejudice and respect for abstract thinking are so great that sensitive nostrils will begin to smell some satire or irony at this point; but since they read the morning paper they know that there is a prize to be had for satires...*

... and yes, just as in Hegel's time, a mere glance at the front page of today's tabloid provides ample testimony to Hegel's claim. Every day headlines trumpet out the most abstract of abstract catch phrases for consumption of the masses. Hegel provides an example of such high abstraction, the Murderer.

> *A murderer is led to the place of execution. For the common populace he is nothing but a murderer. Ladies perhaps remark that he is a strong, handsome, interesting man. The populace finds this remark terrible: What? A murderer handsome? How can one think so wickedly and call a murderer handsome; no doubt, you yourselves are something not much better! This is the corruption of morals that is prevalent in the upper classes, a priest may add ...*

Having taken in Hegel's little gem of wisdom we are now able to answer the question, "What does a radio shock jock and a theoretical physicist have in common?" The answer, of course is - abstraction.

However, this doesn't answer the question as to the alternative to abstraction. Our Western universities have become abstraction factories. Is there an alternative product? The purpose of this book is to present the natural sibling to abstraction, the generic. Instead of thinking abstractly, think generically. However, what is the generic? Equally, what is abstraction for that matter?

Two Fundamental Questions

Our aim is to move towards a formal knowledge of knowledge. There are two kinds of knowledge. On one side, there is what we call left side knowledge, which is dependent on *a priori* information. On the other hand, right side knowledge expounds on what can be known, without any *a priori* information. Each kind of knowledge answers a different question. Thus, two very precise questions characterise each of the two kinds of sciences. We can simplify much of philosophical and scientific tussling over different answers if we recognise that there are two different questions behind the scene. The questions are in a natural opposition and antonymic symmetry with each other.

The domain of discourse for each question is totally disjoint. The questions are so distinct that they can be imagined as being "orthogonal" to each other. The first question, suitably schematically simplified, was posed by Kant in the *Critique of Pure Reason*:

Q1.

*What knowledge can be achieved **without** reliance on any experimental evidence whatsoever?*

The answer would fall under the rubric of metaphysics. This question is familiar to all modern philosophers but is still waiting to be answered. To some, like Karl Popper, the question is summarily dismissed. The problem posed by Kant is "not only insoluble but also misconceived." (Popper, 1963) We all know from Hume that there is no such thing as certain knowledge of universal truths. Thus, the problem must be insoluble. The only possibility was knowledge gleaned from observation of singular or particular instances. The inescapable truth is clearly "that all theoretical knowledge was uncertain."

According to Popper, the problem was misconceived because Kant, even though he employed the term, was not talking about metaphysics. Kant was really talking about what he didn't mention; notably the pure natural science that had burst on the scene in his day, the science embodied in Newton's gravitational theory. Newton's theory has since been shown not to be the infallible exercise in pure reason that so impressed eighteenth century thinkers like Kant, but rather "no more than a marvellous conjecture, an astonishingly good approximation." With the passage of time, Newton comes crashing down to earth and brings Kant's question down with him. This demonstrates Kant's misconception.

Popper concludes his demolition by replacing Kant's bold question with his own languid alternative.

His question, we now know, or believe we know, should have been: "How are successful conjectures possible?"

In this book, in order to arrive at a refutation, we actually go much further than Popper by bringing in some modern arguments to prove more convincingly that Q1 is insoluble. This is accomplished by showing that it is out of bounds of all formal mathematical reasoning. To answer Q1 no axioms are allowed. Not only are operators of all sorts dispensed with – the commutative, the non-commutative, the associative, the non-associative, even operator composition is declared a no go area. Traditional mathematics simply becomes non-operational in this zone. This is the domain where nothing can be said to proceed or succeed anything else.

In the business world, there is nothing more enticing to the entrepreneur than the accepted wisdom that something simply cannot be done. The proposition becomes even more enticing when learned abstract thinkers like Popper claim to have proven that it cannot be done.

Kant's question Q1 viciously casts us into this apparently hopeless ultimate state of undetermined chaotic ignorance. However, by Popper arguing the futility of the enterprise, the question becomes so well defined that surely there must be an answer. After all, it is only when the prisoner is actually placed in the confines of the four concrete walls of his cell can he really start plotting his way out. You cannot escape until you are locked up. Kant built the prison; a Popper slam shut the door and rams home the bolt. It is time to get out of this hellhole.

Once Q1 is clearly shown to be mathematically insoluble beyond any doubt, we are then in possession of our first truth arrived at from pure reason alone. This is achieved without recourse to any experimental evidence whatsoever. We thus arrive at our first negative fact. We could call it a neg-fact. The exercise then becomes one of building metaphysics out of neg facts in some way. This is obviously not an exercise in formal mathematics but an exercise in another genus of formalisation. We call it formal anti-mathematics. This and the remarkable results flowing from anti-mathematics eventually leads to code, a kind of "DNA of the Cosmos" so to speak. This is the principle theoretical contribution of this work and clearly the most enigmatic.

We now come to the second question, diametrically opposed to the first. It reads:

Q2.

*What knowledge can be achieved with **only** reliance on experimental evidence?*

The question is very brief and needs to be expanded somewhat in order to convey the intent. What kind of knowledge can be achieved under the assumption that only what is measured is real and only what is real is that which is being measured? What knowledge can be obtained by totally excluding counter factual reasoning? Stated this way the answer to the question is probably already apparent as will be seen below. The implication is that the moon only exists if you are looking at it. What kind of knowledge can imply that?

In the first question, the only discernible real was that discerned by all-embracing pure reason – the Parmenidean real, the big picture. Q1 addresses the uppermost confines of the top down reality bucket barrel. On the other hand, this second question, Q2 imposes the opposite sense of what is real. It demands the ferociously materialist atomism and absolutist one to one nominalism that only the Epicureans ever had the audacity to contemplate to the fullest degree. To each sensation, there is something, to each something a sensation. There is nothing else. For the Epicureans, this was the way the world ticks. For modern science, it becomes a particular scientific methodological paradigm. The world ticks the way from a particular viewpoint. It is the view from the bottom up. What kind of knowledge can be achieved within the confines of such a dogmatic straight jacket?

In this case, the answer historically came before the question was ever seriously posed. The ancient answer was the physics of the Epicureans complete with their deterministic atoms moving along Bertrand Russel like causal lines but armed with an occasional, unpredictable, and at that time, indiscernible "swerve." The modern answer is in the form of Quantum Mechanics, Heisenberg's uncertainty principle, and in particular the classical Copenhagen interpretation of Quantum Mechanics.

The Epicurean ontological straightjacket implicit in Q1 limits the knowledge quest downwards to the minute, indivisible "Epicurean atoms" of reality: the elementary subatomic particles of modern physics. The only difference is that the atoms of Epicurus we assumed to have extent. Modern physics is more radical in this regard. The elementary particles have no extent whatsoever. They are assumed point-like. Such particles have nothing in their interior. They simply do not even have an interior. If there is something in the interior, your particle is not elementary. You have not reached rock bottom of the reality bucket.

The brutal minimalism of QM is succinctly expressed in Dirac's razor principle.

Dirac's razor

Quantum Mechanics can only answer questions regarding the outcome of possible experiments. Any other questions, philosophical or otherwise, lie beyond the realms of physics.

This is the declaration that QM is a philosophical desert. QM declares that it is fundamentally a philosophical, metaphysical, epistemological, ontological, theological, spiritual vacuum. This is not a weakness it is a strength. This gives it its rigour and even its vigour.

The Entanglement Problem

A situation arises in QM that there can exist minute particle systems that are non-localised. Consider the case of a pair of entangled photons produced by a photon splitting in two. Pairs of such photons can be produced in experiments. The polarisation of one entangled photon will be the opposite to that of the other. According to QM, the actual polarisation for each photon would be indeterminate until the polarisation of one of the photons was actually measured. The measurement performed on one particle would flip its polarisation to say horizontal or vertical. According to QM, the polarisation of the other photon will instantly become the opposite polarity irrespective of how far away it is.

Einstein didn't like the indeterminacy aspects of QM – "God doesn't throw dice" but it was this "spooky action at a distance" that really bothered him. In the famous EPR paper, written with Podolsky and Rosen, he argued his case. QM conflicted totally with Einstein's classical view of physical reality. According to his view, a theory must allow for the simultaneous existence of "elements of reality" which are independent of measurement. The EPR paper gave a very concise and lucid definition of elements of reality:

If, without in any way disturbing a system, we can predict with certainty (i.e., with probability equal to unity) the value of a physical quantity, then there exists an element of physical reality corresponding to this physical quantity.

The EPR paper then put forward a thought experiment that revealed a paradox in the QM theory of entangled particles. The EPR paper argued that each of the "entangled" photons would possess their own element of reality and have their polarisations determined at the time of the pair's creation, not at the time when one of them was measured. The measurement of the polarity of one wouldn't affect that of the other as its polarisation had already been determined and couldn't be altered by any "spooky action at a distance," as predicted by QM.

Basically, the EPR paper argued for what is sometimes called "local realism." The two fundamental principles are that there exist elements of physical reality or "hidden variables" and that this realism is local. The locality principle demands that theory must adhere to the principles of relativity (causes cannot propagate faster than the speed of light). Thus, the measurement on one member of an entangled pair of particles should not affect any measurement carried on the other member.

The simplified argument is that either the locality principle and with it the special theory of relativity was violated or the elementary particles harboured internal "hidden variables." In the first case, relativity theory is proved wrong. Alternatively in the second case there are aspects of reality not accounted for by QM. QM is not proved wrong but is proved "incomplete."

With the passage of time, thanks to the ingenious theorem of J. S. Bell and the experiments devised by A. Aspect et al and others, it has been demonstrated that the EPR proposed construct of local hidden variables could not possibly explain particle entanglement. This left the possibility that QM entanglement explanation would violate relativity theory. However, that is not a problem either as there is no determinate causal relationship between the particle pairs. The process cannot possibly be exploited for signalling and thus does not violate relativity theory.

Popper on Quantum Mechanics

We have used Karl Popper as a point of reference for the first of our reality barrel questions, the one stemming from Kant. He dismissed the question outright with scant regard to any possible answer. For symmetry, we should consider the other side of the reality barrel where we found an already existing answer in want of a suitable question. We provide the question but what would Popper think of the answer? The answer was in the form of Quantum Mechanics. Would Popper in fact agree that QM was a proper science? As is well known, Popper had great difficulty accepting many of the tenets of QM. For a start, QM would have to abide by his falsifiable criterion in order to be acceptable as a science. Hence, provisionally valid propositions can be deemed scientific provided that the possibility for the propositions to be proven false is still open. To Popper, all that was admissibly scientific was uniquely constructed from such potentially falsifiable propositions.

If one takes the long view of what Popper is saying here, one can easily get the impression that Popper is more concerned in fighting political dogmatism on the campus, than engaging in real science. He was more intent on arguing that what was inherently anti-dogmatic was inherently scientific. But was hard core science itself inherently hard core non-dogmatic?

This question takes on great importance when we consider Quantum Mechanics, the most fundamental of the empirical physical sciences. The difference between Quantum Mechanics and all other empirical sciences is not expressed in the details of the subject matter addressed, but in the fact that it is the only *pure* empirical science. Being purely empirical methodologically pushes its subject down to the very bottom reaches of reality. It means that Quantum Mechanics is the only empirical science that tolerates absolutely no "elements of reality" which exist independently of the actual act of measurement. In order to achieve this goal it must dig down to the bare, nude essentials of reality.

Quantum Mechanics is the only purely empirical science. To put it another way, Quantum Mechanics is *dogmatically* empirical. To put it even more bluntly, Quantum Mechanics is empiricism as an absolute dogma. This dogmatism is most clearly expressed with its Epicurean like dogma of the one to one relationship between the sensation and the real. Quantum Mechanics theory of the real is that only what is measured is real. This science, located at the very bottom of the bucket of reality, where there is nothing is deemed below, expresses itself in empirical tautologies. The measured and the real are two sides of the one thing. As such, this most reliable, accurate, and most dogmatic of the empirical sciences *is inherently unfalsifiable* at its core.

All the same, Popper stuck to his guns and had no alternative but to reject some of the essential tenets of Quantum Mechanics as being, in his terms, "unscientific." In so doing, he ignored one of the two most fundamental questions one can ask concerning knowledge of reality. In the case of Q2, the knowledge is not only true, but measurably so. After all, the Copenhagen dogma declares only that which is measured is real. What is real is only that which is measured.

This has led to a tautology, an implicit "analytic judgment." Kant would have found that fascinating. Moreover, this fundamental tautology appears not on the transcendental side of the equation but on the empirical. Even more fascinating is that this fundamental construct defines the pure empirical itself. The pure empirical is, well…, purely empirical. Such is the fundamental nature of Quantum Mechanics as declared in the Copenhagen interpretation.

Is There a Fundamental Level?

There are two takes on reality. In the empirical perspective, There are two fundamental questions Q1 and Q2 that express the fundamental opposition between two fundamental perspectives on reality. The fundamental opposition reveals itself in many ways. An important consideration concerns whether there is a fundamental level of reality.

Is there a fundamental level? Jonathan Schaffer asks the question and summarises the fundamentalist response:

The fundamentalist starts with (a) a hierarchical picture of nature as stratified into levels, adds (b) an assumption that there is a bottom level which is fundamental, and winds up, often enough, with (c) an onto-logical attitude according to which the entities of the fundamental level are primarily real, while any remaining contingent entities are at best derivative, if real at all.

He lists the physicalist, epiphenomenalist, and atomist variants on the theme. Schaffer finds plausible the hierarchical view of nature in (a) as being compatible with the discoveries of science; Schaffer homes in on (b) as the problem area, which he remarks has been almost entirely neglected. Concerning the primacy of what is real, the fate of (c) is linked to (b) as a reasonable but not inevitable conclusion.

And so, is there a fundamental, bottom of the bucket, level in Reality? In our preceding discussion of Quantum Mechanics we argued, with scarcely camouflaged glee, for a dogmatic interpretation of the science findings which would seem to place us firmly in Schaffer's camp of fundamentalists. We were advocating the bottom of the reality bucket theory. On the face of it, we supported without reservation all three tenets of the fundamentalist argument. At the risk of seeming, or even blatantly being, excessively schematic we identified the ontological approach of Quantum Mechanics as smacking of pure Epicureanism, a natural logical set of conclusions resulting from a pure unadulterated atomistic, uncompromisingly blunt materialism. In addition, there is the one to one nominalism evolving down from Democritus, a thinker not particularly notable for his subtlety and dexterity of thought, (At least Aristotle didn't seem to be very impressed, advocating at one time that Democritus's books should all be burnt.) These Epicureans, and by implication the author, certainly seem to resemble Schaffer's bottom feeding fundamental-ists.

However, the Epicureans should not be treated too harshly. After all, they were primarily engaged in a peaceful quest for happiness in this life. They had identified perhaps the greatest obstacle to leading a happy life, notably fear of the gods and the accompanying troublesome predisposition towards deep, contemplative ways of thinking. An anti-metaphysical, anti-philosophizing, theologically bland, and some would say, anti-thinking creed called Epicurean-ism was the result. Few would have predicted that this creed would one day serve as the ontological stalwart of the successful and accurate modern sciences of today. Modern physics can even mathematically describe, at least probabilistically, the dynamics of Epicure's mysterious micro-physicalist "swerve." The science is strong on empirical scientific prediction and mathematical accuracy on one side, a self-declared philosophical, ontological desert on the other. It aims to describe it all but can explain nothing.

We return to the question. Is there a bottom fundamental level? We have answered in the affirmative. In so doing we have sided with a kind of metaphysic which, as Schaffer points out, is not particularly palatable for the more reasonable and civilised of people. Painfully it appears that we have excluded ourselves from such a community. Self-declared metaphysical pariahs, we must face the dire consequences of our apparently foolhardy *prise de position*.

However, as we have argued throughout this work, there are two takes on reality, not just one. Hence, we have assented to the proposition that there is a fundamental layer. This corresponds to the left side science take on the world, the simple, rather simplistic, abstract, naïvely realistic view of the world.

The right side science take on reality has a different vocation to its uncivilised and rather uncouth partner in crime. Right side science must not merely be content with describing the qualities that a thing *has*, it must explain what a thing *is*.

From a historical perspective, we argue that the ancient exponents of the left side take on reality were the Epicureans. In our sometimes desperate attempt to gain some traction for a right side science, we have singled out the Epicurean's nemesis, the Stoics.

Of immediate concerns to our current discussion is the Stoic view on whether or not there is a fundamental level. The general view amongst the Stoics was that there was no bottom fundamental level. In some way, reality was infinitely divisible, at no matter what level. This was also a position held by Leibnitz who made pains to add the nuance of being infinitely divided rather than infinitely divisible. Our position is that, since there are two fundamental takes on reality, there will be two different answers. For the left side take, there is a fundamental layer. The right side take, because of its generic nature, provides a much more nuanced response. As discussed in more detail in the appendices, right side science is starting point invariant. The theory sates that no matter where you start, or what you start with, you will always get the same theory. However, if you start theorising about the physics of matter, the results should be compatible with those arrived at from an empirical left side perspective. Thus, there will appear to be a fundamental layer. However, if one starts with a different problematic, the structure of biological organisms for example, one will also arrive at a fundamental layer. Moreover, the form will be the same as that of the fundamental layer of particle physics. The way that this can be formalised into a formal right side science is by formalising the generic geometry and "generic chemistry" implicit in the four letter generic code of which the genetic code is an instance.

Finally, the right side answer to whether there is a fundamental layer becomes quite tortuous. No, there is not a fundamental layer, as any so-called fundamental layer is only that associated with a particular point of view. Change point of view and you change fundamental layer. However, yes, there is a fundamental layer, the generic foundation to anything that exists.

As to answering the two fundamental questions Q1 and Q2, we can claim to have dealt with Q2, but the enigmatic Kantian question Q1 remains to be answered. Nevertheless, we are starting to see what needs to be done. Rather flippantly, we can say that all we have to do is to revamp ancient Stoic physics and logic and make it scientific. This is all that is required in order to solve, once and for all, the Kantian question.

10

Formalising the First Philosophy

There are two takes on reality, two kinds of knowledge, and two kinds of science. The traditional sciences, traditional knowledge, we call left side. The left side kind of knowledge relies on *a priori* knowledge. Right side knowledge is unconditioned by any *a priori* and must be developed using pure reason alone. The two kinds of knowledge sit on opposite sides of a great divide. A deep understanding requires a deep understanding of this great divide. The ancients had their say on the matter. Consider, for example Parmenides who came after Heraclitus but before Empedocles.

Parmenides

Parmenides was one of the first to put forward a clear view of the structure of reality from a reasoned perspective. The philosophy of Being, ontology, starts with him. His poem, *On Nature,* recounts the epic voyage of the young man on his quest for knowledge.

The tone is dramatic and urgent. Despite only disparate fragments of the text remaining, one can almost hear the pounding hooves and see the sparks flying as the wise chariot steeds gallop at a furious pace through the black darkness carrying the man on the renowned Way of the Goddess. Streaking towards where Dark meets Day, maidens show the way. There is no room for dillydallying here. This is not the time for fine-spun arguments. It is time to confront the truth. The goddess, with her own hands, unerringly conducts this man who knows through things. Wheels swirling, the chariot axle glows red in the socket and gives forth the sound of a pipe as they approach the gates. The daughters of the Sun, hasting to convey him to the light, take back their veils. Having arrived at the gates of the ways of Night and Day, after much persuasion they pass through the gates. The young man is greeted kindly by the goddess. She welcomes him to her abode, far from the beaten track of mortal men.

She invites him to have an open and critical mind, to learn all things, but above all the unshaken heart of persuasive truth. There are two takes on reality. On the one side there is that taken by mortal men as they blindly stumble about in a world of opinions. On the other hand, away from the world of fickle beliefs, there is another world that harbours unshakable truth, a world where reason holds sway. At the epicentre of this world of reason is the simple truth that there are two possibilities. The goddess declares:

"The first, namely, that *It is*, and that it is impossible for anything not to be. This is the way of conviction."

She then recounts the other possibility:

"The other, namely, that *It is not*, and that something must needs not be." She then explains, "That, I tell thee, is a wholly untrustworthy path. For you cannot know what is not - that is impossible - nor utter it."

The Parmenidean Paradigm

An epic journey is always a good ploy to get the reader into the right frame of mind. Excuse the pun. Parmenides' allegory, like any allegory, can be read in many ways. One could see the tale as taking place on two levels, a logical foreground, and a gender background. The background imagery displays the feminine in all its plurality, culminating in the abode of the goddess. There the ethereal feminine welcomes the singular masculine, he who knows through things. However, he is but a youth still in quest of knowledge. Nevertheless, he knows he has been conducted here for a purpose.

Across the unsaid background tapestry, spreads the ephemeral abode of goddess and teaming maidens. It is far from the world of mortals and untainted by their bumbling subjectivity. This is the pure feminine, so pure that it is totally devoid of attribute. It is completely unknowable. The goddess is not even real. The goddess, the feminine incarnate, explains that "you cannot know what is not." You can only know what *is*. It looks as if the goddess has negated herself out of existence.

This is Ground Zero of rationality. Many thinkers have come here over the ages, each making their own interpretation. Parmenides paints his picture in great clarity. On one side, we have the land of mortals enmeshed in opinions. On the other side is the immortal land of Truth. This other side is what we have been referring to as the right side take on reality. It is in this domain that Parmenides must construct his version of non-dualist philosophy. He must explain the non-duality of *Oneness*. Like all who have toiled in this domain, he must tackle the dialectic of the One and its other. His reasoning is stark even brutal: the One *is*, the Other *is not*.

As we have seen in previous sections, right side reasoning is expressed in oppositions and oppositions applied to oppositions, a dialectical, semiotic form of thinking. Left side reasoning starts from given preconditions such as opinions, traditions, rumours, gossip, innuendos, measurements, experiences, sensations, axioms, stabs in the dark, fabulation, and sometimes wicked self-serving deception, just as the good goddess explained: this is the natural lot of mortals. Right side reasoning has no preconditions. It must start from the primary opposition. It is here that we find the primordial form of Ground Zero. The task for the philosopher, the ontological scientist, is to provide an explanation of Ground Zero in terms of the primary opposition. The understanding of this primary opposition varies, depending on the thinker. In the case of Parmenides, his position can be summarised as follows.

1. What is the primal entity?

Answer: *It.*

2. What is its specificity?

Answer: *It is.*

3. What is the primary opposition?

Answer: The opposition between what *is* and what *is not*.

4. What exists?

Answer: Only what *is* exists, what *is not* does not and cannot exist.

Comment: What *is* determines the knowable. Even more strongly, it determines the known. The deep essence of *It* is known. *It is*, end of story. What *is not* determines the inherently unknowable. The totally unknowable is tantamount to not existing.

5. Where is the origin of the primal entity located, i.e., where is Ground Zero?

Answer: In the eternal present.

The above five points do not cover the complete ontological paradigm of Parmenides. Like any ontologist worth his salt, Parmenides must provide the

enveloping rationale for why the above points are necessary. David Furley has
honed in on a passage from the narration that attempts to explain the core of
the argument, the *raison d'être* of **It**:

> *The last section of the Way of Truth is particularly difficult. Parmeni-*
> *des repeats his assertion that there is no not-being and there are no*
> *different degrees of being; what exists is equal to itself everywhere and*
> *reaches its limits everywhere. From this he concludes that it is "perfect*
> *from every angle, equally matched from the middle in every way, like*
> *the mass of a well rounded ball" (Furley)*

We see here Parmenides' attempt to explain First Classness, the central, all
enveloping characteristic of rational reality. A world satisfying First Classness
must be totally unconstrained in every way, the ultimate in perfection. Here
resides the great challenge for ontology; this is the task that confronts us. We
have to understand such a world that is totally unconstrained with no one
behind the scenes pulling strings, and no kingpin calling the tune, whether
seated on high, low, in the middle, in or out of reality.

If there were to be hidden forces at work behind the scene, then a fundamental science of reality would be impossible. On the other hand, if reality is
left free to be dominated by the draconian requirement of the totally
unconstrained system, then indeed it must be exactly that, a totally unconstrained system (the only constraint allowed). The iron laws of First Classness
(FC) spring into play, the laws of ontological fair play organise fair play. In this
great riddle, there is only one answer and, as the good goddess said, only one
"unshaken heart of persuasive truth."

Reconciling the coming-into-being, the genesis of reality, with FC is not
easy. The very notion of a determined beginning violates FC, as this privileges
the starting point entity from all others. FC does not allow privileged entities
and there is nothing more privileged than coming first. Parmenides resolved
that violation by saying that an entity at the beginning does not exist, as it *no
longer is*. He then argued that it never could have existed, finally ending up
with the formula that the only thing real is the eternal present and nothing
changes.

...

It is interesting to look at how the Stoics resolved the problem in their
cosmology, as they too, we claim, adopted a doctrine based on FC. For the
Stoics, the only immortal was Zeus. They were pantheists, so the universe was
the body of Zeus. The body changes, is born, and dies away in the conflagration, but Zeus stays immortal throughout the process. The universe had a
beginning and eventually ended up in the conflagration. In this more complex

scenario, avoiding the violation of FC becomes even more difficult. The beginning *and* the end of the Cosmos in the conflagration become privileged moments, one preceding all that will exist and one succeeding, thus violating FC. The Stoic solution was that time was circular, with the whole story *exactly* repeating itself the next time around. In this way, there are no privileged points in time and so FC is not violated. This Eternal Return solution retained many aspects of the Parmenidean solution. Bodies in the past did not exist, nor did those in the future. The only bodies that exist are those in the present. By eternal repetition of the cycle, nothing really changed and no state of being was irreconcilably privileged over any other. FC was respected.

It is interesting to note that Stoic cosmology went beyond the notion of a single universe. They postulated multiple universes. In other words, the Stoics were the first to propose a *multiverse*. The Stoic multiverse was an unending sequence of universes. In order not to violate FC, each universe had to be indistinguishable, but not necessarily identical, to the previous. This explains their theory of the Eternal Return.

...

Parmenides' allegory of the young man streaking across the heavens, waved on by the veiled daughters of the Night, heading straight for the abode of the goddess, evokes the image of a spermatozoid streaking to a rendezvous with the unfertilised egg. However, Parmenides does not allow the union to be consummated. There will be no masculine principle uniting with the feminine principle in this scenario, despite the atmospherics being full of it. The closest we get to any explicit such union is at the level of logic: one proposition in the affirmative and the other in the negative and never the two shall meet. Parmenides pitched his paradigm at the loftiest level and really could not embrace any explicit masculine-feminine union, as this would imply an explicit beginning and so violate FC. For a way around this conundrum, we have to wait for Empedocles.

In the meantime, we understand that there are two ways of understanding reality, one is the Way of Truth, and the other is the Way of Opinion. For Parmenides, the only repository of truth was in the One, a pureness of eternal, ungenerated Oneness. It is that unique Being that "neither was nor will be, because it is in its wholeness now, and only now." In truth, only the One *is*. For Parmenides:

> *The only true reality is Eōn—pure, eternal, immutable, and inde-*
> *structible Being, without any other qualification. Its characterizations*
> *can be only negative, expressions of exclusions, with no pretence of*

attributing some special quality to the reality of which one speaks.
(Calogero, 2010)

As for the Way of Opinion, this is the world of appearances, a misleading world of falsehood. It is this Being *(Eon)* and only this Being that truly, objectively *is*. All else is illusory. Parmenides provided a vivid image of how the very deepest reality could be comprehended. This image is useful to carry forward in one's mind in the development to follow.

Ontological Calculus of Empedocles to the Stoics

Ours is not a scholarly work. For example, in writing about Empedocles and the Stoics, we imply that this is the way they thought and expressed their ideas. Clearly, this is not the case. Rather than the scholarly, our approach is to *reverse engineer* the concepts of ancient thinkers. As a consequence, what we write is often more of what the ancients *could* have thought, and sometimes perhaps what they *should* have thought, if they remained true to their doctrines.

Keeping this in mind, we will now repeat the summary of Parmenides doctrine and adapt it to the next phase in the development of ancient physics. In order to develop the theory of the Four Elements, the initial structure of Parmenides must be pushed to the next stage. What follows is a brief summary of this next step, as seen from a modern perspective. The key idea is what we call ontological gender. It is this structure that explains the fundamental starting point for our unifying science. The ancients got there first; but there is some cleaning up to do.

We have already considered these concepts in earlier sections, but here is yet another angle.

The Gender Paradigm

1. What is the primal entity?

Answer: Any entity whatsoever.

2. What is its specificity?

Answer: The entity *has* the attribute of absolute non-specificity.

Comment: This attribute, that of absolute non-specificity, is an entity in its own right, as demanded by First Classness.

3. What is the primary opposition?

Answer: The opposition between what *has* the attribute and what *is* the attribute.

Comment: The entity that *has* the attribute determines feminine gender, the entity that *is* the attribute determines masculine gender. This is the definition of ontological gender. These two entities are different by gender but are indistinguishable.

4. What exists?

Answer: Only what *is* exists.

Comment: This follows Parmenides but adds some detail: only what *Is*, that entity of pure masculine gender, fundamentally exists. The pure feminine entity *has* something (the attribute of total non-specificity), but when it comes to whether it *is*, clearly it *is not*. Having something differs from *being* something.

The feminine *has* being, the masculine *is* being. Both need each other. The pure feminine entity, although different, is indistinguishable from the pure masculine. Parmenides' poem did not mention that point. Anyhow, the indistinguishability only applies to a third party. For this primordial couple, they know who is who without a show of doubt.

5. Where is the origin of the primal entity located, i.e., where is Ground Zero?

Answer: The location is that determined by the primal entity.

Comment: The whole Cosmos gyrates around this location. Thus, Ground Zero can be thought of as *any location whatsoever*.

This summary has added in some innovations that don't belong to the ancients. However, it is in keeping with a strictly generic approach that the ancients were pioneering. The summary represents where they were heading more than where they were at.

Traditional sciences, what we call left side sciences, express all knowledge in terms of attributes of things. This is quite reasonable, because it is impossible directly to know scientifically the thing that has the attribute. The knowledge of things is always indirectly achieved via attributes of things. Note also that in pure left side science, an attribute is not a proper thing, as it is in pure right side science where the attributes of a thing are things in their own right.

In our work, we are endeavouring to develop the right side science that does not rely on any preconditions whatsoever. Harvesting attributes as a precondition for theorising is a fundamental left science activity, but not so for right side science. Thus, all the empirical attributes that abound in left side sciences are forbidden on the right side. If right side science is to have any attributes then it has to *reason* them into existence, not measure them. This reasoning process leads to the attribute par excellence, that of masculine gender.

Then comes the incredible core notion of right side science. This primal attribute is the *only* attribute we need! All other attributes can be constructed from it or with it. We enter into the web of gender intrigue. The only guiding principle at our disposal is that of FC. FC demands that an attribute of an entity must be capable of being considered as an entity in its own right: in this context, it becomes the masculine *qua* feminine MF. What is good for the gander is also good for the goose. The feminine entity must also enjoy the possibility of being considered as an attribute in its own right: it becomes the feminine *qua* masculine FM. Here we are starting to get compound gendered entities, depending on roleplaying. The other two combinations are the pure gendered entities *qua* themselves, notably the masculine *qua* masculine MM and the feminine *qua* feminine FF. In the process, we have advanced from the primordial Parmenidean paradigm through to the Doctrine of the Four Roots of Empedocles and later developed into the four-element doctrine of the Stoics.

There are four elements that have mixed gender MF, FF, FM, and MM entities, corresponding respectively to expansive air, earth, converging water, and fire. These binary terms can be thought of as noun-adjective pairs, where F and M can play the role of noun or adjective, depending on their position in the pair. These terms are the elementary terms of the generic code. The genetic code is the biological instance of the generic code where the four letters, using RNA notation, are A, U, G and C respectively.

The Double Articulation

Two birds living together, each the friend of the other, perch upon the same tree. Of these two, one eats the sweet fruit of the tree, but the other simply looks on without eating. The two birds are the Jiva and Isvara, both existing in an individual compared to a tree. [Mundaka Upanishad]

What interests us about Stoicism is the paradigm. The paradigm does not belong uniquely to the Stoics but stretches back from Heraclitus and Parmenides right forward through Leibnitz to Kant, Fichte, Hegel, Marx, and Jung and to some of the more recent moderns. The paradigm has been referred to as monism, but should more correctly be interpreted as non-dualist. The natural opposite to monism can be seen in the present sciences of our day. These sciences are characterised by atomism, fundamental dualities, and abstraction. These traditional sciences, which include axiomatic mathematics, are aligned on virtually the same epistemological axis as Epicureanism. As we have previously stated, we characterise the Epicurean type paradigm as left side science, as this is the kind of thinking that is privileged by the left hemisphere of the brain. What is of interest to us is the complementary mode of thinking, right side thinking. Much less is understood about this side of brain functioning. Here we are talking about the biological brain and also the "epistemological brain," one being the metaphor for the other. We are not particularly precise as to which, in the final count, is the metaphor. Basically, we are concerned with two generic paradigms, one we call left side and the other right side. The left side paradigm articulates a bottom up form of rationality whilst the right side employs a top down strategy.

In our view, the right side dominant thinkers par excellence were the Stoics. It might seem strange that we have difficulty talking about the Stoics without continuously considering them in oppositions with their opposite number, the Epicureans. Even worse, we keep spreading the discussion across a left side/right side epistemological dichotomy, a dichotomy that has distinct biological reverberations. Not only is knowledge organised along these lines, not only the universe, but the very architecture of our brains also has this shape: a left side, a right side, and even a back side with frontal lobe front side.

The basic message we want to get across is that there are two fundamental rationalist philosophical paradigms. Now everyone knows that already. The Western philosophical tradition has been split right down the middle almost since the year dot. What interests us is the multi-paradigm paradigm. Rather than take a partisan position, we prefer to change hats. We have two hats, a left side hat, and a right side hat. Changing hats changes the way we think. If you are asking the question as to which side is right and which side is wrong, then you are wearing the left side hat. This is the kind of question that left side hat wearer tend to consistently ask. By changing to the right side hat, the pertinence of the question evaporates. What is true and what is false gives way to the search for truth. Truth is not obtained by a *prise de position* but rather by taking up all positions. The reasoning is via oppositions between positions, and oppositions between oppositions. We have obtained an intuitive handle on this mode of reasoning through practical examples of the semiotic square.

And so we finally come to the main point. We believe that one cannot arrive at a deep fundamental understanding of Stoicism without simultaneously understanding the nature and way of thinking of its opposite, which in the occurrence, is incarnated in the Epicurean dogma. In addition, this great epistemological divides should be looked at from a higher perspective than merely its historic occurrence. These two doctrines are instances of a greater generic dichotomy that has further instances in Cosmology right across to brain architecture.

Right side thinking expresses itself in oppositions. Thus, even to promote a particular doctrine such as the basic paradigm of right side thought itself, that paradigm must place itself in opposition against its opposite. Left side reasoning can ignore the right side. It doesn't understand it anyway. Right side reasoning cannot do this. All players must be present. Besides, without the left side, the right would have nothing to criticise.

Perhaps one should note in passing that Chrysippus himself seemed to practice this technique. As Diogenes Laërtius wrote that Chrysippus, early on in his studies, "got the habit of arguing for and against a custom." (Yonge) Chrysippus was also known for his prolific and lengthy citations. This is an important aspect of the critical nature of right side thinking. The demise of a vibrant critical tradition in politics and the social sciences is a lamentable fact in present day society. In the case of the mathematical sciences, such a tradition is totally absent: the only critique is that of peer review. What we need is a totally different kind of science to put the cat amongst the pigeons.

...

At this present time, our Western culture and our education system place all the emphasis on left side style thinking. This is particularly apparent in the sciences. There the fundamental left side paradigm goes right back to the first atomists, Leucippus and Democritus, forerunners to Epicurus. The present day traditional sciences and modern formal mathematics continue the development. Some of the key aspects of this tradition are:

- Dualist and atomist, tolerance of absolute dichotomies
- Abstraction
- Language: strong on syntax, weak on semantics
- Bottom up methodology
- Labelling technology.

All of these characteristics are interrelated and are expressions of a single underlying paradigm. From a neuropsychological perspective, these are also fundamental characteristics of left hemisphere brain function.

Some key characteristics of right side thinking are:

- Monist, no rigid dichotomies, everything relative.
- Generic (Non-abstract)
- No Languages, just bare meanings – semiotics and a semiotic code.
- Top down methodology
- Likes expressing knowledge in terms of oppositions, not labels.

Language Difference

On the biological front, a most dramatic differences between the two hemispheres for humans is that the left hemisphere can speak and the right hemisphere is mute. A subject with the left hemisphere immobilised will be mute. With the right hemisphere immobilised the subject will be able to generate syntactically correct speech, although it may be severely impaired with regard to its meaning.

Hemi-Neglect

Another dramatic difference between the two hemispheres is that the left hemisphere, on its own, suffers from hemi-neglect. This means that the left hemisphere appears to be only conscious of one half of reality. It will only recognise and be conscious of one side of the body, the right side. It will not be conscious of its left side or of anyone or anything on its left side. It may even only be able to see the right side of a clock. The right hemisphere exhibits no hemi-neglect, recognises both sides of the body as its own, and sees a whole world.

Biological versus Epistemological Brain

In brief, the left hemisphere can have language whilst the right side is mute. The left hemisphere sees a half world; the right hemisphere sees a whole world. These are differences easily observable in the clinic. However, these are differences between the *biological* hemispheres. A persistent theme in this work is that the biological brain and the structure of knowledge, the "epistemological brain," share a common generic structure. If indeed this were the case, then the phenomenon of hemi-neglect for the left hemisphere and language muteness for the right hemisphere should have some epistemological analogues.

It finally comes down to the fact that there are two distinctive modes of thinking, a left side, and a right side. If we admit of a second dichotomy, as indeed we must, then we end up with four modes of thinking. This secondary dichotomy is the front-back dichotomy that we have been talking about. We argue that this four-cornered playing field provides the common ground for organising knowledge. For right side science, all knowledge is defined relative to this playing field, relative to the generic subject.

Biochemistry and the Double Articulation

The architecture of the generic subject was worked out starting from the Parmenidean question: What *is*? Answer: Reality **is**. This entity that really *is*, this Parmenidean incarnation of the real that is the eternal *now*, this thing can be thought of as the *impersonal subject*. It is from the point of view of this subject that situates the God's eye view. Such a subject merely *is* with respect to its Other, which *is not*. However, the way we see it, instead of conceiving the Other as that which *is not*, we see it as that which *has*. This is an opposition between the pure masculine, which *is,* and the pure feminine, which *has*. In this equation, One entity *has* but *is not*, and the other *is* but *has not*.

There are no absolutes. This Other, that *is not* (relative to the Subject), this other entity corresponds to the *impersonal object*. The impersonal subject and the impersonal object relate to each other in the most intimate and profound way. The Other, from the perspective of the science, is totally devoid of any specificity whatsoever. However, so pure and profound is this lack of specificity that the lack of determined specificity becomes a character-ising principle. This lack of attribute, this utmost purity, can actually be thought of *as* an attribute. This Other *has* the attribute of such purity and lack of specificity.

We have already discussed this dialectic of **to be** and **to have** and cast it into a relative typing system called gender. The pure feminine was character-ised as the entity having the pure attribute and the masculine, considered as an entity in its own right, becomes that attribute. The feminine **has it** the masculine **is it**. What ***has*** the attribute can only be known via the attribute, via what ***is***. This fundamental dichotomy between the feminine and the masculine is that between the object world and the singular subject, what we have been calling the left-right dichotomy.

This is the dichotomy between reality and the impersonal subject, the result of the "view from nowhere." In order to arrive at a tractable science, a tractable view of the world, a more determined purchase is necessary than the view from nowhere. What is required is a view from *somewhere*.

The view from somewhere is the view you get when you are *at* this loca-tion, be it in time and space or whatever. In order for this to be the case, you must take on the determination commensurate with a subject sufficiently qualified to do the job. No longer are you an impersonal subject, but you become a personalised subject. The same gender qualification will do the job, but it is a qualification relative to you and not reality. The qualification must provide the singularity of you. This is the masculine side of the qualification. The feminine is the other side of the qualification. The side that is not a singularity in itself but has the singularity all the same – as an attribute, but not as its essence. The essence of the feminine remains a total mystery in this metaphysic: the mystery *is* its essence in fact. The feminine remains the wildcard at all times. On the other hand, the masculine is an open book. One can only know the world via its attributes. Knowledge of the world is fundamentally a process of masculinising the world, that is to say, by interpreting the world in terms of attributes. Left side reason harvests the attributes from nature. Right side science constructs its attributes, its typing system, through a process by building knowledge out of bricks of ignorance (the feminine) and the total certainty of the ephemeral masculine. It sounds rather poetic and far-fetched, but according to us, the genetic code is built precisely on these principles. However, to become a science a much more precise development is called for. Some preliminary work along these lines is included in the Appendices.

The Indian Hindu thinkers, particularly the Advaita Vedanta school, understand this metaphysic very well and in detail. Many interpretations are of a religious and spiritual nature, which illustrate these difficult and allusive ideas in vivid colourful detail. Sometimes the concepts are adorned with ancient mythical stories dating right back to Vedic times. At the more technical level, the absolutely unqualified entity, which even escapes any subject-object dichotomy, is the *Brahman*. The Brahman is the cosmic self and is aligns with *ātman*, the self of a living being. At the first level of qualification, the pure subject-object dichotomy associated with the view from nowhere corresponds to the Nirguna Brahman, the non-manifested form of the Brahman. The next determination leads to Ishvara, considered as both a transcendent and immanent entity. By coming into being, he creates the world and thus ends up living in it. In fact, he is the world in the sense that "this world is covered and filled with Ishvara." After these two determinations, the world is divided up into four. This is ground zero, and like the North Pole, all roads from here lead in a single direction South. The interpretations of this original semiotic square abound. In all cases, there is the usual Three-plus-One form of the square. Since we know the most generic labelling of the four elements of the square, we can consider different variants using the gender typing as a constant frame of reference. In other words, we use a gender-labelling scheme, noting that the labels are not arbitrary, as for left side methodologies, but ontologically *constructed*. One semiotic description of the Ishvara semiotic square consists of the triad made up of Ishvara, all sentient entities, and the World. The triad is complemented by the One, corresponding to the Brahman. Other accounts explain creation in gender terms resembling the Stoic version. The feminine is unqualified, formless substance. The masculine expresses the Individualisation Principle that fertilises the feminine, bringing about individualisation.

From all of this, a universal structure emerges based on two oppositions and the mutual opposition between them, a metaphysical binomial, so to speak. One opposition, the impersonal, is between Reality and its Oneness. The other is the opposition between an individual being and its Oneness. Oneness translates as the masculine, the Other as the feminine. We adopt the left-right polarity convention for the first opposition and front-back for the second opposition. The masculine Oneness sides of the convention correspond to the right and front sides of the resulting semiotic square. This square becomes the generic reference frame for all that follows. An intuitive interpretation is to see this structure as the architectural layout for the generic mind. In terms of gender typing, the four parts of the square are typed MF, FF, FM, and MM. These binary typed entities can be labelled with four single letters. We have chosen the convention of the four letters A, U, G, and C respectively, using RNA coding. A central tenet of this work is that these letters correspond to the four letters of the genetic code. We call it the *generic* code as we claim it can code *any* material being whatsoever, not just the biological.

From this generic Ground, the development must take into account the generic Figure. Ground, relative to itself, is obviously *static*; that is the nature of any ground, the nature of anything relative to itself for that matter. On the other hand, Figure, relative to Ground, is *mobile*. Figure itself can be understood in terms of a semiotic square structure. The structure of Ground is a Three-plus-One structure, where the Oneness part is typed MM and denoted by the letter C, the singular singular. The same applies for Figure. Figure can be determined relative to Ground, where the AUG triad of quarters of Ground act as placeholders for the mobile quarters of Figure. So mobile are the quarters of Figure that for a determined Figure any triad of combinations is permissible, including even repetitions and duplicates. This is like a dynamic Rubik Cube of prodigious complexity. Everything might appear jumbled and all over the place, but there is a higher plan to produce a Oneness from this apparent chaos.

Figure and Ground are the two moments of Self. Relative to Self, both Ground and Figure are mobile. Ground determines *where*, Figure *what*. Ground determines angle of point of view, Figure the view.

This kind of discussion can easily lapse into either poetry or abstraction. In an attempt to escape from this temptation, we provide the following easily understood example from biology as illustrated in the diagram below. The diagram is a schematic, an artist's impression, not a detailed functional diagram. It shows schematically the relationship between the genetic code and the biological body that it is coding. The body is coded in a sequence of triads called codons. Each codon will consist of three letters from the alphabet A, U, G, and C.

To each codon there is an implicit semiotic square where the fourth element corresponds to the body being encoded, looked at as a whole. The first codon is AUG and is the start codon. It determines the body to be coded as Ground, the initial reference frame for all that follows. Each succeeding codon articulates a succession of holistic prescriptions for biological synthesis. Although lacking in determination and qualification to begin with, this sequence of holistic views increases in qualification. The specification becomes increasingly more precise as requirement on requirement compound and stack up. However, this code is nothing like the recipe for a cake, the sort of thinking characteristic of left side science. By continually viewing and recognising the organism as a whole, no matter from what viewpoint as the specification progresses, the wholeness of the whole becomes an invariant in this kind of language.

This generic language is hardly the kind of language that you would want to use to order your lunch. For that problem environment, a left side language is adequate. However, if the objective is to create an autonomous unity of Oneness, you need a code whose very essence is based on Oneness. For that you need the generic code, the genetic code, and as some call it even in its genetic version, the God Code.

Now the important point to note in the diagram below is that this biological body language of the genetic code illustrates what some French linguists call, the *double articulation*. The first articulation is what the biochemists see. Firstly, they see transcription of the Code to amino acids. At a more macroscopic level, they see the encoding of genes. The genes subsequently interact and even exhibit algorithmic type behaviour. Things happen. Current knowledge even allows the prediction of gene interactions. The combinatorial gives way to the algorithmic. This left side science perspective starts seeing the mechanics of what looks like a machine in action. To use a term of Dawkins, one of our favourite left side science exponents, it starts to look like the clockworks of a blind watchmaker. This is all part of the First Articulation. The machine ticks aimlessly on and will soon give birth to other equally aimless offspring. Some will be destined for success, the others for genetic oblivion.

This left side First Articulation certainly is based on blindness. Absent from the scene is the organisation principle that unites all this disparate paraphernalia into a single, unified being. This organisational principle is not comprehensible from the left side science perspective. As we know, it is based on gender and the principle of First Classness. Any living being can be characterised as being a successful autonomous entity. There is no puppeteer behind the scene. From our analysis, the only organisational principle that accomplishes sure success this venture of the self-supporting organism out there in its world is the principle of First Classness. Any individual organism will have its own distinct disposition. However, one thing than no organism can do it to circumvent the hard facts of life. These hard facts, the implicit laws, are spelt out in the Code. The Code even has moral dimensions. The Code, as an algebraic expression of First Classness, transmits an inherent desire to achieve the same perfection as Nature. Nature and any other organism, all aspire to First Classness. This First Classness is articulated in the gender calculus implicit in the Code. This is the Second Articulation where by the organism, viewed from any angle, appears as a whole. The Code is the algebra of wholes.

Looking at the highly schematic diagram in Figure 25, we see that there is an implicit *second* articulation at work. The opaque four-letter code harbours an underlying generic structure based on ontological gender. Rather than three letters per synthesis event, the diagram implicitly refers to six as each letter is double gendered. Rather than a dumb sequence of three opaque letters, we see revealed the seeds for a kind of generic geometry. This code is no longer just transcribing, it is building a highly elaborate structure of generic interiors and exteriors and so forth that make a Klein bottle look like child's play.

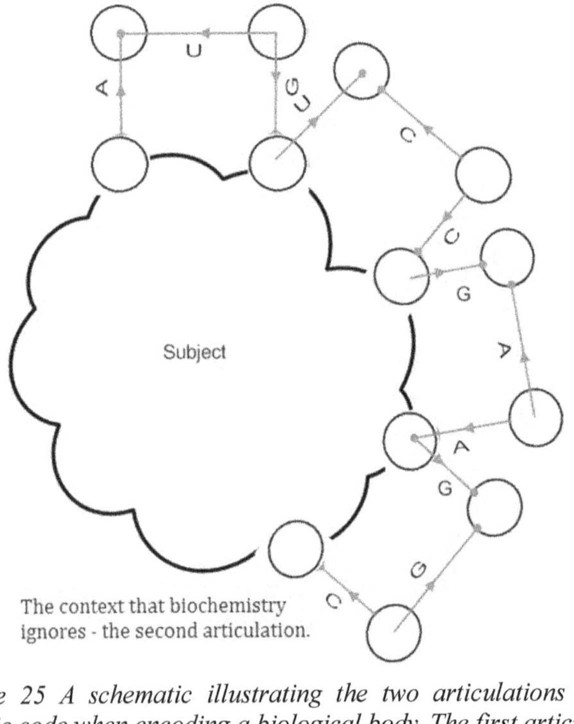

Figure 25 A schematic illustrating the two articulations of the genetic code when encoding a biological body. The first articulation is biochemical transcription. The second articulates the specification in terms of a sequence of wholes. The genetic code becomes the generic code.

The big challenge before us is to understand this second articulation. Until we make important inroads along this direction, we will have to downgrade the incredible achievements of the biochemists. They need provocation. Our finger points to the moon and the biochemist only sees the finger.

In the next section, we look at the generic logic behind the generic code. In a later section, we will look at generic geometry.

Generic Logic and Aristotle

We have already intimated that present day biochemistry, as a life science, only fires on one cylinder. It is now time to turn to the logicians. In this case, the problem is to prove that present day logicians only use half a brain. Our task is considerably simplified by the fact that the logicians have already proved that this is indeed the case. For this achievement, they must be congratulated. The only thing that perturbs the author is that the logicians seem very proud of their feat, even to the point of boasting. The going might not be as easy as we expected.

When it comes to logic, the best place to start is with Aristotle. As well as being the greatest philosopher of all time, Aristotle was also the greatest fence sitter of all time. With him, our neat dichotomy between left side and right side thinking meets a blank. This man has a foot firmly placed on both sides. Nowhere is this more apparent than with his categorical logic and in particular his square of oppositions. In this section, without going into too much detail, we summarise the aspects that immediately concern our project.

Name	Symbol	Term
Universal Affirmative	A	Every S is P
Universal Negative	E	No S is P
Particular Affirmative	I	Some S is P
Particular Negative	O	Some S is not P

Figure 26 The four kind of terms. The Scholastics later labelled them with four letters.

We end this chapter by presenting an alternative approach of breath taking simplicity and elegance, that pioneered by Chrysippus, considered by many in his time to be the equal of Aristotle. However, to understand the full significance of the Chrysippus approach, we will need to add some important clarifying innovations on our part that will not be found anywhere else. This is quite important. If the reader has progressed this far into our exposition, it will be well worth the effort to understand what follows. Great beauty lies ahead.

What follows is not rocket science. Each step is easy to understand. The hard part, as always, is to grasp the full significance of the matter.

Aristotle's Organon

Aristotle's syllogistic logic occupies a central place in what is nowadays called classical logic. This was the logic studied by the learned peoples, mainly monks, of medieval Europe for a thousand years. During this time, the notation was refined and elaborated, but the essence barely changed. Even by the Enlightenment, Kant was known to exclaim that the only logic that one needed to know was that of Aristotle.

Aristotle's syllogistic logic was a central component of what he called the Organon, Greek for tool or organ. Syllogisms are logical arguments made up of three parts, a major premise, a minor, and a conclusion. The most famous is the very familiar:

Major: All men are mortal

Minor: Socrates is a man

Conclusion: Socrates is a mortal.

Aristotle's syllogistic logic is sometimes referred to as term logic where each syllogism has a major, minor and conclusion term. What interests us is how many kinds of term are necessary for such a logic. This also interested Aristotle. He argued that there were four distinct kinds of term. During the Middle Ages, the Scholastics gave each term a letter as shown below.

The Four Terms and the Left Side

Aristotle's syllogistic term logic was half modern and half ancient. We will suspend judgment on which was the better half. The modern half is exhibited in two ways: it relies on abstraction and its logic is static. The abstraction can be seen in the use of the existential qualifier "All." "All men" for example, means every man. By referring to "all men" or every man, one is referring to an abstraction, a generalisation. As the Stoics pointed out, abstractions and generalisations do not exist as real entities. In addition to abstraction, there is the fact that the logical representation of these syllogisms can be covered by Venn diagrams as shown below. The terms can be said to have "Venn Diagram" semantics.

Both of these aspects, the abstract and static nature of the logic, are characteristics of left side thinking. By default, left side thinking has become synonymous with the modern.

The Four Terms and the Right Side

However, what is *not* modern in Aristotle's logic is that his infrastructure of the four kinds of terms is not determined by a set of axioms, but rather by a pair of oppositions and the opposition between these oppositions. This is exactly the approach we have been using to construct our semiotic squares. Firstly, obtain a pair of oppositions. Employ one opposition to define a left-right dichotomy and the other opposition for the front back structure.

A Every S is P

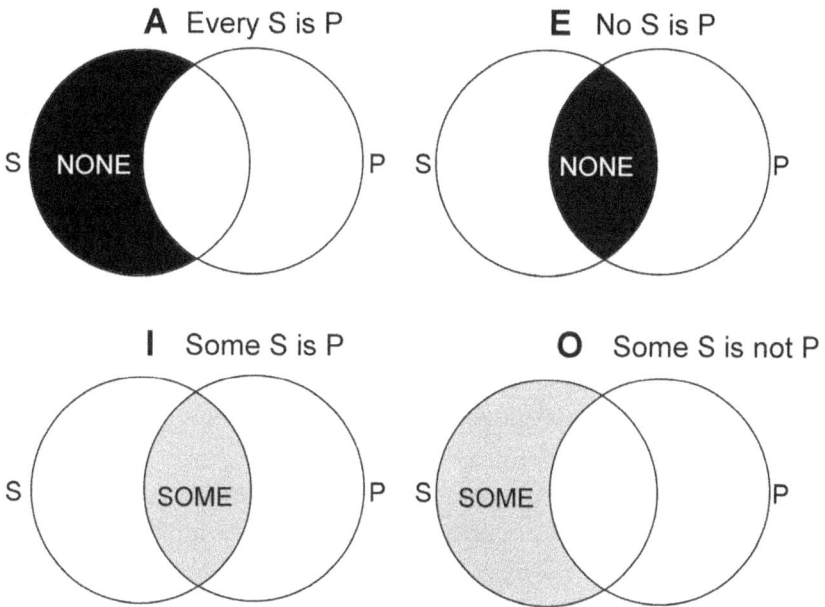

S NONE P

E No S is P

S NONE P

I Some S is P

S SOME P

O Some S is not P

S SOME P

Figure 27 Venn diagrams for the four terms of Aristotle

In Aristotle's case, the left-right dichotomy is a strict logical opposition between the affirmative form and the negative. The second opposition is between the universal and the particular. Both these oppositions must be true dichotomies in order to construct a non-trivial semiotic square. This is a technical point, but a very important one and will be discussed later when considering Aristotle's square of oppositions. It turns out that there can be certain cases where an opposition is not a true dichotomy. This can occur when the subject of a term has no existential import. In other words, when dealing with empty sets such as "All centaurs."

	universal		
affirmative	A Every S is P	E No S is P	negative
	I Some S is P	O Some S is not P	
	particular		

Figure 28 The semiotic square for the four terms of Aristotle's Syllogistic logic. The square is formed from two oppositions, the negative/affirmative, and the universal/particular.

Term Logic

During the middle ages, the scholastics labelled the four kinds of terms with the four letters A, I, O, and E. Syllogisms consist of three propositions, a major, a minor, and a conclusion. Each syllogism could thus be labelled by a triplet of letters taken from the four-letter AIOE alphabet. This fascinated the Scholastics and, many years ago, entertained the author's curiosity for some time. The reason for the author's interest was that such a system did have some resemblance to the triadic structure of codons in the genetic code. With a bit of effort, one can make some kind of rapprochement between the AIOE alphabet of the scholastics and the genetic-cum-generic AUGC alphabet, but the effort is probably not justified, as there are richer pickings elsewhere, notably in Stoic logic.

The genetic codon structure only has 64 combinations. What we have ignored for the Aristotle's syllogism is the detail of how the three propositions in each syllogism hook together. We have ignored the fact that there are four different figures of the syllogism. Thus, taking into account the four figures, instead of 64 possible syllogisms there will be 256. Only nineteen of these syllogisms are regarded as leading to a valid conclusion.

Aristotle's syllogistic logic provides a logical tool that is applicable to the contingent world. Unlike modern logic, it also brings with it some nontrivial semiotic infrastructure, the square of oppositions.

The Square of Oppositions

Aristotle described how the four kinds of terms could be placed in a square illustrating the various oppositions between them. He then went about characterising each kind of opposition, although the subalterns were not mentioned explicitly. The oppositions between universal statements are contraries. Contraries have the property that both cannot be true together. One may be true and the other false. It is also possible that both can be false together. On the other hand, subcontraries involve oppositions between particulars. In this case, both cannot be false together.

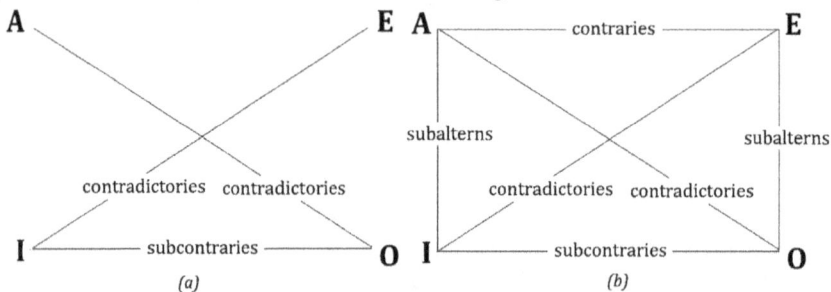

Figure 29 (a) The modern logic version of the oppositions. (b) Aristotle's square of oppositions.

The Modern Square of Oppositions

Of great interest to us is an opposition at a higher level altogether, the opposition between Aristotle's syllogistic structures and modern logic. The dramatic difference between the two approaches was clearly illustrated by George Boole, in what has become the modern version of the Square of Oppositions.

Modern logic differs from the ancient logic by simply replacing the universal with the general, in other words with the abstract. This can be achieved by using labels and the logic becomes *symbolic logic*. Thus, the term 'All men' is replaced by the abstract version 'All X'. The thing gets replaced by a label and introduces different semantics. One could say that the semantics go out the window and are left trivialised. The label becomes simply a placeholder and as such, like any placeholder, may be empty. The logicians explain this as relaxing the requirement of existential import. From a classical mathematics perspective, the generalisation introduced by modern logic is to allow sets to be empty. This allows modern logic to talk about things that are known not to exist, a characterising feature of abstraction.

Once the reasoning becomes abstract, the logical difference between yellow centaurs and canaries evaporates. Not only that, but all the oppositions except the contradictories have also evaporated. For example, both sides of the contraries opposition 'All centaurs are yellow' and 'No centaur is yellow' are true. The contraries opposition has evaporated.

Figure 29 (a) shows the resulting modern logic version of the square of oppositions. The square has virtually collapsed and only the contradictories and the subcontraries survive. We have deliberately drawn the modern version on the left side relative to Aristotle's square to illustrate that this is the left side variant of logic. The other variant is Aristotle's seed for the right side version. The left side involves abstract, symbolic logic. The right side in the diagram represents Aristotle's version of elementary generic logical structure. In practice, the modern symbolic logic approach boils down to a simple bipolar nominalism where the basic opposition is between two particulars, *I* and *O*. The letters *A* and E act as pure label signifiers for the *I* and *O* respectively, acting as the signified. The contradictory oppositions *A-O* and *E-I* model the relationships between signifier and signified. In essence, the system becomes a simple two letter system labelled by *A* and *E*. Thus, although we have not shown that modern day logicians only use half a brain, we are starting to see that they reason using only half of Nature's alphabet.

This is our first exploit into the differences between abstract, symbolic logic, and generic logic. We can do better. This will be our task in the next section where we investigate Stoic logic and discover great beauty in the land of Chrysippus.

Generic Logic and the Stoics

In this section, we are going to look at the Stoic version of Square of Oppositions. This will undoubtedly upset the scholars. There is no record that the Stoics ever proposed an alternative to Aristotle's Square of Oppositions, at least not to our knowledge. However, we are not constrained by historic

Stoicism. If something is missing from the puzzle then we must endeavour to reverse engineer it. We attempt to follow in the tradition of Chrysippus himself, of whom it was said:

> ... *in many points he dissented from Zeno, and also from Cleanthes, to whom he often used to say that he only wanted to be instructed in the dogmas of the school, and that he would discover the demonstrations for himself. (Laërtius)*

Stoicism, particularly the Early Stoa, is a very tightly integrated body of thought. It is much tighter than what might be imagined, especially after Chrysippus had a hand in the matter. Traditionally, Stoic philosophy involves a tight integration of physics, ethics, and logic. Likened to an egg, the yolk was physics, the white ethics, and the shell was logic. Logic protects and holds it all together.

Stoic logic differs dramatically from that of Aristotle. There is no static classificatory apparatus. There are no species and no genera. There is no extension or comprehension of terms. The figures and modes of the syllogistic evaporate into thin air. To the Stoics, Aristotle's syllogistic logic was "useless." (Chénique, 1974) In contrast, Stoic logic is starkly oriented to the particular. As such, it incorporates one aspect that might entitle the logic to be considered a *generic* logic, a logic free from abstraction. It is this kind of logic one needs to construct and deconstruct a *real* world, not an abstract world. However, what precisely is a generic logic? Our immediate task is to answer this question.

Now proficiency in logic demands a certain dexterity and agility of the mind. In this respect, the author has been blessed with a mind as nimble as the Titanic and just as infallible. This helps explain his reaction when thirty years ago he first came across Stoic logic. Almost nothing is left of Stoic texts in modern times. Nevertheless, a rough sketch that can be coagulated into less than a page or so has survived. It was Chrysippus' *Ground Zero*, viewed from a logician's point of view of course, a great logician's point of view. The author was quite excited: it is not every day that one comes across an explanation of the structure of the Cosmos spelt out in hard-core logic, all on one page to boot. There they were Chrysippus' five logical undemonstratables. According to Chrysippus, all reasoning stems from these five logical gems. The author stared at the five gems like a stunned plover. If this is it, he could not see it, even though somehow he knew it.

Over the years, the author came back numerous times to the five unde-monstratables in an attempt to really 'get it'. A favourite reference was *Éléments de Logique Classique* (Chénique, 1974). The pages were getting quite dog-eared. Each time brought about the same stunned plover reaction. He still did not get it. However, Chrysippus' third undemonstrable stuck out a bit from the other four. It looked very much like the modern logic operation called the Sheffer stroke, named after Henry M. Sheffer. It is called a *stroke* because that is the way it is symbolically written, as a vertical stroke. The author would look stony eyed at the third undemonstrable and ask himself, "What does it mean?" He knew that if he asked someone trained in logic they would patiently, and perhaps condescendingly, explain that it means "NOT this AND that." In plain English, it means not both. It is sometimes called the NAND operation. It is important in logical networks because any network can be built uniquely using NAND gates. In other words, any other logical operation can be built up uniquely using NAND. They would then go on to explain that Charles Sanders Peirce had discovered a dual result earlier than Sheffer. In an unpublished work (Peirce, 1880) come up with the mathematical dual, which is now called the Peirce Arrow, written as a vertical arrow. This is the logical NOR operation. Such thoughts would make the author's eyes glaze over. He would ask himself, "But, what does it *really* mean? What did it mean to Chrysippus?" Superficially, it simply looked as if Chrysippus was the first to discover the propositional calculus. Granted he had made the discovery several thousand years in advance of the moderns, but this is the kind of thing one would expect from a master logician. If that was all there was to it, then Chrysippus would have nothing much more to offer the moderns. The undemonstratables could be simply seen as an early attempt to systemise and even axiomatise the propositional calculus. Chrysippus could be brought into the modern camp and branded as one of them. Surely, there must be a deeper message here.

For a long time, Stoic logic remained as a kind of lurking nemesis in the author's mind. The five undemonstratables, where did they come from? What is the underlying principle? For a philosophical system as tight and unified as the Stoic's, the logic must have the same basic epistemological and ontological signature as their physics and ethics.

Our approach so far has been based on an intuitive interpretation of how the physics and ethics could be constructed from a fundamental ontological dichotomy. The dichotomy can be understood linguistically as the difference between the verbs *to have* and *to be*. Two fundamental entities were proposed that expressed primary difference free of any accidental, empirical attributes. In this scenario, one entity had the *attribute* of being devoid of any specificity whatsoever whilst the *entity* was this attribute. One entity *has* an attribute; the other entity *is* this attribute. The difference between these two entities was said to be a difference in *gender*. Using this gender construct, the primary attributers of reality can be constructed from first principles. The attributes are not harvested empirically, but synthesised, calculated.

We note in passing that our gender terminology is more reminiscent of Indian and Chinese philosophy traditions. Rather than explaining the beginning of a creation cycle in terms of a union between the feminine and masculine, the Stoics tended to restrict their vocabulary to the masculine register. It seems that only Zeus and his seed seem to feature. For the Stoics, the two principles translate to the active and the passive principles.

Our development starts with the masculine and feminine opposition. The approach leads to the four elementary letters of the ontological alphabet based on the binary gender typing MF, FF, FM, and MM. For reasons that will later become apparent, we allocated the letters A, U, G, and C respectively from the genetic code for this four-letter ontological alphabet. The detailed algebra of this ontological code based on this four letter alphabet has yet to be determined. This code is capable of describing and proscribing *any being whatsoever*, including the universe, itself a being. Any being must have its own ontological DNA, so to speak. It is in this way that a being can be sure of what it *is*.

Using this gender construct, the theory of the four ontological elements can be explained in terms of four kinds of substance typed by the binary gender typing MF, FF, FM, and MM. This corresponds to the ancient terminology of air, earth, water, and fire respectively. If it starts with M it's light stuff, if it starts with F it's heavy stuff and so on, according to the ancients.

The four elements have multiple instances, are mobile, and mix. The four binary types also apply to something that is not mobile and is located at the centre of the universe; or rather, at the centre of *its* universe. This is the generic *subject*, what can be thought of as any being whatsoever. This is the generic template of mind. The whole universe gyrates around this entity. Not only are the four typed bodies fixed at this location, they are fixed in relation to each other. The two bodies with a binary gender starting with M are located on the right and the F on the left. The two bodies with gender typing ending in M are located in the front lobes, the F in the back. The question regarding telling the difference between left, right, front, and back can be resolved by looking at the gender typing. This might seem a bit tautological, but that is the way things work in this world. Here relativity is not only endemic it is generic.

At this point in the game, there is also the question of whether the four kinds of mobile elements circulate to the exterior of this generic mind or in the interior. This question cannot be answered at this stage of the ontological development, as what is interior to subject and what is exterior is still unqualified.

Finally, we come back to the main question in hand, the Stoic logic question. What is the ontological interpretation of Chrysippus' five undemonstratables?

The Fifth Element

In order to answer this question, we have to start thinking in terms of quintuplets rather than just quadruples. Before we tackle the logic quintuplet, it is worthwhile looking at Stoic physics. What is the fifth element?

Aristotle argued for a fifth element in his physics, which he called *aether*. A fifth element was necessary to fill the heavens above the terrestrial world and to explain the constant, unchanging rotation of the stars. The Stoics also added a fifth element to their system, calling it *pneuma*, an ancient Greek word meaning 'breath'. In this perspective, the four elements air, earth, water, and fire were considered passive, whilst the pneuma expressed the *active* principle. Unlike Aristotle's aether, the pneuma permeates everything and expresses the Logos for both the Cosmos and the body.

Some accounts say that pneuma is created from the fire and air elements. From our previous analysis, we know that the fire and air elements of antiquity have the gender coding MM and MF respectively, which indicates a primary gender of masculine for both. For the Stoics, the masculine gender was interpreted as embodying the *active* principle, which would explain fire and air being associated with the active principle. The other two elements earth and water are gendered as FF and FM respectively and so are primarily feminine and hence considered as embodying the *passive* principle.

So far, we have provided the fundamental ontological justification for the ancient four element based physics that was adopted by the Stoics. However, there was no trace of any fifth element in our development. A clue to the missing fifth element can be found in Chrysippus' five undemonstratables, in particular, the third undemonstratable. As for the other four, we will use them to resurrect a square of oppositions for the Stoic logic of Chrysippus. The Stoic square of oppositions will be comparable to that of Aristotle. Just as the medieval scholars gave four letters to the four terms of the Aristotelian square, we will do the same for the logic of Chrysippus. However, instead of the medieval AIOE lettering we will use the AUGC lettering of the generic cum genetic code that we are developing. As for the fifth term, it has no letter. There is no fifth letter in the genetic code. There is no fifth letter in the generic code.

In the biological genetic code, the AUGC lettering based on the codon triplets of letters, code amino acids that go to making up protein. There is no sign of any fifth player in the scheme of things. Moreover, there is no need to try to find a fifth player as it has been there right from the beginning of our development. It forms the core of the very essence of gender, the generic building construct of anything that aspires to be.

To be what one *is* does not come easily. Being is not something handed out on a plate. To *be* requires Oneness of the being in question and it is up to that being, and that being alone, to maintain and express its own Oneness. The masculine entity expresses Oneness. As attribute it is pure singularity, pure Oneness. The remarkable thing with the resulting science that can be built upon this initial union of the feminine with the masculine is that entities that are more qualified can be constructed, based on compound gender typing. The first compound types are the four binary combinations of the feminine and masculine and can be considered as the four fundamental "letters" of the generic code. Thus, the four compound genders MF, FF, FM, and MM are allocated the letters A, U, G, and C respectively. The arguments for this particularly allocation will be advanced later.

Ontological gender appears, albeit informally, in the cosmologies of the many different civilisations from the West to the East and Far East and beyond. In this work, we present a formalised version of the construct. The formalisation is not an abstract axiomatic, but a formalisation of a different kind, the *generic* formalisation. The generic formalisation leads to a new kind of science that we could call *generic science*.

The essential ingredient in such a science is the rapport between the feminine and the masculine. These two gendered entities are different, but indistinguishable. They are indistinguishable because there is no way to compare them: they both share the one single attribute between them. The attribute that they share is that of *Oneness*: one *has* it the other *is* it. There is a tension between these two entities. If ever the bond between them were broken, then that would spell the end of the world, at least for it.

The tension between the two genders expresses itself at the microscopic level between all the individual masculine-feminine compounded gender typing of a complex organism. Threats to Oneness of the Organism abound at all levels. This complex compounding of tensions throughout the organism is probably what the Stoics referred to as the *pneuma*.

We will continue our discussion of the pneuma, this mysterious "fifth element" further on. In the meantime we take another interlude to hammer home the intuitive understanding of the simplest but most profound concept of all, that of ontological gender.

A Light Interlude

Gender is not an abstraction. Any being is it and has it. Any individual being is gender typed. Once again, this is not an abstraction. For example, every cell in an animal's body contains a copy of its chromosomes consisting of long strings spelling out the genetic coding of the individual. The coding is built up of words consisting of triplets of the four letters AUGC (using the RNA convention). From a gender perspective, the genetic coding has a deeper structure than that determined by biochemistry. The four letters can be represented in terms of binary gender typing MF, FF, FM, and MM respectively. Thus, if the reader wants to know his or her gender typing then all they need to do is to translate their AUGC based genome into the gender equivalent.

Gender can be illustrated more concretely than even this biological version. Consider the following two rather tongue in cheek examples of gendering. The examples may help to overcome the bad habit of always thinking with a left brained mental disposition. Each example aims to prove once and for all, which gender is superior to the other. The reasoning is only vaguely inspired by the Hindu Naya five-step syllogism and so lacks some of the rigor.

The Masculine is Superior to the Feminine

Proposition 1

The masculine gender is superior to the feminine gender.

Example:

Take the case of a bird in the hand and the bird in the bush.

Analysis:

The bird in the hand is of masculine gender as it is in possession of the subject, a fact that is absolutely and even tautologically true because the bird is in the subject's hand. Conversely, the birds in the bush are of the wildcard species, not in possession, and undoubtedly hard to catch. These birds are obviously of feminine gender.

Proof:

In order to prove the proposition we invoke the age-old proverb:

A bird in the hand is worth two in the bush.

Conclusion:

This goes to show that the masculine gender is superior to the feminine (by at least two times).

Not to be out done, there is another argument, which demonstrates the converse.

The Feminine is Superior to the Masculine

Proposition

The feminine gender is superior to the masculine gender.

Example:

Take the grass on your side of the fence and the grass on the other side of the fence.

Analysis:

The grass on your side of the fence is of masculine gender as it is in possession of the subject (which is you). This fact is true for tautological reasons. However, the grass on the other side is not in the subject's possession, and remains tantalising out of reach, a real wildcard. That grass is obviously of feminine gender.

Proof:

In order to prove the proposition we invoke the age-old proverb:

The grass is always greener on the other side of the fence.

Conclusion:

This goes to show that the feminine gender is superior to the masculine gender (and much more desirable).

There is another variant of this proposition that employs the notion that the spouse on the other side of the fence is more desirable than the spouse on this side of the fence. We will not go into a detailed analysis of that case, but the variant has some value as it goes to show that gender is only obliquely related to sex. A spouse of one sex can be gender typed masculine or feminine, depending on which side of the fence they might be.

The Gender Algebra of Oneness

As we have said often, left side sciences are based on abstraction, dualism, empirical attribute harvesting, and employs a labelling, categorising, taxonomic epistemological technology. Right side science replaces the abstract with the generic, dualism becomes a non-dualist monism and the attribute harvesting from the field and laboratory is replaced by attribute construction from one single attribute based on the generic gender construct. As for epistemological technology, labelling, the linear descriptions, and essentially rhetorical approach of left side science give way to the dialectical where concepts and semantics are expressed in purely relative terms. Concepts and constructs are expressed in the form of oppositions. Moreover, whilst left side science specialises in the tunnel view of reality, right side science must always address the whole.

Right side science must always take the holistic view. This is the essence of monism; nothing can be left out of the picture. This means that right side science always deals in wholes. A whole is totality viewed from a particular point of view, the point of view of the present subject. The subject is always present in right side science. In the traditional left side sciences, including mathematics, the subject is always absent. As such, left side science specialises and only recognises one-half of reality: it is half-world science.

Right side science must be a monism. There can only be one such science. The Stoics were pioneers in this area with their unified version of the monism, and we follow in their tradition. However, we make no attempt at doctrinal orthodoxy.

The incredible thing about right side science is that it is independent of scale. One still sees the universe as a whole from even the most apparently microscopic point of view. The science is in fact starting point invariant. It does not matter where you start; you always get the same science. There is one and only one science with this unique property. As a theory, the theory is its own invariant. The science is independent of scale and independent of starting point. It does not matter where you start; you always get the same theory: That is the theory.

Generic Coding

Left side sciences are dominated, even swamped by an ever-increasing avalanche of attributes. Contrast that with the attributes in Generic Science, the right side science. There is only one attribute for the whole science! This single attribute is the attribute that the pure feminine F *has*. It is the pure masculine M. The feminine *has* an attribute. The masculine *is* that attribute. That defines what gender is all about. One entity *has* it; the other *is* it. All other attributes are simply built up from different combinations of the masculine and the feminine.

It might be thought that the feminine F is also playing the role of attribute, which of course it is, but only by role-playing it, not *being it*. And so here is the difference. We can be absolutely sure of what the masculine attribute means. We understand the masculine as it means one and only one thing. It means pure *Oneness*, pure singularity. It represents pure *certainty*, because we know what Oneness means. The situation with regard to the feminine is the opposite. It represents the opposite of certitude. It represents total *ignorance*. We do not have a clue of what the feminine *actually is*, not a clue. This is the absolute expression of the Uncertainty Principle. It is also the secret of Generic Science. Generic Science is the only science that can talk about something that it knows absolutely nothing about, and talk about in absolutely certain terms. Moreover, it can express all this in algebraic form.

Now the reader may have had a similar experience to the author as in the following. The author remembers vividly, many years ago in junior secondary school, when his mathematics teacher started the day's lesson with great drama, something he was prone to do quite often. His name was Harry

Sermon. With great drama, Sermon wrote up on the blackboard one letter. It was the letter x. Now x was just like a number. Apparently, you could add it, subtract it, multiply by it and so on. Remarkable. Even more remarkable was that you could do all this without having a clue of what the actual value of the number x was! Remember that? The lesson of course was an introduction to elementary algebra. Now having gone through what we now realise was not a complete waste of time, we find ourselves faced with the prospect of learning the algebra of the Cosmos, a worthwhile enterprise surely. Instead of the letter x, we have the ultimate unknown of all time and possibly even outside time, the letter F. This is the wildcard of the cosmic algebra. Its value can be *any entity whatsoever*. In the case of x in our first algebra lesson, at least we knew it corresponded to a number, even if we did not know its value. In the case of F, we do not have a clue about *anything*. We don't even know whether it is a value or not. It might be perhaps a placeholder for a value, a fruit cake, or a neutrino, or god knows what.

Our project is to develop a way of describing our pure ignorance of the world. Now some people might say that if you do not know what you are talking about then you should shut up. However, we take the high road, the road of Socrates' confession of ignorance. He is reputed to have said that the only thing he knows with absolute certainty is that he knows nothing with absolute certainty. In addition, this is our position; and our task is to develop algebra up to the task of expressing such wisdom.

One of the main points made in this work, is that we have apparently been beaten to the gun. Instances of the algebra, the algebra of Socrates' confession of ignorance, are everywhere. In your body, repeated in every cell is the same description over and over again, of what you are and are to become. All that has to be done is to "solve for F," based on the situation on hand. The question, 'Why bother?" may be raised by those not motivated to solve things, particularly algebraic things. It is a good idea to solve these Socratic gems of wisdom, as a failure to do so may challenge your very existence. If your body gets its Oneness equations tangled up, you could be in real strife.

...

Only two letters are needed for this code, the letter M and the wildcard F. Combined in pairs they make up the four letters of the alphabet of the generic, pardon, the genetic code. These are the letters A, G, U, and C, all binary compounds of F and M. The macroscopic organism is organised as an immense compound articulated by these four letters, an immense compound of F and M gendered entities. M expresses the Oneness of the organism, F the unqualified, the unknown, "Solve for F," and perhaps you have your life in a nutshell.

However, solving for F may take a lifetime and could depend a bit on what crops up along the way.

Stoic Pneuma as a Body

The generic code applies to any body that exists. According to the Stoics, a material body that exists can act upon or be acted upon. Conversely, what can act upon or be acted upon is a material body that exists. Bodies are made up of what can be described in terms of the four-lettered generic code. In the case of the biological body, any biological body, the body is coded in terms of the genetic code, the same code as the generic. In the case of the elementary entities of physics, as we shall later, there is no discernible difference between code and entity as there is in the more qualified and determined biological realm. The stuff that codes is indistinguishable from the stuff that is coded.

Now, according to our generic science approach, any particular body that exists can be coded in terms of the highly nominalist four-lettered generic code. This seems to be in accordance to the ancient four-element system of physics as, for example, adopted by the Stoics. However, the Stoics would claim that the four elements are passive relative to another additional element, the pneuma. The pneuma is active, permeates, and organises the whole system. If, as the Stoics claim, this fifth element can act upon and be acted upon, then it also must be a material body.

Now clearly, this fifth element is of a different kind from those that can be coded in terms of the four-lettered generic code. For a start, there is no fifth letter to code for this fifth element. The fifth element cannot be coded *explicitly*. The only alternative is implicit coding; the so-called fifth element, this fifth player in the game, must be implicit in the coding. Indeed, this is the case. The fifth element is not coded in terms of gender, as are the other "passive" four elements. It is not a substance, which is encoded; instead, it is a substance that *encodes*. It is "active" stuff.

Stepping aside from all the intricacies of an argument without end, the important notion to grasp is the primordial role of this active stuff, this stuff of the fifth kind. This fifth kind of stuff, this coding stuff, codes the essence of any particular individual. It encodes you, the reader, for example; however, it does not code you to be an accountant or vice president of the tie department of the local department store. It codes you to be that Oneness which is your particular Self. In brief, it encodes *Oneness*, together with a broad strategy of how to achieve and maintain this Oneness in a particular way, your way. The code is oblivious to accountants and department stores and knows nothing about ties. It only knows the iron laws of Oneness as expressed, hopefully without abhorrent defects, in your particular particularity.

Any violation of Oneness is a serious business for any organism. It calls for immediate corrective action. Oneness is the system invariant *par excellence*. It must be.

The generic code for any organism, be it biological or something else, expresses a particular *disposition* of that organism to excel or flounder in particular competencies. For example, the genetic makeup of some people may make them very susceptible to a certain physical disease such as sugar diabetes or a mental disease such as depression. Such afflictions may also be compounded by environmental factors. Such susceptibilities could be termed, in Stoic terminology, as dispositions. Other examples of dispositions might be a disposition to be a high achiever, athletic, studious, indolent, and lazy and so forth.

The Chrysippus Iron Law of Difference

Our immediate task is to determine what the Stoics were talking about with their mysterious *pneuma*, this fifth kind of substance, the expression of the Active Principle that permeates all the other elements and organises the essence of the organism through its tensions and tenor.

We have noted that right side science is starting point independent. Generic Science can be thought of as the science of the starting point. Having started here, you must get the same theory as if you had started somewhere else. Einstein's Special Theory of Relativity states something by demanding that the laws of physics be the same, irrespective of reference frame. Not only that, but it seems that Nature, not only obeys this rule, but also actually writes it down in her Code. In the Appendices, we claim to show how implicit geometry behind the start codon in the generic cum genetic code is none other than the spacetime kind of geometry of the Special Theory. Of course, there is no notion of the invariance of the speed of light, just the requirement that the organism is based on a theory that is starting point invariant. This is quite a remarkable claim. In other times, it would probably be enough to get oneself certified or burnt at the stake. But we make the claim anyway.

The starting point in generic science must formalise the generic requirement for right side science that the subject is always present in any whole. In any whole, there will be substance that belongs to the subject and substance that does not. This is a most fundamental dichotomy. In order to thrive in the world, our generic subject must very quickly realise this distinction. If you lose track of what is *you* and what is *not*, then you are in big trouble.

In order to function in this world, what we need, and what any subject needs, is an iron law that can be applied in order to resolve this vexing question of what is subject and what is not. Once again, we come back to the minimalist starting point, where we find the pure feminine entity whose only specificity is its total *lack* of specificity. This entity is qualified by the unqualified. Its attribute is the attribute of non-qualification. Demanding a reality based on First Classness, the attribute that this primordial entity *has* must be an entity in its own right: here we find the pure masculine and it is here that the problem arises. We have two entities: Which is which? We know that the feminine entity is totally unknowable, a pure wildcard. The only thing certain in this Socratic apology of ignorance is the absolute certainty of the ignorance itself, just as Socrates would have claimed. Here we find the good old tried and trusted masculine, an open book with nothing to hide. What you see is what you get.

It is in an initial minimalist ontological scenario that we gain our first snippet of hard-core generic wisdom. The next step is not so obvious. We know that the only specificity of the masculine entity lies in the fact that it is an attribute of pure singularity, nothing more, and nothing less. With the union of the masculine with the feminine, the mysterious feminine can at last claim to have Oneness. She *has* it he *is* it. Oneness is synonymous with subject. Lose Oneness and the subject is lost, the dichotomy is lost. Any hope of *being* is lost. Thus, in this scenario, determining which of these two entities is subject and which is not comes down to determining which is masculine and which is feminine. We need an iron law to resolve this question. If it is possible to sex chickens, then it should be possible to gender a couple of primordial ontological entities.

The iron law must be generic in nature. This means it must be applicable to any being whatsoever. Fortunately, we do not have to look far to find the generic resolution of our primordial entity gendering conundrum. The Stoic master logician Chrysippus comes to the rescue. Yes, it is in the form of his third undemonstrable; and that provides us with the generic iron law that we are seeking.

The scenario is that of two entities, one of feminine gender and one masculine. They form a whole. However, which is the masculine gender?

The law effectively declares that they both cannot be masculine at the same time. Thus, if one is masculine gender, the other is not; i.e. the other is of feminine gender.

Chrysippus' third undemonstrable is usually stated as:

> You cannot have the first and the second quality at the same time
> You have the first
> Thus, you don't have the second

The one you have is the masculine. The one you don't have and can never have is the feminine. The situation is black and white. There are absolutely no greys. This is the iron law of difference.

Superposition of quantum states

The generic answer to the gendering problem can provide a generic answer to the chicken sexing problem. The question is posed: 'Is this chicken a chook or a rooster?' Frustratingly, the answer comes back: 'It must be one or the other and can't be both at the same time'. Frustrating perhaps, but note that the answer is quite generic as it even covers the case of the chicken having a sex change later in life. Apparently, some fish can change sex, so it is not off the cards.

At this point one might think that the author is not being serious. Sometimes he is, sometimes he is not. You cannot be serious all of the time. However, in this case he is deadly serious. This third undemonstratable of Chrysippus' logic is applicable on both sides of the epistemological divide. Interpreted from the traditional left side logic point of view it becomes the Sheffer stroke. As Sheffer showed, all of the logical operations of the propositional calculus can be expressed uniquely in terms of the Sheffer stroke alone.

However, that is not the exciting bit. What really matters is the interpretation as a fundamental *right side* scientific law, not a logical law, but an *ontological* law. To avoid lapsing into abstraction, consider a biological being. Take the chook that we have been talking about, for example. We will assume it is a chook, but it really does not matter. It could be even be a toad. All that matters is that it is a being of some kind. As a biological being, we know that this chook is coded through and through with its particular version of the chook genome. The biochemist sees this genetic coding in terms of the AUGC alphabet. The ontologists (that's us) see the chook coded in terms of the more detailed gender alphabet MF. From a biochemical perspective, the chook is a seething mass of biochemical activity. From an ontological point of view, this chook is intent on obeying the iron laws of ontology motivated by the desire to *be;* in this case, to be a chook.

Although this chook is genetically determined by its particular version of generic coding, when launched into *being*, it becomes a seething mass of MF interactions in such a way that when viewed as a whole, from any perspective, Chrysippus' third undemonstratable holds sway.

Perhaps we can look at this from a Quantum Mechanics viewpoint. Here we are talking about Quantum Mechanics where Chrysippus' third undemonstratable does indeed start to raise its head as an iron law at play in the material real. We are talking here about the phenomenon of superposition, for example. The magic formula for Quantum Mechanics is Schrödinger's wave equation. The customary story is that the wave equation predicts the states of any minute subatomic system. However, the prediction is nondeterministic; it is probabilistic. Instead of the system being in one determined state, it is generally considered to be simultaneously in a superposition of all the possible states. This is what superposition means in Quantum Mechanics. The actual state of the system only becomes determinant *after* the intervention of a subject, for example at the moment of a measurement taking place. The wave equation is then said to 'collapse' and the system takes on determined states within the probabilistic bounds indicated by the wave equation.

Try as it might, traditional physics has not been able to attain its holy grail of becoming a completely deterministic science of matter. The age-old Epicurean swerve keeps coming back to add spice to what would be otherwise a completely mechanical, deterministic world. The basic truth of the matter is that traditional left side science cannot go any further than this.

Another inevitable outcome of the traditional left side sciences is atomistic dualism. One can see this clearly in the Standard Model in elementary particle physics. The situation is even starker than even Epicurus would have expected. The Epicureans had the elementary particles floating around in a void, just as in the modern Standard Model. This is pure atomistic dualism. However, for Epicurus all of his atomic building blocks had extent. In the Standard Model, all of the elementary particles, the quarks, the electron and neutrinos and so forth, without exception, all of them are point-like singularities. This is quite an incredible view of reality. Matter is totally made up of a vast sprinkling of zero sized specks floating around in the void. According to the ancient Epicureans, the motion of these atoms would not be totally deterministic: from time to time the atoms would experience randomly swerve. Modern science has formalised this random behaviour in the form of another duality, the wave-particle duality. The particles behave like waves. The waves can in turn be interpreted as probability distributions for the position and motion of the particle.

However, one should not make the mistake of attempting to criticise the scientists for coming up with such a philosophically unsatisfying model of reality. This is not bad science. This is just the inevitable solution that the left side science paradigm is destined to produce. Even Epicure, over two thousand years ago, intuitively foresaw the general gist of the atomist scenario. Left side science inevitably produces abstract, dualistic, atomist models of fragmentary aspects of reality. This is not because that is way things are, but how they appear from the point of view of the left side scientific paradigm.

On the other hand, right side science does not produce abstract models of reality. This new scientific paradigm is monist not dualist. As such, there can be no duality between abstract theory on one side and the concrete real on the other. There cannot even be a dichotomy between the describing language on one side and the described spectacle on the other. The right side scientific paradigm demands that language *and* substance are both present in the spectacle. Thus, the science is built around a generic language. In the biological world, thanks to the left side sciences, this generic language can be observed in action. It is the four-lettered genetic code. We interpret it as the four-lettered generic code, observable in its raw form even at the level of traditional particle physics.

This four-lettered generic language has a deeper structure based on binary gender typing in terms of masculine and feminine gendered entities. The algebra of gender typing forms the backbone of right side science. It is via the gender construct that a bootstrapping ontology becomes possible.

Right side science reasons in terms of wholes. Reality is always from the holistic perspective with the subject always present. The subject provides the reference frame for the science. The development starts with the minimalist scenario concerning the pure unqualified substance as object of study. This object of study must *have* a subject present. The generic nature of the object is determined by the generic specificity of the subject. In this case, there is no determined specificity. In this configuration, the object of study is an entity of pure feminine gender. This pure feminine entity has one single attribute. The attribute is none other than the subject as an entity in its own right. This attribute entity is said to be of pure masculine gender. Thus, the feminine, an absolute unknown, *has* a subject; the masculine *is* the subject. The subject itself is the *masculine pure*.

The subject in this primordial dichotomy is the *impersonal* subject, the one with the "view from nowhere," the God's eye view. However, this dichotomy violates the most fundamental principle of the Cosmos, that of First Classness (FC). The right side epistemological paradigm must maintain a monism at all costs, the basic requirement of FC.

The great challenge thrown out to anything that aspires to come into being is to find a way around this conundrum. Life in any form is impossible unless the dichotomy between subject and object, between the feminine and masculine pure is allowed. The only alternative seems to be a futureless formlessness: subject and object might still be present, but in superposition. This situation seems not to violate FC. There is no absolute dichotomy splitting the whole in half.

...

Traditional left side science is fascinated by this simple scenario. The most famous example is the Schrödinger's cat thought experiment. Schrödinger provided the thought experiment as graphic metaphor to help understand superposition. Left side dominant thinkers have since tended to take it more literally.

A cat is hypothetically put inside a box with enough vitals to keep it alive for the length of the experiment. A randomly activated device is also placed in the box. If and when the device activates it will kill the cat. A group of scientists sit outside the box, wondering whether the cat has or has not met its maker. One bright spark amongst them, who doesn't believe in makers, comes to the conclusion that the cat is in a superposition state of being both dead and alive at the same time. Once the box is opened, all will be revealed. The superposition state will collapse into either the dead cat or the living cat state.

Schrödinger's graphic cat metaphor illustrates the seductive nature of left side thinking. Taken literally, it reveals difficulties but the difficulties are solely due to the inability and refusal to deal with subjects. The subject is completely absent in the left side paradigm. The only subject allowed is the one with the God's eye view, the impersonal subject. The left side paradigm subsumes the impersonal subject into its version of objectivity and so it too is effectively emasculated. As for the observers sitting around staring at the black box spectacle, they are also subsumed into the impersonal subject point of view and thus become passive objective onlookers.

Formalising the Active Principle

The thing about a whole is that you cannot put it in a block box and sit around watching it. If it is a whole, *you* have to be in it. As Leibnitz said for his monad kind of wholes, they have no windows. That is the essence of wholes. There is only one sitting place for the whole and that is on the inside. There is no room for bright sparks on the outside. Wholes can have no outside.

We started off with a whole made up of a pure feminine entity and a pure masculine entity. This formless futureless couple apparently eke out an existence in some sort of superposition of genders. Each of these two entities is both masculine and feminine at the same time. Of course, if this were the case then we can no longer talk about two entities but merely one entity with gender in superposition, a real hermaphrodite.

The ontological obstacle encountered here comes about due to the requirements imposed by FC. The formless, futureless hermaphrodite entity seems to provide the perfect solution for a system satisfying FC. This solution expresses the *passive* principle, the superposition solution where something can be *and* not be at the same time. However, there is another solution, which can also be compatible with FC. We can call it the *active* principle and formalise in terms of Chrysippus' third undemonstratable, the incompatibility syllogism. This syllogism states the converse of the passive principle, notably that something *cannot* be and not be at the same time. There is a caveat thought. The active principle involves relativity. The incompatibility of superposition is the internal active organisational principle of any individual being, *relative* to that being. This is the way the individual sees itself. It is conscious of itself without ambiguity, provided the active principle prevails. Any onlooker may get a totally different impression. But then, it is not the onlooker's job to organise someone else's daily ontological and existential dilemmas.

At stake here, is a very fundamental principle. Fortunately, by exploiting the gender construct, the principle can be stated more formally and simply. The active principle demands that in any context, the gender typing cannot be masculine and feminine at the same time. Since gender is a relative determination, the active principle is guaranteed to be relativistic.

From a physics point of view, gender does appear to something like a binary valued quantum state. However, the quantum state is something observed by the outside observer, posing as God. From this point of view, superposition will be the norm, provided the observer does not look. As soon as the observer opens their eyes, superposition collapses. This observer, half god, half-confused left side thinker, is in a rather enigmatic predicament.

A gender state is different. Gender is a *relativity* construct, where all gender typing is defined relative to all the others of the organism. The active principle demands that at any time there will be, or must be, no ambiguity of gender typing throughout an individual's scope. Ambiguity in gender typing spells the doom of gender typing of an organism. It spells the doom of the organism as an organism. Gender typing throughout the organism is the coherence maintaining mechanism of any being, a totally relativistic, interrelated system by which the organism knows what is and what is not itself. Time will also be a relativistic construct, relative to the organism. It is the organism's responsibility to manage its own temporality.

The task of maintaining coherence of gender typing throughout an organism is essential to maintaining integrity of Self. In response to all the onslaughts experienced by an organism, it is faces the formidable task of policing and enforcing the incompatibility principle throughout its being. The author believes that this phenomenon, with all its tensions and *tenor*, corresponds to the Stoic concept of *pneuma*. The pneuma seems like dynamic ethereal elastic binding logico-material substance that bonds the organism into a whole. The basic organisational principle of pneuma is the incompatibility of any ambiguity in gender typing. Something may experience different gender (gender state) but never at the same time, relative to all other gender states.

The pneuma of an organism is the systemic expression of the incompatibility syllogism as the fundamental organisational principle of system coherence with itself. It helps to understand Chrysippus' insistence that logical propositions should be bi-valued. This becomes the very essence of the organisational principle of the organism.

Sidenote:

Chrysippus' claim that logical propositions should be bi-valued worried the author for some time. At first sight, this seems to imply a logic based on the excluded middle. Traditional, axiomatic logic is bi-valued and based on the excluded middle. It is generally considered that constructionist logic does not respect the excluded middle and has an extra truth-value, the "Don't know" value. We are interested in constructionist logic and so should Chrysippus. However, once logic becomes embedded in a relativistic schema, where every proposition is endowed with relativistic ally-determined context, the situation changes. Propositions become bi-valued and the law of the excluded middle returns. When framed in our gender calculus, the excluded "Don't know" value is not lost but gets integrated into the feminine, the ultimate "Don't know" wildcard, the only one of which you know with absolute certainty that you know nothing, the purest of Socratic ignorance.

Chrysippus is correct to insist on bi-valued logic when he is interpreted in this relativistic sense. Where mathematical logic differs is that its propositions are devoid of context. It is a context-free logic, a context free zone. The amazing thing about Chrysippus' logic is that it can be interpreted both as a bread and butter version of traditional propositional logic and, on the other hand, as the core reasoning of a fully-fledged ontological logic. It seems that the ontological interpretation is not at all appreciated or even discerned by the moderns.

Effect and Affect

From a causal point of view, left side sciences see Nature in terms of chains of causes and effects. There is no mystery there: life is just like playing billiards. What should be the right side science take on causality? Before looking into the Stoic version, consider the two takes of the biological brain. The analytical left hemisphere would accord with the cause-effect view of traditional left side sciences. As for the right hemisphere, its reaction to events is more in terms of emotions. Rather than rationalising about A causing B, the right hemisphere is more concerned with what *that* means to the subject, how to accommodate *that*.

The chook cocks its head around so it's right brain, connected to the eye on the other side, can check out that shape lurking in the shadows. There will be no attempt at classification but simply an engendering of emotion, perhaps an icy feeling that it is high time to get out of here or one of annoyance because it is that pesky rooster again. Intuitively, we are starting to see that, from a right side perspective, the importance of effects arising from causes gives way to emotions and affections engendered in the subject due to the interaction with another body. This is an entirely different scenario, but can it be scientific?

On the right side, rather than talking about effects, we will use the term *affects*. An affect is engendered by another body acting as cause of the affect experienced by the subject. Thus, unlike the binary cause-effect structure of left side causality, the right side take involves three players. There are two corporeals, the subject, and the body that engenders the affect experienced by the subject. The third aspect is the affect itself, which is incorporeal. Fairly naturally, we have come up with a causal format that seems to accord with that of Chrysippus as explained by Susanne Bobzien (Bobzien, 1999). The main difference with Bobzien's version, other than our "affect" terminology, is that the body experiencing the affect is not just a body "out there:" It must be the *subject*, because all right side reasoning is relative to the subject. The subject is the centre of the universe for right side science, and the same must apply to Stoicism. Not only must this causality be determined relatively, as Bobzien points out, it must be determined relative *to the subject*. Moreover, the subject must be present in any scenario. This is the central requirement for the right side paradigmatic alternative to the *science sans sujet* of classical left side epistemology.

Our comments here cannot be taken as criticism of Bobzien's presentation. Bobzien is constrained by the rigours of scholarly discipline based on historical evidence, whereas our kind of interpolations and extrapolations might be seen by many as quite hair-raising. In this work, when we talk about the Stoics, we are talking about *our* Stoics, not the strict historic version. We talk about how they *should* have thought. How they actually thought should be left to the scholars, but our ideas might provide some useful insights nevertheless.

Emotion and Nature

According to the Stoics, the art of living happily is to know how to control one's emotions and passions. The guide to leading such a life is to live according to Nature. This raises the question of how Nature controls *her* emotions. Can Nature have emotions? If so, what is an example of Nature actually displaying emotion?

Questions like these are enough to make many a traditional left side scientist reach for his revolver. Unperturbed, we will fearlessly continue with these questions and endeavour to answer them in a coherent manner. To begin with, we must give some precision to what constitutes emotion. To the traditional scientist, emotion is something totally anathema to his science. Maybe that is why he is threatening us with his revolver. However, as we have explained, his science is based on the duality between what he sees as the empirical real and the abstractions he employs to understand the empirical real. He is oblivious to any other alternative epistemological framework. He is a specialist. He doesn't do epistemology, or ontology or any of that kind of stuff; he does his particular scientific speciality. Outside of his speciality, quite often, his cultural level is little different from that of his motor mechanic, and may even be inferior. This fact is not considered a detriment to his scientific capability. The simple fact is that traditional left side science does not require anything but the most minimalist amount of philosophical adornment. It can function with almost no philosophical infrastructure at all. It goes it alone, so to speak.

The polemic aside, what is of particular importance is not just the cognitive style of the left side scientist, but also the discursive style. The traditional scientist has learned to explain his science in the form of a persuasive, well-reasoned monologue fortressed by empirical evidence. He carefully defines his terms and proceeds in a way to which the traditional scientist has become accustomed. The resulting scientific production will never be definitively truth, as Karl Popper pointed out. It will always be tentative. However, is will be rationally presented with suitable citations, evidence based, and persuasively cast. This is the left side science discursive style.

The ancients, including the Stoics, studied this particular discursive style as a discipline in its own right. They called it rhetoric. In addition to rhetoric, there was a second discursive style. Instead of being conceived as a monologue, it was presented as a dialogue. They called it dialectics. Plato's *Dialogues* provide a famous example of this discursive technique. In the classic dialogue, two protagonists argue opposite sides of the argument in an attempt to arrive at understanding. What matters here are not the two protagonists, but the two opposing points of view. A single interlocutor is capable of pitching one argument against another and so argues dialectically. Kant used this technique and came up with his antinomies made up of two contradicting propositions. He claimed that any fundamental attempt at a bootstrapping kind of meta-

physical philosophy would end up being confronted by unresolvable antinomies, and he was quite right.

To the left side rhetorical thinker, antinomies do not present a problem as
they only arise from deeper metaphysical investigations, something that such
thinkers avoid. Antinomies do not present a problem because they do not
present. However, with the *right* side take on reality, antinomies cannot be
avoided. A dialectical approach is required to grapple with these double-
headed slabs of rationality.

Now the dialectic is probably the least understood and most abused rational artifice of all time. There is no time to go into all the grotesque details
but suffice to say that whenever there is an opposition of two diametrically
different sides, there will always be simplistic formulas for resolving the
situation. For example, one way of resolve the squabbling between opposite
sides of the parliament is to replace it with a one party state. This is an example
of the bad dialectics, the dialectics of Evil.

Good dialectics takes the high road and recognizes that truth does not
reside in what is true or not true. The truth resides in a system that recognises
the relativity of what is true and not true. Such a rational system, such a
Logos, must be built from oppositions alone, and oppositions between
oppositions. Such a system must obey the requirements of First Classness,
where there is no absolute entity or proposition that dominates or is antecedent to all others.

Explaining such a dialectic system demands not a linear, monologuing
rhetorical discursive style, but a dialectical discursive style where concepts are
expressed in terms of oppositions rather than from predefined definitions and
labels. In this work, we have consciously attempted to avoid the rhetorical as
much as possible and employ the dialectical discursive style.

Now our immediate topic of interest is emotion. What does the dialectic
have to do with emotions? Dialectics expresses concepts in terms of oppositions. Thus, consider the opposition between dialects itself and it opposite
number, rhetoric. The rhetorical discourse has a particular stance to such
things as emotion, sin, sex and sensuality. The rhetorical talks about such
matters clinically, dispassionately, rationally, abstractly and often with a great
deal of apparent authority. However, it seems that these concerns come from
another planet. The rhetorical discourse is at best, that of the voyeur.

In contrast, the dialectic discourse is one of being immersed in a world in
discourse with itself. It is the discourse of the participant. It is body language.

Here, we are starting to get immersed in the poetic side of the situation.
To come back to basics we will now look at an important aspect of Nature
known to present day science. We will articulate the scientific content first as a

traditional rhetorical exercise and then repeat the process from a dialectical perspective. In the process, we intend to illustrate the difference between effect and *affect*. The *affect* side of the coin can have interpretations of an emotional nature. We are going to look at the special theory of relativity from the perspective of traditional modern science, and then from the point of view of Chrysippus' perspective, seen through the fog of history.

Special Relativity

Slipping into a simplistic description of the Special Theory of Relativity, we can state that the theory concerns the propagation of energy or information in the form of radiated electromagnetic waves. This concerns radio waves, X-rays, infrared rays, light and any other form of radiated energy. A central tenet of the theory concerns the speed of propagation of such waves, usually considered to be in a vacuum. The is usually called the speed of light and is the same for all forms of radiation. The Special Theory declares that the speed of light will be invariant for all inertial reference frames. In other words, the speed of light measured in a spaceship travelling at great speed relative to a static observer, will be the same as that measured by the static observer.

Now everyone has heard about the Special Theory in one way or another, and that the speed of light stays the same independent of reference frame. The more technically inclined would know that the mathematics relating the two reference frames is called the Lorentz transformation. We should remember what this transformation is called, as we will be coming back to it.

What is of interest to us is that the Special Theory provides us with an easily understood example of an *affect*, as distinct to an *effect*. First of all, consider the case of an effect. Effects are the result of causes. First, create a cause by running a plastic comb through you hair. Then create a corresponding effect by positioning the comb over a scrap of paper. The electrical charge on the comb will be the cause producing the effect of the scrap of paper being attracted to the comb.

This is an example of electrostatics and can be explained in terms of Coulomb's Law, which predicts the forces between electrical charges. Electrical charges produce an electrical force field. The predictions of Coulomb's law for electrical force fields are extremely accurate. However, the law becomes inaccurate for the case where electrical charges are in motion, relative to each other. The phenomenon becomes very evident when the speed of the electrical charges starts to approach the speed of light, which is the case for electrical current flowing through an ordinary copper wire. This divergence from the static case can be explained in terms of the Special Theory of Relativity, which adds a correction force to the static electric field case. This correction force field is called a magnetic field.

Putting aside the technical details, what we have here is that magnetism is a necessary relativistic correction that Nature makes in order to maintain the invariance of a physical constant in all reference frames. Magnetism is not an effect brought about by a cause, but an *affect* that is necessary in order to maintain a physical invariant of Nature. It is here we see a very good example of the difference between effect and *affect*.

However, the idea that Relativity Theory is based on the invariance of a scalar number seems very opaque. The natural question is "For God's sake, why?"

We need a more dialectical approach. The reality is that if the Special Theory of Relativity were violated, by particles bodies travelling faster than light for example, then a very nasty situation would arise. It would result in causes from the future producing effects in the present. This is tantamount to uncaused causes. Chrysippus, back in ancient Hellenistic times, declared that a single uncaused cause would spell the end of the universe (Gould, 1970).

Relativity and Causality

The mathematician Erik Christopher Zeeman sheds more light on the situation by effectively showing that the fundamental invariant behind the Special Theory was causation itself (Zeeman, 1964). The principle of causation demands that any affect event must be antecedent in time to any consequent effect event. Zeeman showed that in order for this principle not be violated in any reference frame, the mappings between reference frames must be the same mathematics as for maintaining the speed of light as a constant. In other words, the transformations must be Lorentz transformations. Zeeman's causality interpretation is a weaker, more generic version of relativity than the Special Theory. The constancy of the speed of light could be seen as just an implementation detail for assuring the non-violation of the causality principle in physics.

The causality principle is essentially a polarity condition: causes come first; effects come second. Any violation of Relativity Theory is a violation of the causality principle, which in turn is a violation of the causality polarity condition.

Chrysippus and Generic Relativity

In physics two masses of electrical charge that move relative to one another produce a magnetic force field affect. This is about as close as one can get to a display of "emotion" in Nature. The process involved is that of maintaining the causal coherence of the universe by assuring the polarity of cause/effect relationships. Cause must precede effect. Here we see that the details of cause and effect are secondary. What is at play is the polarity condition between the two. When there is a risk of violation, the magnetic field affect ensures.

In this example, there are the two bodies and the affect in the form of a magnetic field. Whether this in any way corresponds to the Stoic concept of causality is a matter of speculation. At any rate, there is an apparent anomaly as the affect, in the form of a magnetic field, is not an incorporeal as it is capable of acting on bodies, something that the incorporeal are not supposed to be able to do.

In summary, the left side version of causality involves events separated in time. Antecedent events are the causes of the posterior events, which are the effects caused by the causes. The effects then become causes of other effects. The causal construct is binary in nature. The Stoic causality paradigm, a right side version of causality, is a *ternary* construct. It involves the interaction of two bodies; one is active and the other passive. The third entity in the triad is not an effect but an *affect*. Linguistically the word effect is a noun and the word affect is a verb. Thus, the active body affects the passive body. This is a rather more subtle notion of causality than the left side version. The notion that affect is incorporeal starts to make sense.

Our example of affect was the magnetic field affect resulting from the interaction of two masses of charge. The example appealed to the Special Theory and the underlying necessity that the integrity of causality itself must be maintained, a requirement that necessitated the magnetic field affect. A Stoic example cited by Bobzien involved a scalpel as the active body, the flesh as the passive and the cut as what we have been calling the affect.

Trying to compare the thinking of the ancients with modern physics is fraught with danger at the best of times. However, it is interesting to make these brief remarks along the way.

Zeeman's causality interpretation of Relativity hints at the possibility of a more generic form of relativity than modern traditional physics, relativity as an expression of the causality polarity condition. In our case, we are interested in a much more profound polarity condition than that of cause and effect. We are interested in the most fundamental polarity of them all = that of gender, the masculine-feminine polarity. Rather than a calculus of cause and effect, we propose a calculus based on the two genders, an age-old idea going back to pre-Socratic times. This gender calculus goes way beyond the physics of matter and includes the algebra of life itself. The most visible form of this gender calculus is in the guise of the genetic code. We call it the generic code and claim that it is based on a four-letter alphabet constructed from the semantics of the binary gender construct. Having said this we now wish to shed more light on the semantics of this gender calculus by interpreting it in the perspective of Chrysippus' logic based on the five undemonstratables.

Zeeman's causality version of relativity effectively states that an event cannot be active and passive at the same time. The active, as cause, always precedes the passive, as effect, independent of reference frames. Moving over to the *right* side science version, we must talk in terms of *bodies* and not events. Events are not bodies. From a right side perspective events cannot be acted upon or act upon. It is here that we encounter Chrysippus' incompatibility syllogism, the third undemonstratable, that declares that a body cannot play two fundamental roles at the one time, that is to say, a body cannot be active and passive at the same time. This condition underpins causality and the necessity of temporal ordering. Causality demands temporality.

In what follows, we will endeavour to sketch out an alternative approach to relativity than the causality based versions than underpin traditional physics. The traditional approach is based on enforcing and maintaining a particular form of First Classness, notably that the laws of physics should be applicable to all possible reference frames. The reference frames are first class in the sense that no reference frame is superior to any other. The most important condition for realising this requirement is that the principle of causality is maintained for all reference frames.

Traditional relativity theory only provides a limited from of First Classness. The traditional approach violates First Classness by tolerating a rigid dichotomy that splits the Cosmos into two absolutely separate parts. One side is the world of objects floating around in a reference frame. On the other side is a swirling mass of reference frames in which objects can float around. A reference frame is not an object and an object is not a reference frame. Never shall the twain meet. In brief, traditional relativity theory is fundamentally wracked with dualism.

The only way to achieve true First Classness is with a monistic worldview. This is the right side worldview and leads to Generic Relativity. The monist version of relativity replaces all of the possible reference frames by a single entity, the *subject*. All of the objects that float around in the various reference frames are also replaced by a single entity, the same entity in fact. This is what a monistic worldview means: there is only one entity and this entity is the subject. This may seem like sounding the death knell to any further possibility of meaningful discourse, let alone a science. However, there is one escape clause: the subject involved is not just your ordinary everyday run of the mill subject (even though this possibility is not explicitly excluded); the subject is none other than the illustrious *any subject whatsoever*. This can be called the generic subject. Right side science is the generic science of the generic subject. Any fundamental law of physics must be applicable, without fail, to the generic subject. It is this draconian requirement that forms the Cosmos and the world around us into the shape it is today.

Unlike traditional relativity theory, generic science does not attempt a science applicable to every point of view. Every point of view is a meaningless abstraction. Such a purchase point cannot exist anywhere in the Cosmos. Rather, generic science concentrates on the point of view of one unique subject situated right at the centre of the Cosmos. This is the point of view of the *any subject whatsoever*. The whole Cosmos gyrates around this entity. There is a huge difference between the *every* and the *any*.

To develop a science of the generic subject is to solve the Kantian problem of developing a science from reason alone. As we have already discussed, the first principle involved is that of First Classness. FC is the organisational principle for any subject whatsoever. The science is expressed in terms of oppositions and oppositions between oppositions. Any entity must be determined relatively to any other. Absolute determinations violate FC. Predefined labels and definitions, for example, violate FC.

The most fundamental opposition is that of gender, where the feminine opposes the masculine. These two primordial gendered entities are determined relative to each other. The feminine is characterised by the sole fact that it *has* an attribute. The masculine entity *is* this very same attribute.

There are thus two entities with only one attribute between them. Gender provides the relativist typing mechanism for any subject. The generic subject knows entities in terms of compound gender typing. In other words, the subject articulates its knowledge of the world in terms of a calculus based uniquely on gender typing. We can call this the gender calculus. Another name is the generic code. If the four binary gender combinations are appropriately interpreted it can be interpreted as the AUGC lettered genetic code of the biological kingdom. In the context of generic science, if the kingdom is widened to be the kingdom of the generic subject, and not just the biological subject, the AUGC code becomes that of even the Cosmos. Of course, in this latter case, the Cosmos is taken as a particular instance of the generic subject. The life principle permeating the biological world has a much wider scope than thought.

So far our explorations of the gender calculus has been limited to the examination of static semiotic structure. This semiotic analysis perspective is rather wooden and static. In this world of oppositions that we are exploring, there is another opposition than that immediately based on gender. This second opposition is that between the active principle and the passive principle, so popular with the Stoics and particularly Chrysippus. The active-passive opposition is really the gender opposition, but in a different perspective, a fine point that we can blithely ignore in this particular discussion.

It is known that the specificity of any biological organism is overwhelmingly determined by its genetic code. Biochemistry only considers how the code transcribes into the protein building blocks of a living organism. Biochemistry ignores the deeper ontological semantics of such a language. This is where Generic Science comes into the picture. The role of Generic Science is to explain this "second articulation" of the genetic code, the articulation of the ontological semantics that govern the organisation of living organisms. Generic Science claims that the role of the genetic cum generic code is to describe and proscribe the building blocks of life in terms of a totally relativistic typing system based on gender.

Such is the wooden static perspective of the generic code. The code articulates an ontological, semiotic recipe for the organism, no matter what the scale, be it biological or cosmological. However, a recipe on its own is not sufficient to bake the cake. It needs a chef. According to Chrysippus, the chef is the active principle. However, this chef is unlike any other. It is purely an expression of logical necessity. If this coded recipe remains true to the rational constraints of logical necessity, this cake will bake itself.

The Stoics claimed that this logical necessity that guarantees systemic coherence was a material body. The called it *pneuma*. Perhaps an example of Stoic pneuma was the magnetic force field in our example discussed above. The magnetic field came into existence due to the dictates of the Special Theory of Relativity. Looked at more generically, the magnetic force field was an effect produced in order to maintain not just the invariance of the speed of light in all reference frames, but to ensure that the principle of causality was not violated. This example gives us a glimpse of the Stoic Logos in action. It certainly does give the impression of tension and *tenos* in some sort of ontological cosmic membrane like the pneuma. The Stoics claimed that all beings, including humans, worked on the same principle, no matter what the scale. Any such being is an instance of the generic subject, the unifying core of the science.

The problem of eventually formalising generic science starts to gain some traction when expressed via the gender calculus. The language of this calculus is the four-letter AUGC generic code. The genome for any particular being is expressed in this code. One half of the semantics is expressed in terms of the feminine wild card, the complete unknown. The other half is expressed in terms of masculine, the expression of absolute certainty, even though it expresses nothing less than pure singularity, pure Oneness. It is this intertwining thread of pure Oneness that prescribes how the totally unknown can be organised into the fold of the subject, a subject with clear corporeal ambitions. In this perspective, the masculine is an expression of the active principle.

The ontological, semiotic concept of a being is different from the chemical and biochemical. It involves a prescription of the dynamic generic geometry of the beast and a myriad of other organisational structures, each contributing to one unique objective, the coherent conscious knowledge of what is and what is not the subject. To know what it is, the subject must know what it has and *vice versa*. The language for expressing this "is-have" dialectic is the gender calculus based generic code. Right at the core, the wildcard feminine has an attribute; the masculine is that attribute. Together they create the world.

This generic coherence of self is synonymous with the coherence of the gender typing of each and every part of the subject. One way of understanding the gender typing is to realise that the process is *dynamic*. These typings are analogous to dynamically changing quantum states. The whole organism appears as a dynamically changing mass of gender based quantum states. Viewed from a third party point of view the states seem to be incoherent and in continual superposition, a great amorphous mass of perplexing ambiguity. But these "quantum states" are not like those proposed by traditional physics. They are not states defined relative to the view from nowhere of traditional physics. These are gender states each determined relative to each other. Moreover, they are in total coherence with and determined relative to a singular point in time. This point of time is called *now*. According to any subject, this is the only time when the organism really exists. Any subject must be continually conscious of this fact.

Chrysippus provides us with the fundamental principle which starts to formalise the concept. This is his incompatibility syllogism, his third undemonstrable. Interpreted from a traditional left side logical point of view, it is the Sheffer stroke. There, it is of little interest except that all other constructs of the propositional calculus can be constructed from this operator alone. However, viewed from an *ontological* viewpoint an entirely different picture emerges. The syllogism declares that there cannot be any ambiguity or superposition of states. Its premise declares either the first quality or the second, but not at the same time.

Interpreted in the light of the gender calculus, the first and second "qualities" become the feminine and masculine genders. In any context where the question of gender typing is raised, there can be no ambiguity, the laws of the excluded middle must prevail. The gender cannot be masculine and feminine at the same time.

In the classical Special Theory of Relativity, the generic concern is not to violate causality. This can be assured by demanding that at any determined time, an event is either a cause or an effect, but not both at this same time. The event is either an effect arising from a cause in the past, or a cause for an effect in the future. It cannot be both. This is an example of Chrysippus' third undemonstratable. This incompatibility condition on the present effectively forbids violation of causality, such as the looping of the causal chain back from the future to the present.

Demanding temporal gender coherence is a more generic and far-reaching requirement than just demanding temporal causal coherence. However, an understanding of the traditional causality based relativity can help to understand the totally generic variant based on gender typing rather than just causal typing.

Reason with and without a brain

When it comes down to brass tacks, the most important issues involving the sciences all centre on a critical political debate. The debate is about infrastructure: Who needs it? The first thing that a conservative government does on coming to power is to start making deep cuts into infrastructure. They cut back on public servants and teachers, deregulate the stock exchange, deregulate the labour market, and try to destroy the union movement. They then try to put likeminded people into the judiciary. Their motto is small government or, in the limit, no government at all except that determined by market forces. A progressive government does just the opposite. Where the other side deregulates, it regulates. It is for big government and justifies its position by staking a claim to serve the integrity of the society as a whole rather than that of a handful of well-situated individuals.

Nowhere is this difference in attitude towards infrastructure more stark than in in the field of logic. Already, we have seen the difference between modern logic and the syllogistic logic of Aristotle. For Aristotle, science needed an Organon, a tool for reasoning. It needed infrastructure and this applied to logic itself, well-illustrated by Aristotle's square of oppositions. Reasoning involved employing not only different kinds of oppositions, but relationships between oppositions. The square of oppositions provided a quasi-formal way of illustrating the logical infrastructure that regulated the world of reason.

In modern times, there has been a change of government. Led by George Boole and others, the rational superstructures of the ancients were abandoned and even ridiculed. Everything was deregulated and the whole infrastructure from Aristotle and medieval times was dispensed with. From now on there were no dialectical oppositions ravaging the world. Everything was black and white. The quest for truth was abandoned. What mattered was merely the true and false. There was no longer any need for infrastructure. Other than the opposition between truth values, all other oppositions were pronounced trivial hangovers of an overindulgent past.

The purpose of this book is to start the way back to another change of government. The infrastructure is to return. Instead of reasoning with our bare brains, we go back to reasoning in the context of the Organon. The Organon will not be that of Aristotle, but that of those that followed him. It will be inspired by the Logos of the Stoics. From our perspective, the new infrastructure will feature the generic subject, that most important ingredient accompanying any discourse involving fundamental reasoning.

In what follows, we will attempt to construct a Stoic version of Aristotle's square of oppositions. This will form part of the infrastructure of this new science that we call generic science.

Flat Earth Theory

There are two theories of the world. One is the Flat Earth Theory and the other is the Round Earth Theory. Space engineers and astronauts opt for the Round Earth Theory, but most people on the planet prefer the Flat Earth version. Atlases, maps of all kinds, and GPS applications on phones, all show the earth as flat. People simply don't need the constant hassle and overhead of trying to get their mind around the fact that they might live on an oversized basketball. They want to go from A to B. The fact that if they kept on going they would end up at A again, is usually more of a hindrance than a help. All in all, it must be said that the Flat Earth Theory is well and truly alive today. Most people feel that the exotic Round Earth Theory is best left to philosophers, romantic poets. and space walkers.

A similar situation can be seen in the sciences. On the one side, the left side, are the sciences based on the *terra nullus* theory. According to this doctrine, the world is totally devoid of inhabitants. The whole Cosmos is no man's land. The doctrine has enjoyed massive popularity over recent times. This kind of science was even used to send a spaceship to the moon, in so doing adding further credence to the *terra nullus* doctrine: no one was there.

However, on the other side, there is another kind of science, yet still in its infancy. This science is based on the doctrine of *terra plenus*. According to this controversial doctrine, the world is actually inhabited: it is inhabited by beings such as human beings and other assorted species. One of the most contentious claims of the doctrine is that scientists themselves actually live in this world and not somewhere else. The scientist is even present when taking measurements; or so the doctrine claims.

Just as for the Flat Earth Theory, the *terra nullus* doctrine is the most popular amongst the common folk. Most people find it just too irksome to burden the mind with all of the infrastructure necessary to constantly remind them that they actually live in this world. They prefer to define themselves by the things they might have, rather than what they actually might be.

Most scientists and mathematicians, being pragmatic people, opt to go with the masses and so base their research on the *terra nullus* doctrine. These highly skilled researchers have mastered the art of detaching their minds from their bodies and go through all sorts of contortions in order to prove the *terra nullus* doctrine of the uninhabited Cosmos.

Nevertheless, there are some diehards, like the author, that continue to support the *terra plenum* doctrine and stubbornly insist that there still *are* some inhabitants left in this world. Moreover, these stalwarts of *terra plenum* also claim that the inhabitants of the world do matter and should be taken seriously.

The Five Undemonstratables

There are two kinds of logic, logic with infrastructure and logic that can be carried out with the bare brain, the *terra nullus* logic. We first consider the bare brained version. This variety of logic is virtually infrastructure free. The logic is abstract and makes extensive use of symbols that don't mean anything in themselves. It is often referred to as symbolic logic. At the base of symbolic logic is the propositional calculus.

What is interesting about the Stoic logic developed by Chrysippus is that it can be interpreted as a left side symbolic logic as well as a right side logic, all decked out with dialectical infrastructure. Thus, Chrysippus' logic has both a *terra nullus* as well as a *terra plenus* interpretation.

The logic of Chrysippus distinguishes between simple and molecular propositions. Examples of a simple proposition are "its is day" and "Dion is walking". Molecular propositions are constructed from simply propositions using logical connectives. According to Chrysippus, there are only five kinds of logical connectives. He described these connectives in terms of his "five undemonstratables." The five undemonstratables can be stated as five three step syllogisms as follows (Chénique, 1974):

1 Conditional

 If one has the first quality one has the second

 but one has the first

 thus one has the second

2 Contraposition of the conditional

 If one has the first quality one has the second

but one has not the second

thus one has not the first

3 Incompatibility

One has not at the same time both the first and the second quality

but one has the first

thus one has not the second

4. 'OR exclusive' or alternative

One has either the first quality or the second quality

but one has the first

hence one has not the second

5. 'OR non-exclusive' or disjunction

One has either the first quality or the second quality

but one has not the second

hence one has the first

All of these syllogisms can be interpreted from the symbolic logic perspective of propositional calculus. As such, it can be said that Chrysippus was the first to discover the propositional calculus. Also, the first and second syllogisms can be interpreted as definitions of *modus ponens* and *modus tollens* respectively. This is all familiar ground for traditional logic.

The third syllogism deals with the incompatibly paradigm. In the propositional calculus context, this corresponds to the Sheffer stroke. In this context, the syllogism loses its explicit temporal nature and flattens down to the simple formula:

NOT (a AND b) is true

Note that the "at the same time" part of the formula has been dropped. Traditional modern logic has no notion of time. To entertain a notion of time, one needs a brain. The brain of the logician does not count, because that is not a formal part of the logic. Modern logic has no such infrastructure. It has virtually no infrastructure at all. Totally brainless, this is truly the logic of the *terra nullus.*

Sidenote:

We intend to use Chrysippus' logic to develop a *calculus of gender*. The elementary simple propositions will not involve particulars like "It is day." Instead, the simple propositions will be generic. In fact, there will only be two simple propositions, both involving gender. These two elementary propositions will be:

It is masculine

It is feminine

The molecular propositions will be constructed according to the five undemonstratables provided by Chrysippus. The translation from the

Chrysippus to the gender calculus is a bit subtle and works work as follows. The simple Chrysippus proposition translate into gender specific propositions as follows:

"One has the quality" translates to "the quality is Masculine"

"One has not the quality" translates to "the quality is Feminine"

Four of the undemonstratable syllogisms specify a compound gender. For example, "One has the first but not the second" specifies the compound quality determined by the compound gender MF (or Masculine as Feminine). The first and second undemonstrable translate to the compound gender MM and MF. The fourth and fifth translate to FF and FM respectively. These four compound types determine the four elementary building blocks of the calculus. These are the Four Elements of nature that the ancients referred to as Air, Fire, Earth, and Water.

The third undemonstratable does not specify a determined element as do the other four, but rather defines the basic organisational principle. The principle states that there can be no ambiguity in gender typing. A type cannot be masculine and feminine *at the same time*. We believe that the Stoics interpreted this as the allusive fifth element, Pneuma.

According to our gender calculus, these four elementary forms combine in triads to make up 64 possible molecular forms. These are the "codons" of the gender calculus. We claim that this is the underlying algebraic structure of the genetic code. These triads can be interpreted geometrically to provide a new kind of geometry. This explains the underlying semantics of this genetic cum generic code. The code encompasses a generic algebra based on gender, a geometric geometry, and a generic logic based on Stoic logic. This is what generic science is about.

In a nutshell, that is what this book is trying to explain. Hopefully, all will become clear in the long run.

Building the Logical Brain

Logic combined with integrated cognitive structure goes from being logical to being ontological. In other words, it starts to become a science of *being*. Integral to a science of being is the science of the generic subject. The formal presence of the generic subject in the science provides a fundamental point of reference. All propositions become relative, relative to the subject. As we have said before, the subject, *any* subject, is the centre of the Cosmos. This means that you are located at the centre of the universe. Since you could be anybody located anywhere, the centre of the universe can be literally anywhere. This identity of the generic centre of the universe and the individual centre of the universe is a most important principle. (Elsewhere in this book we show that any spatial reality with this property is equivalent to the Special Theory of Relativity)

As we have seen, the generic subject is endowed with particular brain architecture. First of all, the impersonal, undetermined subject is based on a left-right dichotomy, with the subject on one side and its kingdom of objects on the other. The usual polarity convention is right and left sides respectively, but this does not have to be the case.

The above paragraph has a certain apparently outrageous dimension. It's probably enough to make some readers choke on their crumpet. However, it just takes time to become comfortable with the generic viewpoint. The situation can get even more untenable when we move on to the next paragraph. Written in italics in an effort to ease the reader's pain, it reads something like this:

The brain architecture of the personal subject, in addition to the left-right dichotomy of the impersonal, has a front back determination with the polarity, subject in front and kingdom in the back. These left-right, and front back determinations can be explained in terms of gender, where the singular Oneness of the subject corresponds with the masculine and the non-singular wild card Otherness corresponds to the feminine. The configuration at this stage is that of a square divided into four quarters. Reading from left to right, starting from the front, the quarters are gender typed MF, MM, FF and FM respectively. Any subject whatsoever will have this configuration.

Perhaps one redeeming point is that we are not the only ones to have ever argued along these lines. Apparently, the ancients, going back thousands of years, have passed by here many times before. One thing to keep in mind is that we have not yet distinguished between the form of the world and the cognitive structure needed to comprehend it. Our basic thesis on this matter is that:

1. The form of the world and the cognitive structure are different,
2. The form of the world and the cognitive structure are indistinguishable.

This constitutes the basis of generic science and is why it is only necessary to study cognitive structures. Just put yourself into the position of that electron over there, the one that is peering at you at this very moment. It is a subject just like you. It might surprise you to know that, in this context, it has a cognitive structure indistinguishable from yours. One might argue that electrons do not have this kind of cognitive capability. The point is that this does not matter, at least as far as you are concerned. You, as an intelligent, conscious being, have that incredible capability of being able to put yourself in an electron's shoes. Moreover, you can do this even if electrons do not have shoes. You are capable of seeing the world just as the electron does, could, would, or should see it. Whether the electron makes the effort or not, doesn't really matter. It is hard enough to know what anyone is thinking, let alone an electron.

In the final analysis, the two points above apply perfectly to the two basic building blocks of generic science. These are the pure feminine entity and the pure masculine entity. They are both different whilst being indistinguishable. The masculine, in this sense, is the ultimate embryonic cognitive structure; the feminine corresponds to the ultimate embryonic world. These are the two basic building blocks of any entity endowed with consciousness, any entity that really exists.

Bridging Laws of Consciousness

David Chalmers characterised what he called the Hard Problem (Chalmers, 1995) as the problem of explaining the relationship between a physical account of reality and conscious experience. As he saw it, solving this Hard Problem required determining the "bridging laws" that related physical reality and conscious experience.

The "bridging laws" solution to the consciousness question is a natural response of traditional left side scientific thinking. Such thinking is naturally dualistic where dichotomies abound between Mind and Body, the abstract and the real, and in this case, between the realms of the physical and the conscious.

Chalmers posed the Hard Problem over fifteen years ago and it has played an important role in crystallising attention on the problem of developing a science of consciousness. However, one must admit that absolutely no progress has been made. The problem remains as intractable as ever. The problem slides off the table and slips into the same box as the problem posed by Kant, and Aristotle well before him. This kind of problem has remained unresolved and intractable right from the very beginning. The real problem is to find the real problem. We claim that that the implicit epistemological stance of Analytic philosophy espoused by Chalmers is the actual root of the Hard Problem. Just as it is a very Hard Problem to eat soup with a fork, it is an impossible undertaking to understand consciousness from the dualist Mind Body assumptions of such an epistemology. It is like taking a chook, chopping off its head and trying to work out the "bridging laws" to put the body and the head back together again.

The Hard Problem is like after the fall: How do you put Humpty Dumpty back together again? How do you bridge the broken? Chalmers is looking for an abstract solution to a problem that is a direct consequence of abstract thinking itself. Explaining abstractly how to bridge the abstract with the real is definitely a very Hard Problem, reserved only for the most courageous of abstract thinkers. For the less courageous, an alternative approach is to avoid abstraction and think generically.

From the monist right side viewpoint of the generic, there is never any need for a bridge as nothing was ever broken apart in the first case. At the ontological foundations of the generic, the very first spark of consciousness stirs with the pure unqualified feminine that *has* the pure attribute of Oneness. The masculine entity *is* this attribute. The embryonic physical unites with embryonic consciousness: One *has* an attribute; the other *is* this attribute. The bridging here is more like how some of the Hindus poetically describe it, as a coital embrace. This couple have no need for a prosthesis, bridging or otherwise.

According to our embryonic Generic Science based on the generic algebra of gender, any being is coded and organised in accordance with this generic code. In the case of the biological, the generic code becomes the genetic code. The four-letter code is really based on binary valued gender. Any life form is coded in this gender algebra and organised through it. The original gender construct of the masculine and feminine now becomes a massive complex entwinement of gendered entities. The overall coherence and survival of the organism absolutely depends on maintenance of the coherence of the gender typing that runs throughout every nook and cranny of the organism. If there is failure of coherence then no bridging Band-Aid will ever bring this organism back into consciousness. The organism would be well and truly beyond the Hard Problem stage. It will be a true cot case.

When viewing the healthy gender typed organism from the perspective of a third party, everything appears to be in ambiguous and chaotic superposition. Gender states are dynamic and something like quantum states, except that they are relative to each other and the organism, not absolute. Unlike quantum states, the subject sees its states quite differently from any third party. Viewed from the perspective of the organism, these states are in coherence with its own being and articulate its being. There must be no ambiguity whatsoever in gender typing.

We propose that the formal mechanism of gender regulation can be articulated in the form of Chrysippus' third undemonstratable, that of incompatibility. The premise of the syllogism states:

One has not at the same time both the first and the second quality.

In the context of the generic coded organism, this becomes:

One has not at the same time both the masculine and the feminine gender.

It is by the implementation and maintenance of this principle that any living being maintains coherence of Self. This solution demands a dynamically gendered system with a global mechanism for the maintenance of gender coherence.

For a cosmological system, the mechanism is that of pure rational coherence, including the non-violation of the causality principle. As a science, it will present as a much more generic version of present day relativity theories. The geometric aspect of the mathematics (or anti-mathematics) will however need a substantial overhaul. In fact, a new geometry is needed. It will be a more generic version of what is now called *geometric algebra.*

In biological systems, the genetic code, although material, is a different substance from the proteins it codes. In the realm of pure physics, the code and the substance entities are possibly the one and the same. However, the same generic principle is at work in any realm.

The Chrysippus Square of Oppositions

Chrysippus' remarkable logical system can be naively interpreted as a simple left side version of the propositional calculus. There the incompatibility paradigm can play a pivotal role. Interpreted from the abstract point of view of symbolic logic, the other four syllogisms can be derived from the third syllogism. In this sense, the other four syllogisms are not undemonstratables, but demonstratable from the third. However, the Stoic system is not abstract logic, but logic with an ontological bent. Abstract logic has no semantics. This is not the case for Stoic logic. Looking at the semantics of the Stoic syllogisms, one can discern how they match perfectly with their Four Elements theory in physics. Figure 30 illustrates how each of the syllogisms matches with the gender typing and thus with the Four Letters of Empedocles. Provocatively, we have added in how these map to the generic code letters of A, U, G, and C.

Chrysippus developed Stoic logic following in the footsteps of Aristotle and his Organon. Here we have *our* version of Chrysippus' Logos, the forerunner to the semantic cracking of the genetic code.

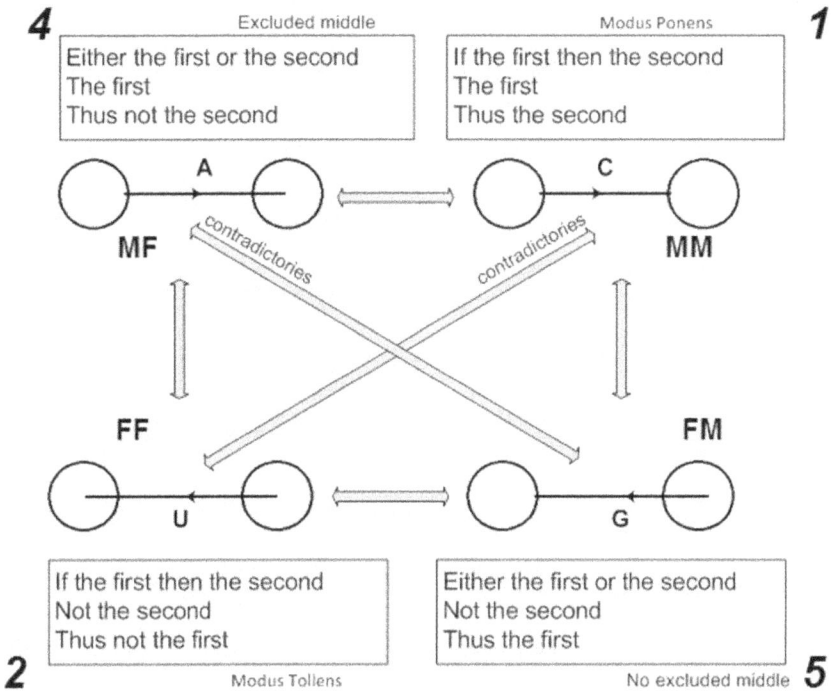

Figure 30 Author's hypothetical reconstitution of Chrysippus' Square of Oppositions. The four syllogisms shown, match up with the Stoic Four Elements

Chrysippus and Ground Zero

By Ground Zero, we mean the centre of the Cosmos. Ground Zero has a certain shape, the shape of the entity located at the centre of the Cosmos. As we know, this entity is none other than any subject whatsoever that takes the pain to reflect on its particular spot in the universe. Without fail, this subject, like any other subject, sees itself as being located at the aforesaid location, notably the centre of the Cosmos, the centre of *its* Cosmos, at least.

As for the shape of this entity, the generic subject, it has a left, a right side, also a front, and a back. This is the structure we have been referring to as the semiotic square. It is a structure that can be interpreted in many ways, as a blueprint for epistemological organisation of knowledge, for example. There are also ontological, and of course many biological interpretations. On the biological front, this structure can be thought of as the structure of a *whole*, as coded by a chromosomal codon. Any biological organism is organised as an entity viewed as a whole from a myriad of points of views. To each codon, there corresponds a holistic point of view. The genetic cum generic code is the language that articulates the geometric algebra of this holistic view of the organism. Another interpretation of this highly generic semiotic square is that it offers a schematic for elementary cognitive structure.

So prodigiously generic is this structure that it can make the head spin. To top it off, we now have Chrysippus joining the fray. Now Chrysippus was conceded by the ancients to be the equal of Aristotle. In above, we have organised four of his five fundamental syllogisms into a form that falls quite naturally into the elementary structure of the generic semiotic square. As can be seen, the premises of the first and the second hypothetical conjunctive syllogisms make one diagonal of the square and the two disjunctive forms mark out the other diagonal. It is becoming clear that we have here, a structure that resembles the Square of Oppositions of Aristotle. The Scholastics added the AEOI four lettered labelling to Aristotle's system and spent over a millennium probing into its delights. Not to be outdone, we have added our lettering to our reconstructed version of the Chrysippus Square in the hope of preparing it for its reinvigorated role in the present millennium. Suffering from a lack of creativity, we have borrowed the RNA version of the biologist's genetic code. Why invent when you can steal, is our motto. It took the author a little while to get the right fit, but he is reasonably confident that his allocation of the CAUG lettering is spot on. He would be very miffed if this was not the case.

Chrysippus and the Grand Unification

The ancient Stoics have been the historic mentors for the material presented in this work. They developed the most successful and diverse form of monistic philosophy that the Western world has ever seen. Zeno provided the intuitive and informal core elements of the doctrine. Chrysippus logic marked the first tentative steps towards the formalisation of a unifying science. The full significance of Chrysippus' contribution has been little understood by the moderns, blinded as they are by the achievements of the current day sciences. Despite these achievements, the present day sciences are lacking in any kind of cohesive unifying discipline. The unifying science pioneered by the Stoics, will provide such a unification.

Of critical importance is to learn how to reason in a different way from what is customarily taught in modern schools and universities. The moderns only have a partial grasp on rationality. Modern science and mathematics only understand the notion of the true and the false. What lacks, is the understanding of *truth*. However, the very mention of this word, truth, can seem off putting. After all, probably more people have been burned at the stake because of an allegedly incorrect understanding of truth, than for any other reason. However, buried amongst the historic debris of lost causes lurks indeed the rusted hulk of truth.

Nevertheless, as any philosopher knows well, truth of this kind must be self-justifying. For many, such as Karl Popper, the notion of a self-justifying truth is synonymous with the blind faith of religious zealots and doctrinaire extremists, something anathema to science. Popper is content with the kind of knowledge where each proposition is forever condemned to the judgment that it might be false. Even worse, at the same time the proposition must accommodate the stark reality that this judgment might indeed be true. Then again, it might not. Modern scientists are a brave lot.

Sidenote:

Popper did eventually nuance his views on this matter in the light of the self-justifying biological organism notion. In so doing he implicitly admits that the biological organism is obsessed with self-justifying its continual existence in the world. As such, biological organisms seem to have ontologically more in common with the logic of religious zealots and political fanatics, than with the cool, dry head of the analytic philosopher.

The stark truth about truth is that it must be *relative* and never absolute. Only in this way can it become an absolute truth. In other words, it becomes an absolute truth *relative to itself*. This is the essence of monistic philosophy: It is the rationality of the self-justifying Self. Relative to this subject, there is only one truth.

We have already made inroads into the science of the subject. Unlike the analytic rhetorical type reasoning of analytic philosophy, the reasoning of this right side, monist philosophy, is expressed in terms of oppositions and oppositions between oppositions. It is in this way that the reasoning becomes a relativistic form of reasoning. Rather than rhetorical, it becomes *dialectical*. The nuts and bolts of the reasoning deals with the dialect of two entities, one which *has* and the other that *is*. These entities differ by gender, the first corresponding to the feminine gender, the second to the masculine. The dialectic of *to have* and *to be*, constitutes the core essence of the monistic, right side form of reasoning.

This is the dialectic of the subject minimally conscious of itself. It leads to a particular kind of knowledge. It leads to the generic truth that reality, viewed from any particular perspective, is the reality viewed from the point of view of the generic subject, the *any subject whatsoever* kind of subject,

The elementary form that arose from our investigations was the semiotic square. This structure arose from the opposition between what the subject *is* and what the subject *is not*, that is to say, what it *has*. This opposition was formalised in terms of the gender construct. This leads to the four distinct parts of the square being gender typed MF, FF, FM and MM.

This very generic quadruple structure is highly lacking in determination. The edifice is so undetermined that it is not even clear whether it corresponds to the semiotic structure of knowledge of the world, or the structure of the world itself. Is this epistemology or is it physics? Is this the structure of Mind or is it the structure of Body? Is it the structure of a generic language or that of a generic world?

Finding an answer to these kinds of questions is key. It is here that we find the great enigma of this science. Unlike the analytical thinkers who want to understand the relationship between Mind and Body in terms analogous to that between horse and cart, the synthetic monist thinker must take a different tack. The horse will not be separated from the cart, but treated as an organic whole. One cannot have one without the other. We came across the very essence of the monist solution in the form of the gender construct. Rather than plucking attributes from a predefined definitional framework or harvested from empirical measurements, we *constructed* the one single fundamental attribute from which stem all other attributes of our science. This was the attribute possessed by the pure feminine entity. The attribute, an entity in its own right, was the masculine entity. These two entities are different. They differ by gender. However, they are absolutely indistinguishable. Two entities are distinguishable if they have different attributes. Here there might be two entities, but there is only one attribute between them: two entities; one *has* an attribute, the other *is* the attribute.

This gender construct provides the generic formula for all of the science that follows. The dialectic of the masculine and the feminine provides the generic base for all other seemingly dyadic structures such as the popular Mind-Body duality of the analytic philosophers. The relationship between the pure feminine and masculine is a generic form of the same relationship between Mind and Body.

Not everyone will agree with this assertion. Certainly, an analytic philosopher or anyone reasoning from a Cartesian viewpoint would take the abstract road, abstractly arguing that Body is like a machine and Mind is an intelligence that drives the machine. The two are linked together by some kind of "bridging laws" perhaps. There is no dialectic here, as the notion of a bodiless mind and mindless body, is considered quite respectable. They can conceivably go their separate ways: put the brain in the bottle and the brain dead body on life support, should do the trick.

Such a surgical separation is impossible for an organism constructed from the gender construct. The organism is constructed according to a four-lettered code. According to our gender calculus version of this code, each letter is made up of one of the four binary gender typings, MF, FF, FM and MM. On the face of it, the organism might be just a highly complex assemblage of hydrocarbon-based compounds. However, from an *organisational* point of view, it is a seething mass of intertwined, gendered entities. It is this gender typing of content and form of the organism that ensures systemic coherence. It is in this way that the One can be constructed from the inseparable and indistinguishable Two.

The Stoics saw this dynamic systemic organisation of the organism in terms of the tensions and *tenos* of a fifth kind of substance they called *pneuma*.

> *The pneuma is in constant motion. It is a process into itself, and from itself. The inward process produces unity and substance, the outward process dimensions and qualities. The pneuma is a disposition (hexis) in process. As a disposition, the pneuma holds the Cosmos together, and accounts for the cohesions of each individual entity. The pneuma is the cause of the entity being qualified: for the bodies are bound together by these. [Chrysippus views on the pneuma (Reesor, 1989)]*

The coherence, the very being of an organism, is synonymous with it maintaining Oneness. The mechanism for achieving and maintaining Oneness is through the establishment and maintenance of gender typing. The organism must know, without a shadow of doubt, what it *has* and *has not* and what it *is* and *is not* in all cases. These are the key determinants of consciousness. In addition, the determinations are purely relative. They are purely subject-ive. This, one must admit, is truly a beautiful, self-referring system.

Beautiful indeed, but how does it work? With profound beauty, one would expect an accompanying simplicity, a profound but simple principle. Seeing that everything involved in this kind of self-organising organism is relativistic, there should be some fundamental relativistic principle at play. In the traditional sciences of our day, the only relativistic principle known is in physics. There is no known equivalent in biology. In physics, we see relativity theory expressed as demanding that the laws of physics remain invariant from one reference frame to another. Perhaps more pointedly, as shown by Zeeman, the principle of relativity is intimately bound up with the non-violation of the causality principle. It is here that one can grasp the simplicity and elegance of the theory. System coherence demands the coherence of causality. The claim of generic science is that this is not enough. A much more demanding form of relativity is we call *generic relativity*.

If the work presented in this book is to be more than the usual exposition of inconclusive philosophical prose, then we should be able to advance an equally simple and elegant formulation concerning the essence of generic relativity, the cornerstone of the generic science we are trying to develop. Fortunately, we do not have to look very far. The principle is located at Ground Zero and there is no one who knew this spot in the Cosmos better than Chrysippus, the Stoic logician par excellence. Ground Zero is the location of the Logos, the reasoning faculty of any subject whatsoever. The form of the Logos can be understood in terms of the dialectic of having and being, a form expressed by the semiotic square. Chrysippus provided the logical framework of the Logos semiotic square in the form of four of his five undemonstrables. We have resurrected this structure as an alternative to Aristotle's Square of Oppositions, discussed previously. We have named this the Chrysippus' Square of Oppositions. The fit between this structure and the four undemonstratables is comfortable and reasonably self-evident. The structure effectively provides an additional logical impetus to the thrust of our argument. The four undemonstratables provide a logical dimension to the interpretation of the four-element theory and the corresponding four letters.

Absolute Incompatibility

Five undemonstratables minus four leaves one. The missing syllogism is the third undemonstrable, the incompatibility syllogism: *One cannot have one quality and the other at the same time.* We now come to the fundamental tenet of generic science. It is founded on the premise that there is nothing more incompatible in this world than the masculine and the feminine. This premise does have some intuitive appeal and so we will stick with it. This is not a bad idea, as it appears that the whole Cosmos hinges on the concept. It is the incompatibility principle that holds not only the Cosmos together, but any being whatsoever that exists.

In the case of biological organisms, the concept should be relatively easy to grasp. A stumbling block might be in accepting that the genetic code is more than a mere transcription language. One should keep in mind that curiously, and apparently accidently, the code became a convention adopted by all living organisms since the year dot; without exception. Accidents do happen, but this accident does seem a little bigger than most. Life might be subject to evolution, but the *language* of life seems absolutely impervious to change. The gamble of Nature seemed to have hit the jackpot absolutely spot on, right from the start.

The reader may rest with that interesting accident hypothesis or move on to considering that the code may be based on a generic semantic and ontologi-cal structure. According to our take on the question, this structure is based on the dialectics of *being* and its naturally orthogonal counterpart, that of *having*. This can be formalised in terms of the gender construct and leads to a four-letter code based on the four possible binary combinations of the two genders. It is generally accepted that all biological processes are coded by the genetic code, what we claim to be the *generic* code. Moreover, in multi-celled creature, the same code is repeated for each cell. We say that this code expresses a relative typing on all aspects of the organism. At the very ground roots level, the typing is in terms of complex combinations of gender typing. We claim that the organism relies on this form of organisation in order to arrive at knowledge and consciousness of itself. It is via this absolutely relativistic gender typing that the entity knows what pertains to it or what does not. This is the most elementary and most essential feature of life.

Moreover, the basic health of the organism will be placed in peril if this typing mechanism starts breaking down. The cohesion of the system demands the constant maintenance of the integrity of gender typing through the organism. The Stoic picture of a *pneuma* permeating every aspect of the organism is very helpful. The pneuma is constantly attracting and repelling, constantly maintaining the equilibrium of the organism.

The Stoics claim that there are two primary principles working through the pneuma: the active principle and the passive principle. This terminology is also helpful, as long as we recognise that the active and passive ultimately refer to the masculine and feminine, in a particular configuration. For example, we refer to the *feminine as active* by the mixed gender term FM.

The *masculine as active* becomes MM and so on for the passive MF and FF variants.

Maintenance of the integrity of gender typing throughout the organism is paramount. Since the system is changing and reacting to its environment, this integrity must be synchronised. This brings us back to the key logical ingredient that guarantees such coherence: the coherence principle.

The Gender Coherence Principle

The organisational coherence of an organism is regulated through gender typing. The maintenance of organisational coherence is synonymous with maintaining the integrity of gender coherence. This can best be expressed in the form of Chrysippus' third undemonstratable, the incompatibility syllogism. The premise can be restated in the form:

In no single moment can an entity be both masculine and feminine at the same time.

We will call this the *gender coherence principle,* the fundamental organisational principle of Nature.

Note in passing that an entity can have *multiple* gender typing. However, it cannot have two different gender typings at the same time. This raises interesting question regarding the degeneracy of the genetic code. Take the amino acid asparagine, for example. It can be coded by the bases either AAU or AUC. In gender terms, this translates to the gender typing MFMFFF or MFMFMM. According to the gender coherence principle, such an entity has two possible "quantum" gender states. At any time, it can be functioning as either MFMFFF or MFMFMM, but not both at the same time. Remember that gender typing at any instant of time is not absolute and cannot be measured deterministically by a third party. Gender typing is *relativistic* and dynamic and in coherence with the organism so typed.

Note that the so-called superposition of states addressed by Quantum Mechanics disappears if they are considered to be more like relativistic gender states. Any observer that deterministically tries to measure a relativistic gender state of an organism will encounter superposition. For the organism in question, there is no superposition whatsoever. Relative to its integrity system, the gender coherence principle demands that the very *opposite* apply at each and any instant.

As for the organism, in the life sciences the organism might be a cat on a slab in the lab. For the physicist, the organism might be a much smaller or much larger creature. However, it is still an organism based on the same generic organisational principles.

Physics Interpretation

In an appendix attached to this work, elementary particle physics will be interpreted from a generic point of view. This leads to elementary entities like quarks and leptons being gender typed in terms of codons reminiscent of biology. In this way, any being in nature codes itself in terms of the generic code based on gender typing. This includes the Cosmos itself, as a dynamic self-organising being.

In traditional relativity theory, one can discern an elementary organisational coherence that can be stated in a form comparable to the gender coherence principle. In this case, it becomes the *principle of causal integrity*. The principle states the dialectic of cause and effect:

The cause event is always antecedent to the effect event.

This is the most fundamental organisational principle known to traditional physics. The law must not be violated in any context (i.e., in any reference frame) and so demands a system that obeys Einstein's Special and General Theory of Relativity.

One can see that the form of Einstein's relativity has a certain resemblance to the generic form expressed, not as a causality coherence, but as gender coherence. There is also a fundamental difference. Einstein's relativity demands coherence across time: causes must precede effects in time. In other words, Einstein's relativity is diachronic in nature. In contrast, the generic version of relativity demands coherence at the *same* time. In other words generic relativity is *synchronic* in nature and, up until now, has been totally ignored in physics.

Computer Science Interpretation

It is important to keep in mind at all times when dealing with the generic that it is not an abstract science. Generic science is capable of formalism but not as an abstraction, which is necessarily dualist. Generic science is monist and non-abstract. Some effort is required to become accustomed to this totally different paradigm. Interpreting some of the concepts in a Computer Science setting can help, in this regard. Unlike axiomatic abstract mathematics, Computer Science is a constructive science and naturally synthetic in nature. The science also enjoys a natural tendency towards monism in the sense that the theory of code can be expressed *in* code.

Generic science is a discipline, which has for its vocation the task of articulating the structure and organisational principle of any living being. The science is naturally constructionist. This raises the question of how to construct an organism based on generic science principles. Such an organism would have to be based on gender typing and be organised on the gender coherence principle. In addition, the whole system must not violate the principle of First Classness. Is this possible?

This is a silly question as our very own presence on this globe is at least some kind of feasibility proof of the concept, a living proof in fact. What we wish to do in this section is to provide a very simple example of how Computer Science, unknowingly, has already started to go down the path of Generic Science. Our example is the very computer itself, the Von Neumann computer.

Before Von Neumann, there already existed programmable calculating devices. However, they all had one thing in common. They were based on an absolute dichotomy between data and program. For example, the program might be hard wired into the device and the data fed in via paper tape. If we want to put some gender typing into the mix, we could say that the program was masculine stuff and the data feminine. With this arrangement, the gender coherence principle could be satisfied because at no time is any confusion possible between what was program stuff and what was data stuff. Data was always on the paper, and program in the machine. The only problem was that such a device violates First Classness. First Classness is incompatible with such a blatant and absolute duality. First Classness cannot tolerate a world cut up into two, one made of paper and one made of the other stuff.

Von Neumann started the process of moving the calculating machine into the realm of a generically organised entity. He made two innovations. The first innovation was *shared memory* where there was no longer to be any absolute dichotomy between data stuff and program stuff. They were all loaded into shared memory in the same format as small chunks of information.

Von Neumann was then faced with the problem of how the computer could tell the difference between program stuff and data stuff. It was here that Von Neumann decided to invoke his version of Chrysippus' incompatibility principle. The principle was that:

No chunk of information in shared memory could be both data and program at the same time.

In order to implement this principle, he came up with his second innovation. It was called the Program Counter. The Program Counter is a pointer into the shared memory of the computer. The rule was that a program instruction was the chunk of memory pointed to by the Program Counter at a particular instance in time. All the rest of the chunks were considered data. Having executed that instruction, the Program Counter would be incremented to the next memory location, and that would then be considered a program chunk and no longer potential data. In the case of a JUMP instruction, the Program Counter would be moved to some other distant place in memory and the process continues. The computer was born.

Like practically every major advance in computer science, the Von Neumann's computer was an exercise of eliminating violations in First Classness; in this case, eliminating the fixed dichotomy between data and program. Henceforth, the distinction became *relative* to the dynamically changing Program Counter. What was program and what was data depended on context.

However, such a device is far from freeing itself from violation of First Classness. The Program Counter itself becomes a rigid privileged memory location, totally estranged from the run of the mill information chunk in shared memory. That is yet another dichotomy to be eliminated by generic engineering principle. There is a long way to go.

...

The Von Neumann computer needed a few further innovations in order to become operational. However, not many other innovations were needed. Add a stack, interrupts, and a few input/output ports and that is about it.

The Semiotic Logic of Chrysippus

Before reaching an understanding of our reconstruction of the Chrysippus semiotic square, we need to know a bit about semiotics, or at least, our version of it. We provide here a summary of our approach.

The author's first acquaintance with the semiotic square came from following the courses of Greimas back in Paris, many years ago. The term "semiotic square" is nowadays generally associated with his name. The big weakness in the Greimas approach was his failure to come to terms with the subject. His semiotics is *sans sujet*. We will sketch out here a more fundamental approach to semiotics and the semiotic square that *does* include the subject.

To begin with, there are two kinds of semiotics, one associated with Ferdinand de Saussure (dyadic, arbitrariness of the sign etc.) and one associated with Charles Sanders Peirce (triadic). In our view, the approach of de Saussure is not semiotics, but General Linguistics. Like Greimas, the approach of de Saussure is *sans sujet*. If there is a subject, it is part of the Spectacle, not the Spectator. It is merely what Hegel referred to as the empirical ego. In this perspective, the de Saussure approach is like that of the traditional sciences and mathematics. All of these sciences are *sans sujet*. We call all of these traditional science left side sciences. Left side sciences claim to be objective, which is another way of saying that they only concerned with a reality of objects where any reference to the subject has been excluded. They are all *sans sujet*. As such, these sciences look at the world from a very specific point of view. This point of view has been described as the "view from nowhere" or the "God's eye view." This is a general characteristic of science *sans sujet*. It is a general characteristic of all the sciences and mathematics of today.

The other possible scientific paradigm goes in the opposite direction. It demands that the subject is always present. In other words, if there is a spectacle there must also be an accompanying spectator. You cannot have one without the other. We call the science based on this paradigm, right side science. The right side science becomes, in fact, the dialectic of the Spectator *and* the Spectacle, the Subject and its kingdom.

Unlike the many left side sciences, there is only one right side science. This is because its focus is on the science of the subject and this is quite different to the science of objects. It is the science of the Self. For a Stoic logician like Chrysippus, it is the science of the Logos. This generic entity, the Self, the Logos, the Ego, has a generic form. This form can be worked out from pure reason.

Now Charles Sanders Peirce was more inclined to the right side paradigm but, like everyone else, had difficulty gaining much traction. He also despised the Stoics, which didn't help. Thus, we have to start from scratch. Starting from scratch means that we start with a subject and its kingdom. Alternatively, we start with a kingdom and its subject, the same thing. Both spectator and spectacle must be present in the same moment.

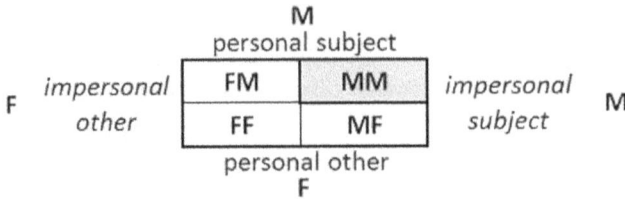

Figure 31 The generic semiotic square is constructed from the feminine masculine opposition applied to itself.

This is where we have to put our thinking caps on. The relationship between the Subject and its Other is a very particular kind of relationship. They each determine one another. The Hindus sometimes see this as a coital relationship. The subject corresponds to the masculine. The mysterious other is the feminine where gender gets interpreted as sex, poetic licence *oblige*. The Stoics saw the relationship as that between the Active Principle and the Passive Principle. Vedanta philosophy often refers to the Active principle as the Principle of Individualization, the Spiritual Principle, or simply the masculine principle. We have here the building block for right side science. It is getting a bit steamy, so here is one way to arrive at a dispassionate view. It involves the gender construct.

The main role of the subject in this right side science is that it does provide a *determined* point of view. As such, it is a pure singularity. What is non-subject is non-singularity. This can be formalised with the concept of gender. The gender concept is very ancient, in both the West and the East. First, there is the unqualified substance totally devoid of any determined specificity. Such an entity is typed as the pure feminine. One might say that the pure feminine is devoid of specificity and so has no attribute. This is not the case. It is only devoid of a determined specificity. It has an *undetermined* specificity. That is its attribute. This attribute, using the argument of First Classness, must be an entity in its own right. (Note that the Stoics always claimed that the property of an entity is an entity in its own right). This attribute entity will be said to be of masculine gender. Two entities; one *has* an attribute, the other *is* the attribute. The first entity corresponds to the feminine, the second to the masculine. These two entities provide the building blocks for the right side science paradigm.

The first thing to construct is the semiotic square. One way of understanding this square is as the architecture of a whole. Totality can only be understood from a determining point of view of the subject. Instead of comprehending the totality in any moment, which is impossible, it is understood as a whole. A whole is totality looked at from a particular point view. There are as many wholes as there are points of view. This requires that the subject must be present *in* the whole. Right side science always understands things in terms of wholes.

Thus, the semiotic square, as a generic understanding of a whole, is a map of the subjects conscious understanding of the whole, any whole. The first moment of understanding is "Wow, here I am, this is me and the rest is not me." We thus draw a square, cut it down the middle and adopt the convention that the right side corresponds to subject and the left side to what is not subject. The right side is masculine typed and the left side is feminine typed.

However, the subject in this particular configuration is not you or I. It represents the *impersonal* subject. In fact, it is this subject that corresponds to the "view from nowhere," the "God's eyes view" of the traditional sciences. These sciences, in their quest for objectivity, remove all reference to subject from consideration. They even remove this impersonal subject from consideration, as they have no need for it. They demand a godless science, a pure science *sans sujet*. Thus, the semiotic square for the left side sciences is the same as for the right side science, except that the right side is blacked out. Left side sciences thus suffer from a symptom well known to the psychiatrist. It is called hemi-neglect. Right side science knows about the left side, left side science wings it alone, content with half a brain, so to speak. Curiously, in passing, the human brain exhibits exactly this same bi-lateral specialisation. The right hemisphere does not exhibit hemi-neglect and sees a whole world. Only the left side exhibits hemi-neglect.

This is now where left side and right side science part company. Not content with just the presence of the impersonal subject, right side science must find a way of introducing a more determined subject, the *personal* subject. This is constructed by applying the first feminine-masculine opposition to itself, an opposition of two oppositions. It might sound complicated but is easily visualised with the semiotic square. The second opposition is orthogonal to the first and so instead of a left-right dichotomy, the dichotomy is front-back. We use the convention of masculine in front, feminine at the back. It appears that we are not the only ones to adopt this polarity convention.

The end result is that we end up with a square shaped kind of placeholder for dealing with knowledge. The first kind of knowledge involves an elementary consciousness of self, a knowledge of what is and what is not. This is expressed logically in our reconstruction of the Chrysippus square. For the moment, note that the four parts of the semiotic square have been binary typed with gender. For example, the left front part is typed as MF. This reads that, from the impersonal subject perspective, it is typed as feminine. From the *personal* subject perspective, it is typed as masculine. Thus, the first letter in the binary gender typing is that of the personal subject, the second letter is that of the impersonal.

The semiotic square is a placeholder, the architecture of the generic mind, so to speak. The semiotic square is static and unique, for the purposes of the science. You only need one brain, it can be said. In addition to the placeholder, there are values relative to it. These values are mobile. There are the four kinds of elementary substance that can be binary typed by the four binary gender types. The binary typed substance corresponds to MF, FF, FM and MM. The ancients called them air, earth, water, and fire respectively.

We now come to the semiotic square constructed with four of the Chrysippus undemonstratables. Note that one diagonal is constructed from the conjunctive syllogisms. These are known to logicians as Modus Ponens and Modus Tollens. The other diagonal is constructed from the two forms of the disjunctive. The diagram can be gender typed by matching the *is* copula with the masculine and the *is not* with the feminine, as shown. This matches perfectly with the semiotic square gendering shown above.

What is interesting is that the logic of Chrysippus has introduced yet another dimension into the semiotics, a vertical axis. The square becomes the "Chrysippus cube"! We have used the convention of the implication arrows in the diagram going left to right to signal the upwards direction, and the downwards for the right to left. Talking intuitively, this indicates that the top two entities have an "upward flow" and the bottom two entries have a "downward flow."

One should note that the gender coding of the top two elements correspond to the "elements" of air and fire. These are the "light" elements, being predominantly masculine and less substantial than the feminine bottom two elements of earth and water. Such reasoning is not very rigorous, as we are not talking about the same kind of elements as in the left side, traditional science. The logic of Chrysippus however adds a different complexion to the matter.

These principles must have been part of core Stoic teaching, as Marcus Aurelius wrote in Meditations.

> *Your aerial part and all the fiery parts which are mingled in you, though by nature they have an upward tendency, still in obedience to the disposition of the universe they are overpowered here in the compound mass. And also the whole of the earthy part in you and the watery, though their tendency is downward,*

The Stoics claimed that theirs was a unifying science that integrated logic, physics, and morality. Some people are attracted to Stoic values whilst thinking that their science has been completely eclipsed by the modern day sciences. However, how antiquated is the science of antiquity? Consider the following.

In our diagram, we have added in the four letters CAUG matching up with the gender typings MM, MF, FF and FM respectively. This is part of another story in this book. These are the four letters of what we call the generic code. We have taken them from the RNA version of the genetic code. The genetic code is a standard code that codes all living beings, without exception. This is an established fact. The generic code is impervious to evolution and has remained unchanged since the year dot. By extending the notion of the living to that of the universe, itself considered as living by the Stoics, this same code takes on a generic vocation. In this book, we explore its application to understanding elementary particle physics from a new angle (see Appendix). We use the generic code to code quarks and leptons. These claims may test our short-term credibility. However, in the longer term that is the way it will pan out once we have properly digested this new science, a science with such ancient roots.

11

Science without Attributes

It is a remarkable fact that the one single language, the genetic code, codes all biological organisms. The code is universal across all autonomous living organisms and seems impervious to evolution, remaining practically unchanged from the very beginning. However, the central dogma of biochemistry downplays the importance of this observation claiming that, linguistically, the language is of little interest. It is a mere transcription code translating genes into proteins. In this chapter we argue that, like any true language, there is a double articulation, As well as coding the means for an organism's life, the language also articulates its ends, by proscribing its generic ontological structure. In other words, this generic code articulates a generic semantics.

The generic semantics applies across the board to, not only hydrocarbon based biological life forms, but to even the world of matter in physics. Coding for biological organisms employs an intermediary process where the code substance and the coded are distinguishable from each other. In realm of the physics of matter, the distinction between the code and the coded vanishes. The components of each elementary substance and the letters that code it are the same thing. Nevertheless, it is the same code no matter what the scale or the realm. Such is the nature of the reality in which we live. We all share the same code, right down the humblest neutrino and gluon.

Our story starts from an unlikely source by tackling the Kantian problem. The problem posed by Kant was that of developing a kind of knowledge totally free of any *a priori* experience or definitional scaffolding. Kant called such science metaphysics. In the book, we call it Generic Science, right side science, or simply the First Science.. Our work must resolve the Kantian problem by developing a science that is free from any determined attributes, the science *sans attributs*. From pure reason alone, the elementary generic structures of the science are developed. We claim the resulting generic language to be the reverse engineered version of the genetic code. We develop an understanding of the code and how it articulates the generic semantics of the missing "second articulation." Central to the development of this new science is a very old philosophy, that of the Stoics. In this respect, the Generic Science presented here is a modernised and reconstituted version of the Stoic paradigm.

In addition to the genetic code implications of the science, this chapter also includes some examples of applying the same generic language to provide new insights into Quantum Mechanics and elementary particle physics based on a new kind of relativity principle, generic relativity (further developed in Appendix B). Needless to say, the material in this chapter goes in the countersense, or is orthogonal to practically all accepted tenets of the present day sciences and philosophy.

Introduction

There are two kinds of knowledge. On the one side, which we will call the left side for discursive convenience, we find all of the traditional sciences including mathematics. These sciences all have a common epistemological structure, which can be summed up by saying that they specialise in conditional knowledge, knowledge that is conditional on such things as empirical data, hypotheses, axioms, and in many cases even opinions. In this chapter, we reopen the age-old case for the other kind of knowledge, a knowledge that does not depend on any antecedent factors at all. A science capable of providing such knowledge we will call right side science. It was Kant that asked the question concerning "the possibility of the use of pure reason in the foundation and construction of all sciences." Turning this possibility into reality requires the development of right side knowledge. The object of this chapter is to argue that the construction of right side scientific knowledge is possible, to show how it is done, and to explore some of the practical repercussions.

Characterisation of left side scientific knowledge

A common characteristic of the left side sciences is that they are all attribute based. From the left side perspective, without exception, all knowledge is expressed in terms of the attributes of entities: attributes can be anything from measured properties, defined properties, just to human attributed labels. Many philosophers claim that this is the only kind of scientific knowledge possible.

They effectively declare that there is no other way to know the "real thing" behind the attributes other than via and in terms of the attributes themselves. Thus, knowledge has to be preceded by attribute acquisition in some way, either by attribute harvesting, as is the case for empirical sciences or by axiomatic proclamation, as is the case in formal mathematics. This is what characterises left side science as conditional knowledge, conditional on having the prerequisite attributes on hand before any reasoning can begin.

The "attributes only" paradigm of left side sciences imposes important limitations including the impossibility of a fundamental explanation of differentiation and distinguishability of objects. The possession of differentiation and distinguishability mechanisms are assumed to be a *fait accompli* before any serious investigation ever even starts.

Right side science is founded on First Classness

Unlike conventional scientific knowledge, non-conditional knowledge cannot rely on being kick started with a bunch of attributes. However, like any science, right side science needs something from which to start. It needs traction. Since this kind of science is to bootstrap itself up from pure reason alone, it will need to predicate, right from the beginning, all the logical development on the very central principle of pure reason. In other words, knowledge must be *conditioned* by the principle of pure reason. Our task is to develop knowledge that is totally non-conditioned and not predetermined in any way; It cannot even be conditioned by its object. A knowledge of astronomy is preconditioned by the nature of stars and galaxies. Unconditioned knowledge subsumes down to knowledge of the totally unconditioned entity. Across the ages, philosophers have come to know this entity very well and give it many names. Kant called it *the thing in itself.* Knowing this mysterious entity is the challenge before us. We scratch our heads trying to think of what we can know about it. Of whatever can be known, we know one thing for sure. There is no one in the shadows pulling strings. This unconditioned entity has to face up to the most severe fact of life. In the final analysis, it has only itself to rely on. It is the study of this autonomous entity that forms the subject of this work, the study of the autonomous entity that is subject to one and only one principle. This pure entity is the most purest of all and bathes in untainted perfection. This is the principle. We call this principle First Classness, the ultimate no strings attached formula.

Such a principle is extraordinarily difficult to describe let alone formalise. However, examples of this amorphous but powerful concept appear in many present day scientific disciplines. Computer Science provides some good illustrations of that concept we are searching. There, it is actually called First Classness (FC). One illustrative example of FC is the Object Oriented (OO) paradigm in Computer Science where the mantra is "everything is an object." An object is defined as being an "instance of a class." In this case, there is an apparent violation of FC as a class is not an object. The OO paradigm answers that a class is in fact an instance of a meta class and so is an object. The same argument applies to the meta class that is an instance of a meta-meta class. Infinite regress is avoided though as it turns out that the meta-meta class is an instance of itself and so are both a class and an object in the same instance. The OO paradigm resolves the dichotomy between specification (class) and implementation (object) and provides a useful, concrete way of beginning to understand FC. Note the objects at the base level, the class level, meta class level, and meta-meta class level construct described here. The overall structure is a Three-plus-One structure, where the meta-meta level is the One. We claim that any serious attempt at a system based on FC, be it a left side or right side science, will employ Three-plus-One structures.

Another Computer Science example of FC is the programming language LISP invented by the mathematician McCarthy as an outcome of his theory of mathematical recursion based on anonymous Lambda functions. In this case, the mantra was that any entity is a list. Procedures, arguments, return values, values, and value placeholders were all instances of lists. In this way, McCarthy's paradigm eliminated the rigid dichotomy between program and data. The Three-plus-One structure in this case is built around atoms, lists of atoms, lists of lists, where the One corresponds to the Lambda functions implemented as lists. "Everything is a list" was the mantra.

Perhaps the most abstract examples of FC comes from mathematical Category Theory which merits a claim to FC by eliminating the rigid dichotomy in Set Theory between the sets of elements and the elements they contain. Instead of sets of elements, Category Theory concentrates on mathematical structure represented in the form of arrows called morphisms. A collection of such structured arrows, together with some axiomatic preconditioning, is called a category. We don't like the axiomatic preconditioning, but we do like the arrows. Many different branches of mathematics can be lumped together by thinking of them as instances of common mathematical categories. In their turn, these mathematical categories reveal higher order structure that can be represented by arrows between categories themselves. This leads to meta categories based on meta morphisms called functors. The abstraction does not end there. Mathematics admits of yet another meta level, what we can think of as meta meta categories with meta functors. These meta functors were first defined by Eilenberg and Mac Lane. They coined the term *natural transformations* for these meta meta arrows. They later wrote that their express aim in developing Category Theory was to study natural transformations. There is no higher meta level above natural transformations. Mathematicians use natural transformations to discover new mathematical objects.

Despite its power in the providing new understanding of mathematics, Category Theory is built on an axiomatic framework itself and so is still only a left side science. However, we will use some of its arrow theoretic thinking to construct right side science and in so doing we will discover some of the generic entities underlying any system whatever as long as it is based on FC and only FC, no axioms allowed. The Three-plus-One structure in the Category Theory case is realised in the form of the category objects, the morphisms, functors, and finally the natural transformations.

The earliest scientific approach to FC can be traced back to Aristotle who developed his mantra that everything can be rationally understood as instances of classifications. Particulars were considered as instances of species, which in their turn ended up being instances of genera. Aristotle argued that any traditional scientific discipline was limited to the study of entities that were under the umbrella of a determined genus. In so doing, he provided a useful definition of what we are referring to as left side sciences. What interested Aristotle in *Metaphysics* was how he could classify something that had no determined genus. He needed some sort of meta genus. He referred to it as Being *qua* (as) Being but left few precise details of what he meant. The study of beings without determined genera, the study of Being *qua* Being, he called metaphysics, a science with a decidedly ontological vocation. Aristotle's metaphysics can be thought of as the first explicit and coherent reference to right side science and its distinction from the traditional left side sciences. Developing the foundations of such a science is the task of this chapter. However, our inspiration will not come from Aristotle but from those that followed him; the little understood Stoics.

Another example of FC can be found in a modern particle physics where the mantra is "everything is a particle." The particle paradigm attempts to resolve the dichotomy between particle and field. The effect at a distance explained by force fields is replaced by a new breed of particle, the gauge boson, which acts as a carrier of force. However, particle physics is a left side science and so can only approximate FC. The reason is that FC is irretrievably violated at the very foundational level where the dichotomy between entity and attribute is not resolved but ignored by only considering the attribute side of the equation. In particle physics this violation of FC reappears in the form of the irresolvable formless, point-like elementary particles floating around in a void; the particle/void dichotomy. What can be measured is particle, what is without measure is void. We will not attempt to discern any Three-plus-One structure in Particle Physics, as there is probably little point in the exercise at this stage.

These examples show that FC is a very powerful and widely applicable principle. Central to the principle is that it abhors rigid dichotomies. However, even more fundamentally and quite surprisingly, as we shall see, the principle abhors symmetries preferring instead asymmetries in the form of Heraclitus style oppositions. In fact, practically everything taken for granted in the traditional left side sciences is torn inside out for right side science,

If particle physics is founded on the doctrine that the only things that exist are pinpoint particles in the void, right side science upholds the antithetical position that the only things that can exist are bodies. Voids have no role in the science. This is the world of bodies. Only material bodies exist. It is by rejecting the particle principle and the inevitable void of left side science and replacing it with bodies, the particle-void dichotomy is avoided, and FC not violated. This mantra dates back to the ancient Stoics as David E. Hahm writes, "According to the Stoics the only things that really exist are material bodies." (Hahm, 1977) He also remarks "For half a millennium Stoicism was very likely the most widely accepted world view in the Western world." Thus, the concept has some pedigree behind it. This doctrine goes part of the way for ensuring the purity of a deeper doctrine. If left side sciences are fundamentally dualist, right side science must be fundamentally monist.

The Difference Dogma

The central dogma of the traditional left side sciences concerns difference. According to the dogma, two entities are determined to be different from each other according to a difference in their attributes. For the physical sciences the attributes may be perceptible or measurable qualities such as, for example, mass, colour, position or velocity. In the case of mathematics, the properties are assigned via the definitional framework, or derived in some way. Of central importance to the dogma is that an entity is an entity and that an attribute is simply not an entity, it is an attribute. The dichotomy between the world of entities and the world of attributes is central in establishing the fundamental dualist nature of this kind of science. We will call this central dogma of left side sciences, the Difference Dogma.

Associated with the notion of difference, is that of distinguishability. In this chapter, we will define that two entities are distinguishable if they have different attributes. From this definition it follows that if the Difference Dogma holds, and so difference is determined by attribute comparison, then the same applies for distinguishability. Thus, for a science satisfying the

Difference Dogma, difference and distinguishability are synonymous. This is understandable as the distinguishability notion begs the question: "Distinguishable by whom"? The "whom" referred to here is the subject. Left side sciences, because of their objective epistemology, are devoid of any determined subject. The perspective is that of the "view from nowhere," the viewpoint of non-determined subject, the viewpoint of what can be called, the impersonal subject. From this 'God's eye', 'mind independent' point of view, it is no surprise that difference and distinguishability are synonymous.

Difference without Attributes

The Difference Dogma appears so obvious and familiar, that few people of modern times seem to question it. However, there is an alternative viewpoint, the monist viewpoint. Monism totally negates the Difference Dogma. For a want of a better name, we will call it the Anti-Difference Dogma. This dogma effectively declares that there is no determined difference between entities whatsoever. Clearly, this is what is required in order to establish a truly monist worldview.

At first sight, the Anti-Difference Dogma, and the implied monism philosophy, appears counter intuitive and diametrically opposed to common sense let alone any kind of scientific enterprise. Even Hegel's tortuous attempt at such a science has done more to reinforce this assessment than achieve the original intention. How can one construct a science based on the premise that there is no ultimate difference between entities?

Unlike the left side sciences, the object of right side science is devoid of any determined attributes. Such an object of enquiry is quite familiar territory in philosophy. Kant called it the *thing in itself.* The Stoics referred to it as unqualified substance. We will refer to it as an entity characterised as being devoid of any determined attribute whatsoever. We then make the observation that being devoid of any determined attribute does not imply that the entity is attribute free. The contrary even, it implies that the entity has a very specific attribute, that of the attribute of being free of determination.

Apparently, this primordial starting point entity is not alone. The entity is in the company of its attribute, the attribute of being free of any determination, qualification, and specificity whatsoever. It is at this conjuncture that the principle of FC most be invoked, FC abhors the rigid dichotomy. In order not to violate FC, there must be no dichotomy between objects and their attributes. The attribute of this primordial starting point entity must be an entity in its own right.

Thus, this generic starting point for the new science is not a lonely "*thing in itself,*" as Kant imagined it, but necessarily two different entities. These entities are of a different kind. We will formalise this difference in kind by recycling some ancient terminology, the terminology of *ontological gender*. The starting point entity, which has the specificity of being absolutely devoid of specificity, we will say, is an entity of *feminine* gender. This entity has an attribute, that of non-specificity. In order not to violate FC, the attribute must be an entity in its own right. The attribute will be said to be an entity of *masculine* gender. Thus, the feminine entity *has* an attribute; the masculine entity *is* this attribute. Here resides the object of the right side science. From here on, the science at its most fundamental becomes the scientific dialectic of *to be* and *to have*, the dialectic of gender. The central task of right side science will be to develop a new calculus, the *gender calculus*. It is via this calculus that one will be able to describe and proscribe any entity whatsoever, any world whatsoever, as long as FC is respected. In the process of developing the gender calculus, we will make a remarkable discovery. We will realise that we are in fact reverse engineering a calculus that already is apparent in Nature, the four-letter genetic code. We are discovering the underlying generic structure of Nature.

Note that these two entities are formally different as they differ by gender. However, they are indistinguishable to a third party. Distinguishability requires a difference in attributes; however, in this case the two entities have only one attribute between them: One entity *has* it the other *is* it. Unlike the left side sciences, right side science does not rely on distinguishability to determine difference. Right side science is a monism where maintaining in distinguishability is paramount. Rather than attribute comparison to determine difference, right side science is totally based on the relative typing scheme provided by ontological gender.

Differentiation

Right side science is the science of the oneness of monism. The first question concerns the compatibility of the unified oneness demanded by the doctrine with a differentiated reality seemingly dominated by multiplicity. The science must provide an account of such differentiation. There are two approaches one objective and one subjective. In this section, we recount the objective approach.

Unlike left side sciences, right side science accounts a reality where the subject is always present. The subject can be present in two ways, either implicitly or explicitly. In the implicit scenario, the subject becomes the Spectator. In the explicit scenario, the subject becomes Spectacle. Traditional left side science and left side philosophy, such as analytic philosophy, only consider the first scenario and ignore the second, as indeed they must in order to remain true to the objectivity paradigm. After all, the whole thrust of traditional science is to be objective and this demands elimination of the subject from consideration. This objectivist *prise de position* of the left side sciences has obvious advantages as witnessed by the spectacular success of the objective sciences over the past few centuries. However, it comes at a cost.

The downside can be illustrated by what appears to be a similar *prise de position* underpinning the organisation of the biological brain. In the biological case, subjects acting with only a functional left hemisphere exhibit hemi-neglect. where they only identify with the right half of their body and even may only be conscious of the right half of a clock. The subject may only eat the food on the right side of their plate and still complain to be hungry. They may only shave the right side of their face and wash only the right half of their body and so on. (Berlucchi G, 1997) (McGilchrist, 2009). Acting with only the left-brain, they are only conscious of half a world. In the converse case, a subject operating with only a functional right brain does not exhibit hemi-neglect but is conscious of a whole world. One could say that the left-brain only *has* a body, in fact only half a body at best. In contrast, if the right brain operates under the sway of the monistic paradigm then it must conceive itself as part and parcel of body. In this scenario, the right brain would not be conscious to having a body, but conscious that it *is* body, all body, all of body.

The traditional left side sciences suffer from a similar form of hemi-neglect. This fact does not have to be proven because, as remarked above, hemi-neglect forms the core of the objectivist stance of the traditional sciences. To be objective, one must eliminate any subjectivity pollution that may be introduced by the subject. The traditional left side sciences and mathematics adopt their own form of hemi-neglect, based on the neglect or refusal to allow the subject into their epistemology.

The hemi-neglect syndrome is most easily observed in axiomatic mathematics where mathematicians even boast about it. Axiomatic mathematics is abstract left side mathematics *par excellence*. Such mathematics necessarily produces a symmetric view of the world where every mathematical theorem, every mathematical object, every mathematical space possesses a symmetric dual, without exception. Any theorems proven valid in one side of this reality will automatically be valid in the dual reality. Pure mathematicians see this as a godsend, often boasting to their students that they get "Two theorems for the price of one!" This must be one of the most popular clichés of abstract mathematics. It gives mathematicians a perfect excuse to practice hemi-neglect and always only work on one-sided realities. Most still eat food on both sides of their plate though.

In contrast, right side mathematics, due to the ever-present subject, will be seen to produce a fundamentally asymmetric world. The structure of right side mathematics, although complementary to the left side version, is quite antithetical. Farther on in this book we will call it *anti-mathematics*.in order to emphasize this fact.

Hopefully, the biological split-brain allegory helps to illustrate an essential difference between traditional left side sciences and the embryonic right side science introduced here. The right side science is based on an anti-dual, monism paradigm. This makes it immune from hemi-neglect and demands that it not only considers reality as a whole but must continuously embrace reality as a whole. From the standpoint of monism, both sides of the equation are always present. The subject cannot be separated from its object kingdom. They form an inseparable whole. This said, the monistic paradigm must come to terms with the same world of objects that the left side sciences so adeptly study. However, it must avoid the half world mentality of left side science. Thus, there are two worlds of objects, that viewed by left side science and that viewed from the monistic perspective of right side science. Left side sciences completely discard the implicit subject and have no need for it. Right side science must retain the implicit subject at all times. Right side science exploits the presence of subject to the full, using it as the reference point for development of a monist kind of knowledge of reality. In the process, it comes up with its own science of matter, for example. Our aim here is to show that such a science will be a more modern version of the four-element theory of the ancients, dating back to Empedocles. The end results are deceptively simple. The reasoning leading to the results is tricky to explain but should be easy to understand.

Left side sciences address a reality populated with composite entities ultimately made up of atomic or sub-atomic particles. All particles and their composites possess attributes that are measurable by a disinterested third party. To prove that the third party is disinterested, all results must be demonstrably reproducible. In contrast, the right side science version cannot employ attributes harvested from experiment. The attributes must be constructible from first principles. As for the disinterested third party, it becomes the subject and plays an integral role in the development of knowledge. We proceed as follows.

In this scenario, the subject is the same impersonal subject as for the traditional left side sciences. However, there is a difference. Rather than being dismissed as irrelevant, the subject is always present and in this scenario provides the reference point for the argument. Attributes are involved in the exercise but in every case, the attribute is determined relative to the subject. One could even say that the attributes are determined *by* the subject. There are no idle players in this game. Unlike left side science, there is no *a priori* notion of stand-alone attributes that are indifferent to all subjects. The only attributes permissible in right side science are those calculable from the unique position of the subject. For the case in hand, the subject is *any subject whatsoever*, the undetermined subject, or what we call the *generic subject* in its totally undetermined guise. Any attributes calculable from the generic subject will be generic attributes. By an ironic twist in the argument, any subject whatsoever will experience the same generic attributes and so there will appear to be, in fact, "stand alone arguments" which are indifferent to any subject. However, they are not *a priori* to the subject. They are part and parcel of what determines subject as subject.

As anyone who has ever delved deeply into this area knows full well, this domain is a minefield. The dialectic of the Spectacle and the Spectator sometimes appears as a morass of hopeless self-referring contradictions. There are many ways of tackling the problem. One way, the one opted for here, is to see it as a problem of choice. The subject, in order to be such, must be capable of choice. It must be capable of choosing between this and that. The immediate problem is to determine what these choice alternatives are. In the left side science, this and that just appear willy-nilly. In contrast, right side science must ignore the accidental and employ an entirely different technology for acquiring and processing attributes. An integral part of this technology is the process of choice. In this highly relativistic game, the very act of choosing is intimately implicated with the ontological status of the object chosen.

We start where the generic Subject is called upon to engage in a thought experiment involving objective choice. We note in passing that this "thought experiment" may indeed be a "real life experiment." At this stage of the development, there is no determined difference between the two. Confronted with Kant's *thing in itself*, the Subject has already reasoned that it is confronted with two entities, not one; on one hand the entity which *has* an attribute and on the other, the entity which *is* that attribute. The Subject must now attempt to differentiate the one from the other.

Note that differentiation is not the same as distinguishing. Distinguishing is difference determination that is *posterior* to an already accomplished differentiation. Distinguishing can only take place when differentiation is already a *fait accompli*. Differentiation is an active process involving the Subject itself. It is the *fait à acomplir*. Differentiation involves an act and the specificity of that act. The act is that of choosing, in this case choosing absolutely objectively. This brings us to the First Choice, the most objective choice of them all.

The First Choice, the most fundamental of all choices, involves two alternatives, one passive, and the other active. The passive alternative is to choose not to choose. This is the "*Let the mountain come to Mohamed*" type choice. The active alternative is to choose to choose. The active case is the easiest to understand, but like any choice, opting for one alternative at the expense of another always restricts the range of possibilities further down the track. In the case of the active choice, the Subject has opted to know the entity by its attribute. Knowing via attribute is the methodology of the traditional left side sciences. The choice involves opting for a masculinising view of the world. Such a world is perceived as being uniquely populated by attributes. What this means is that, viewed through this methodological eye, the two primordial entities, one feminine and the other masculine, will now *both* appear as attributes. In other words, they will both appear *as masculine*. This is a consequence of enacting the active choice alternative.

Choosing to choose is the masculine choice, the active choice. Choosing not to choose is the feminine choice, the passive choice. In the feminine case, the two simply gendered primordial entities do not appear as attributes. Given that attributes are essential for any appearance, they do not appear as anything. In other words, no explicit knowledge can be garnished from these two entities viewed through the prism of the passive choice. This means that in this context the two entities must be considered *as feminine*. However, these two entities can be known, not explicitly, but implicitly. They can be known not be what there *are*, but by what they *are not*.

This is the land of the poets and there are many ways of explaining the self-referring reasoning involved here. In a nutshell, relative to the impersonal Subject, there are not two, but four types of elementary substance making up the generic substance, the original unqualified stuff, or *thing in itself*. In other words, relative to the ever-present impersonal subject there are four kinds of stuff.

1. MM the masculine *as* masculine or masculine active
2. MF the masculine *as* feminine or masculine passive
3. FM the feminine *as* masculine or feminine active
4. FF the feminine *as* feminine or feminine passive.

This construct is the basis of the ancient theory of the four element. Empedocles called them the four roots, associating fire and air with male deities whilst water and earth were associated with female deities. We also note that Empedocles saw the four elements as a Three-plus-One kind of structure that we have associated with FC. In this case, Empedocles associates the One with Fire or Zeus, to which we have associated the binary gender MM, the doubly masculine, the masculine as masculine.

Right side science dispenses with determined attributes and replaces them with self -determining attributes where system entities are determined relative to each other. At all times the subject is present, either explicitly or implicitly. In the present scenario leading to the four classical elements, the subject is implicitly present. Implicitly we are discussing the very substances that constitute this organism that we have being referring to as the Subject. This subject, any subject, is constituted along these lines according to the doctrine of the four elements. The coherence of this relativistic gender typing scheme is known to the subject in question. In fact, the very coherence of the organism depends on the relative coherence of its gender typing system.

The subject in question is none other than the impersonal subject. But the same argument can be applied to any subject whatsoever. Any such subject will be constituted from matter based on the organisational structure of ontological gender. The gendered structure is determined relative to that subject and is in coherence with that subject, and that subject only. To each subject, there is only one centre of the Cosmos, and that is itself. Right side science involves the ultimate in introspectively.

In the final analysis, the masculine is synonymous with subject. As we shall see, the gender calculus that we are introducing provides the algebra of subject and subjectivity via the masculine and its interplay with the more allusive feminine. Both the personal and the impersonal subject are handled by this algebra, depending on the relative position of M in the terms. For the case in hand, there are four terms, the binary gendered MF, FM, MM and FF. In this book, we explain the gender calculus from many angles. A deep understanding of this blindingly simple gender construct takes time, but will be well worth the effort. Remember that this gender construct is a very ancient concept and has never been properly formalised before. Here, we start gently.

Differentiation is based on the right side version of objective reality. In that scenario, the subject is implicitly present where the explicit becomes a material presence. The gender argument leads to the classical four element constituents of material presence. Any substance present will itself be constituted from the four elements. Such an organisation just does not happen by chance but comes into being from a creative act, the creative choice process that endeavours to stay within the bounds of FC at all phases of the development.

Unification

The previous section considered the right side science version of the objective reality that underpins the traditional left side sciences. In the first instance, this leads back to the ancient doctrine of the four elements. In a nutshell, this can be summarised by enumerating all of the primordial choice alternatives confronting the generic subject. With impeccable precision, the science predicts that the outcome of any choice will be one of four possibilities. According to right side science, anything can be constructed from these four generic types of entity. This provides the basis for the right side version of composite structure, a certain kind of "monistic atomism."

We have already seen that even with the right side epistemological version of left side science, the resulting knowledge is expressed in terms of oppositions. The primary opposition, of course, is that between entity and type. In order not to violate FC, the type of an entity must be considered as an entity in its own right. Thus, according to the right side paradigm there are fundamentally two kinds of entities.. This difference in kind can be formalised as a difference in gender, the ultimate and most generic expression of any ontological opposition.

In summary, left side science relies heavily on using labels that do not mean anything. Left side Science is also based on dualism and atomism. Right side science replaces labelling technology and dualism with ontological oppositions. If there are any labels or letters involved, they must be constructed and will necessarily mean something.

As Kant observed, these kinds of oppositions are, in effect, antinomies. Thus, right side science describes and cognizes its reality in terms of two faced coins. The coin itself is a whole, but always has two sides to it. In this way, right side science must express itself dialectically. Plato expounded the practice of dialectical argument in the form of the dialog between two antagonists. Here we see that monism based thinking is monist on the outside but dualist and antonymic on the inside. Such is the nature of dialectic reasoning. This contrasts with the left side sciences to which the dialectic is totally foreign. Left side thinking is rhetorical, speaks with a single voice on a single subject, and so expresses its non-duality in that way. Nevertheless, it always remains totally dualist in regards to its subject matter. The left side sciences are thus dualistic on the outside and monistic on the inside, so to speak.

Instead of dealing with objective reality, we are now dealing with subjective reality. Our attention is drawn to the generic specificity of the subject. What is this specificity and by what mechanism does the subject, *any* subject, maintain its coherence and integrity? By what mechanism does the subject know what it **is** and what it **is not**? According to the monist doctrine of right side science, the coherence of such an organism can be articulated in terms of a relativistic typing system. However, such a typing system is only a means to an end. The end is the organisational coherence of the subject itself. This involves a process of unification. Once again, four binary typed entities come into play but this time there is no multiplicity and the entities are no longer mobile. Rather than values, we are dealing with placeholders. The placeholder *par excellence* for value is the subject.

Relative to object, the subject is active. It is an innovator, a source of creativity and, embarrassingly, a wildcard when it comes to prediction. Subject can be seen as a causal factor of certain events, of certain effects. A subject thus enters into the causal chain and, in so doing, upsets the deterministic applecart. Epicurus was perhaps the first to recognise this problem and find a suitable remedy. Implacable determinist that he was, he had to find a way of admitting the subject into the realm of causes whilst at the same time salvaging a deterministic concept of causality. His was a left side science viewpoint advocating a dualist, atomistic and deterministic world, His solution to admitting some slack into a brutally deterministic system was in the form of the Swerve. All atoms behaved deterministically, but he added a caveat. Atoms behaved deterministically *most* of the time. Occasionally they experienced an inexplicable, imperceptible Swerve. According to Epicurus, it is via this mysterious Swerve mechanism that the universe micro swerves from its primordial state into the highly structured reality we know today. Remnants of Epicurean Swerve theory can still be discerned in modern science, resurfacing in the form of Heisenberg's Uncertainty Principle and Darwin's Theory of Evolution, for example.

Our topic is how to construct a science without attributes, not to explain causality. Nevertheless, when it comes to science of the subject, causality cannot be ignored.

Causality and the Subject

In physics, the Principle of Causality is very much related to the Theory of Relativity. Erik Christopher Zeeman sheds some light on the situation by effectively showing that the fundamental invariant behind the Special Theory was causation itself (Zeeman, 1964). The principle of causation demands that any cause event must be antecedent in time to any consequent effect event. Zeeman showed that in order for this principle not be violated in any reference frame, the mappings between reference frames must be the same mathematics as for maintaining the speed of light as a constant. In other words, the transformations must be Lorentz transformations. Zeeman's causality interpretation is a weaker, more generic version of relativity than the Special Theory. The constancy of the speed of light could be seen as just an implementation detail for assuring the non-violation of the causality principle in physics.

The causality principle can be thought of as a polarity condition: Causes come first; effects come second, not the other way around. Any violation of Relativity Theory is a violation of the causality principle, which in turn is a violation of the causality polarity condition. Relativity Theory guarantees the coherence of causality. It can intuitively be thought of as guaranteeing the integrity of the Arrow of Time, which expresses the irreversibility of time. The Arrow of Time can be thought of as a polarity convention. It declares that, in our universe, time flows this way, not the other way. From the left side science perspective, this is difficult to formalise. Arthur Stanley Eddington introduced the Arrow of Time term stating:

Let us draw an arrow arbitrarily. If as we follow the arrow we find more and more of the random element in the state of the world, then the arrow is pointing towards the future; if the random element decreases the arrow points towards the past. That is the only distinction known to physics. This follows at once if our fundamental contention is admitted that the introduction of randomness is the only thing which cannot be undone. (Eddington, 1928)

The Arrow of Time points in the direction of increasing uncertainty. He then continues to a most important observation:

We shall use the phrase 'time's arrow' to express this one-way property of time which has no analogue in space.

Eddington highlighted the familiarity of the arrow of time to consciousness and how this stood out in stark contrast with physics where all of the fundamental laws are essentially time symmetric. In desperation, he was forced to turn to entropy considerations to get a handle on the direction of temporality. He admitted that this was a poor solution, remarking:

I do not think he [the scientist] would say that the familiar moving on of time is really an entropy-gradient.

Eddington was writing in 1928 and since then little has changed to challenge his observations. Particle physics is dominated by Schrodinger's Wave Equation, which is time symmetric and so completely catholic in regard to which way time flows. What is not time symmetric is the "collapse of the wave function" which is irreversible. Applied statistically we see this irreversible process, just like all others, leads off to increased entropy according to Eddington's time arrow, hardly an illuminating observation.

Eddington pointed out that there was no satisfactory "this way" arrow for time in the four dimensional Murkowski space necessary for Special Relativity. The same applies to Riemann space for General Relativity. An extra dimension has been added for time, but it is just like the other spatial dimensions, totally lacking in orientation.

Eddington's plea for a physics that somehow included consciousness has been revisited in more recent times. Roger Penrose writes:

A scientific world-view which does not profoundly come to terms with the problem of conscious minds can have no serious pretensions of completeness. Consciousness is part of our universe, so any physical theory which makes no proper place for it falls fundamentally short of providing a genuine description of the world. I would maintain that there is yet no physical, biological, or computational theory that comes very close to explaining our consciousness. (Penrose, 1994)

Nobel laureate Brian D. Josephson also joins the fray with his Mind-Matter Unification Project (Josephson) in his quest to find a physics that embraces mind as well as matter. In effect, all of these writers are lamenting an incurable condition of the traditional left side sciences: all sciences that operate under the left side epistemological paradigm are condemned to suffer from hemi-neglect. Such sciences can never accommodate a role for consciousness due to their wired in antagonism to a science of the subject. After all, how can a science *sans sujet* ever produce a non-trivial science *de sujet*?

The right side scientific paradigm must not only render a science that makes embracing consciousness possible but fundamentally necessary. The right side scientific paradigm moves the subject into the explicit realm of study. The subject becomes the fulcrum for all subsequent knowledge which follows and as such, provides the answers to the "which side up" type of questions posed by Eddington. There are no absolute, context free answers. All answers must be relative to the subject. By proceeding along these lines, we will be able eventually to arrive at a fundamental notion of the Arrow of Time. In this kind of scenario, not only does time become asymmetric, but also so does spatiality. The Cartesian spatiality of the left side sciences treats any orientation, any frame of reference as being equally valid. There is no preferential reference frame in Cartesian geometry and so one could argue that this is a good example of FC. The fact that no reference frame is privileged over any other is an example of FC, but a very second rate example. Cartesian geometry violates FC because it is based on a dualism, a dichotomy between reference frames and geometric objects situated in these reference frames. FC demands that these geometric objects should be treatable as reference frames in their own right. This form of FC is impossible in Cartesian style geometry. However, in the geometry of the right side scientific paradigm, the converse applies. This science of the subject treats the subject as the origin and reference frame for any spatio-temporal geometry that might be involved. Such a geometry, dictated by the specificity of the subject, loses its time symmetry and its spatial symmetry. What is on the left and what is on the right, the front and the back, the up and the down, all becomes absolutely determined and not interchangeable. A new form of absolutism enters the scene. The only caveat is that the absolutism only applies to a specific subject. As far as this subject is concerned, it knows with absolute precision, what is on its left and right and any other relative position to it. In this right side scientific paradigm, the subject becomes the privileged entity in the system. As such, one might think that this violates FC, as FC is anathema to absolutely privileged entities. The way out of this apparent conundrum is the fact that, in this science, the subject under investigation is any subject whatsoever, the generic subject. The generic subject is not an abstraction. Understanding the difference between the abstract and the generic, is synonymous with understanding the difference between the left side and right side scientific paradigms. Abstractions cannot exist. This cannot be said of the generic.

In effect, all of these writers are lamenting about an incurable condition of the traditional left side sciences: All sciences that operate under the left side epistemological paradigm are condemned to suffer from hemi-neglect. Such sciences can never accommodate a role for consciousness due to their wired in antagonism to a science of the subject. After all, how can a science *sans sujet* ever produce a non-trivial science *de sujet*?

Shape of the Generic Subject

A classical problem for traditional schools of Western thought such as Analytic Philosophy is the Mind Body problem. Viewed from the left side perspective, this presents as a problem to be tackled. It is however, a fruitless quest as the very problem is a mere immediate consequence of the explicit dualism of the left side paradigm itself. The best that can be achieved is some kind of psychological neuroscience brain theory. Nevertheless, left side philosophy has bravely pushed ahead and come up with a philosophy of Mind. However, this philosophy essentially consists of systematically cataloguing all of the known possible (left side) approaches to the philosophy of Mind. This is like presenting a nineteenth century theory of flight by cataloguing of all of the brave would be aviators who jumped of cliffs with a flight contraption strapped to their backs.

From the right side perspective, the problem vanishes as the monistic paradigm, due to FC, does not allow any rigid dichotomies at all, let alone any Mind Matter kind of split. At first encounter, this concept can be annoyingly difficult to grasp fully. In what follows, we will be constructing the "shape" of the generic subject. In so doing, the subject becomes endowed with a determined form. Now it may be all very well for us to say that this is the generic "shape of the subject," but the nagging anxiety arises as to what constitutes the subject. Is this a subject confronting Nature? Alternatively, is this subject really Nature looking back the other way? As for the specificity of the shape, is this the shape of Mind or is it the generic shape of Matter or of Nature herself? Perhaps annoyingly, right side science does not even attempt to answer any of these questions, as they do not make sense in the monist framework. The difficulty can be traced to the difference between abstraction, which is a left side science technology, and the generic, the right side sibling. Abstraction is naturally dualist. On one side, there is the abstract understanding of the thing in the form of a theory, opinion or whatever. On the other side is the thing pure and simple. There is no mix. This is a harsh dichotomy.

The dualist epistemology of left side science is easy to understand. Going from the abstraction formalism of left side science to the generic formalism of right side science is much more delicate. The existence even of such an alternative formalism has long been contested. Suffice to say at this point, is that the dominant characteristic of right side epistemology is the absolute intolerance to rigid dichotomies as dictated by the demands of FC.

Even though the absolute dichotomy is anathema to right side science epistemology, the generic structures from which generic knowledge is constructed are built on what appears to be absolute dichotomies. This is only an apparent absolute contradiction as all dichotomies are relative to the subject. The subject needs to know its reality in terms of absolute dichotomies. It needs to know what it is and what it is not and know so absolutely and even urgently. However, viewed by a third party, such dichotomies, such determinations, so vivid to the subject, will be virtually undiscernible. After all, the third party has its own boat to row.

Other than expressing structure and knowledge in terms of oppositions, the second cardinal characteristic of right side science is that knowledge is expressed in wholes where, at no time, can any aspect of the whole claim autonomous and separate existence. Any aspect of the whole is necessary for the existential coherence of any other aspect.

The third characteristic is that the subject is always present.

The totality of reality cannot be conceived or perceived. Reality can only be known in the form of a whole. Each whole is reality considered from a particular pint of view. There are as many wholes as there are points of view.

Any whole must include the subject. Thus, due to the presence of the omnipresent subject, any whole can naturally be considered as a dichotomy, the subject on one side and what is not subject on the other. We will adopt the convention that the subject is on the right side of the dichotomy and the other on the left. Certain subjects may be based on the opposite polarity convention, but that does not affect the argument. To each its own polarity convention, as long as it lives by it.

Any subject is characterised by its own specificity. In the case of the left-right dichotomy, the subject involved enjoys the specificity of being totally devoid of any determined specificity. Such a subject is customarily referred to as the *impersonal subject*. The impersonal subject is what it is, nothing more and nothing less. It inevitably finds itself in the company, of what it is not. We will informally refer to this Other of the subject as its kingdom. However, we keep in mind that there is no construct that determines in any way that the kingdom may be of a mental or physical nature, or of any other kind. Relative to the totally undetermined impersonal subject, there is the presence of the equally undetermined kingdom.

As we have already remarked, left side science also is involved with this kind of dichotomy. However, the left side science ends up dispensing with the impersonal subject and concentrating on getting to know the great unknown kingdom via various forms of pragmatic techniques, the details of which we are more than familiar.

At this point, right side science definitively splits from its *suck it and see* sibling and starts its own distinctively dialectical journey. The step in the quest for the conscious subject was based on the realisation that "I am not alone." The second step involves being conscious of this realisation. This is achieved by the presence of a more determined subject than the first. Entering into this reality, already split asunder by the presence of the impersonal subject, is the personal subject. This subject is neither on the side of the impersonal subject nor of its kingdom. It must straddle both and in doing effectively produces a dividing incision orthogonal to the first. This results in the whole ending up being cut into four quarters. One might argue that the notion of an orthogonal incision does not make much sense here, as there is no *a priori* notion of orthogonality in the first place. This objection has some merit; however, the objection comes too late. In the wake of this dichotomous operation, a primordial form of orthogonality becomes a *fait accompli*. Having achieved this tautological operation of effectively applying the first opposition to itself, we will adopt the convention that the personal subject is located at the front lobes of the resulting four part square whilst its kingdom is located at the back. We retain in mind that the primordial rapport between the front back axis and the left-right axis is some equally primordial kid of orthogonality.

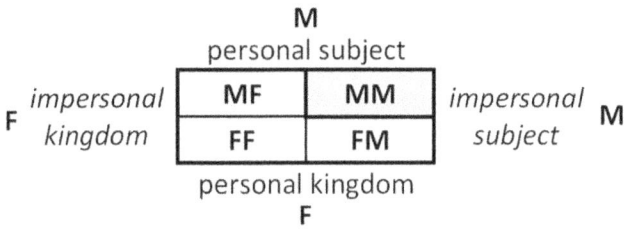

Figure 32 A semiotic square illustrating the generic form of the subject.

Keeping in mind that both the personal and impersonal subjects are singular relative to their kingdoms, we can say the both subjects are of masculine gender relative to their respective kingdoms. The kingdoms themselves are of totally unknown specificity and hence of feminine gender. What we now have constructed is relative typed form of a generic whole as illustrated in Figure 32. This corresponds to the generic form of the subject as a generic placeholder of value. There are four placeholder positions, each binary gender typed. The gender typing relative to the personal subject is defined by the first letter whilst the second letter conveys the gender typing relative to the impersonal subject.

The Semiotic Square

There are two fundamental ways of organising knowledge, the left side and the right side way. The left side technique is the easiest to understand as it is based on a taxonomic hierarchical form of organisation involving different levels of species and genera. This kind of structure can usually be represented in tree diagrams. For example, Chomsky's transformational grammar for representing the syntax of natural languages is based on tree diagrams and transformations between tree diagrams. The essence of the left side tree-diagram approach is that entities, and knowledge of them, can be broken down into increasingly smaller and refined components.

The right side organisational approach is the converse to the atomist and atomising left side approach. In this scenario, any entity, viewed from a particular point of view, can be known in terms of a whole. The innovation comes from the fact that a whole, any whole, has a generic form consisting of four gender typed parts as shown in Figure 32. We will call this generic form, the *semiotic square*.

The semiotic square conveys that in any whole there are always four players in a Three-plus-One structure. Each of these players has an attribute that distinguishes one from the other. This attribute is not of the empirical kind but of the ontological kind, being based on ontological gender. With the semiotic square, no labelling is necessary, only the polarity conventions. The semiotic square can be thought of as a generic cognitive structure. The semantics behind the gender typing formalism can be grasped more intuitively by using natural language; The semiotic square shown in Figure 33 replaces the gender typing with natural language generic attributes. In this way, we start to see that our science with attributes starts to take attributes on board by actually constructing them.

The taxonomic, tree structure classification systems of the left side sciences are simple to understand and apply. For the right side version, it is the semiotic square that first springs to the fore. By applying a simple semiotic analysis to practical scenarios viewed as a whole, one can start to comprehend the semiotic square and its extremely generic nature. Algirdas Julien Greimas developed his own version of the semiotic square, inspired by Aristotle's Square of Oppositions, Based on the semiotic square, he applied semiotic analysis to many different areas such as the semiotic analysis of text. (Greimas, 1991).

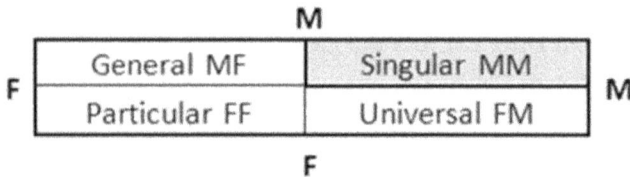

M

General MF	Singular MM
Particular FF	Universal FM

F   M

F

Figure 33 Intuitive natural language interpretation of the generic semiotic square.

The semiotic square provides an intuitive approach to formalising the Three-plus-One structure we mentioned in the introduction. The One is the Singular MM typed player with the triad making up of the rest. This singular MM typed entity corresponds to that which is simultaneously impersonal and personal subject, a singular singularity. Over the ages, many minds, from different cultures have mused over the nature of this being.

The Theological Semiotic Square

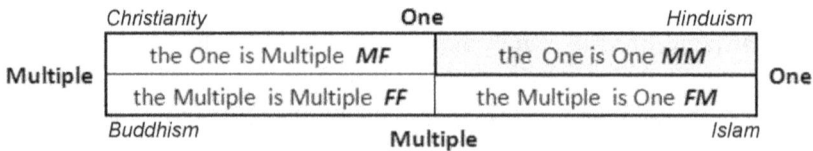

Christianity   One   Hinduism

the One is Multiple **MF**	the One is One **MM**
the Multiple is Multiple **FF**	the Multiple is One **FM**

Multiple   One

Buddhism   Multiple   Islam

Figure 34 On a higher note, the generic nature of the semiotic square is well illustrated by applying it in the theological domain in order to study generic theology. There are of course four generic religions and it is interesting to see where they fit into the semiotic square and relate to each other.

Semiotic analysis based on the semiotic square has been applied to many different areas by practitioners of semiotics. It can even be used in advertising to settle such questions as "How many fundamental ways are there to advertise underarm deodorants? It is not an abstract way of thinking, but a generic way of thinking. There are, of course, four strategies to advertise underarm deodorant, the singular, the general, the particular, or the universal approach. It is left to the reader to fill in the details.

On a higher note, the generic nature of the semiotic square is well illustrated by applying it in the theological domain in order to study generic theology. There are of course four generic religions and it is interesting to see where they fit into the semiotic square and relate to each other.

The four generic religions make up a whole. The whole can be informally constructed as follows. The generic structure of a whole is constructed from one single opposition applied to itself, leading to the semiotic square structure. For the generic semiotic square, the primary opposition is between the subject, necessarily masculine and its feminine gendered other. A simpler to understand version of this opposition is to talk of the masculine subject as the One and the feminine as the Other. The primary opposition thus becomes that between the One and the Multiple. Such an opposition can qualify as an opposition satisfying the Socratic Uncertainty Principle, as defined previously.

When the opposition is applied to itself this leads to four kinds of entity as shown in Figure 34. This semiotic square illustrates the four fundamental answers to the theologically impregnated question, "What is the relationship between the One and the Multiple? Each answer corresponds to a fundamental generic paradigm that can be interpreted in theological terms. Each paradigm corresponds to one of the four world religions. The Providence paradigm of Christianity declares the rather individualistic stance that the One is Multiple. Islam takes the converse position declaring that the Multiple is One, where the collectively is decidable under the hammer of the One. Hinduism, according to the Advaita doctrine of Non-Duality declares that the One is One, and to add a phrase of Sankara, "Everything else is illusion." All three of these religions incorporate the masculine subject into the paradigm in one way or another. The exception is the fourth paradigm that we associate with Buddhism. Buddhism, according to its doctrine of Non Self, declares that the Multiple is Multiple. There is no One. This doctrine is without subject, pure FF.

Christianity

Father	Jehova
Son	Holy Ghost

Hinduism

Vishnu	Brahmana
Buddha	Shiva

Buddhism

The Four Noble Truths

Islam

Faith	Allah
Divine Koran	Divine Justice

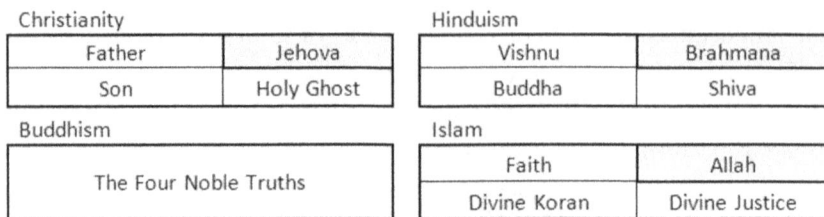

Figure 35 Each of the four world religions can be considered a whole and thus has its own semiotic square. The Buddhist semiotic square would probably be based on the Four Noble Truths, but no details have been included in this diagram.

Here is not the place to attempt an exhaustive semiotic account of theology, however it the importance of religion studies cannot be underestimated. Before moving on, we note that each of the world religions themselves can be viewed as a whole. As such, each will be characterised by its own semiotic square. For example, the semiotic square for Christianity will be based on the Trinity and have God in the singular position, God the Father in the general, Christ in the particular and the Holy Ghost in the universal slot. The Three-plus-One structure for each of the world religions is shown in Figure 35. In this way, one ends up with a semiotic square of semiotic squares.

In the diagram, no details are provided for the Buddhist Multiple-is-Multiple paradigm but it would probably be based on the Four Noble Truths. However, one must tread carefully as this paradigm is devoid of any explicit subject.

Aristotle's Square of Oppositions

We will now start considering the right side version of logic. To make it clear that there are two quite distinct forms of rationality involved for the two sides, we will refer to the left side version of formalised reason as being *logic*, pure and simple. The right side version of formalised reasoning we will refer to as *ontologic*. Ontologic will become the formalised reasoning component of ontology. To emphasize the difference between these two forms of rationality, we can exploit a distinction what was very popular amongst the Stoics, the distinction between the *true* and the *truth*. Rather flippantly, we can say that logic is the science of the true and ontologic is the science of the truth. It is now time to start to explain what it all means. We start with Aristotle.

Aristotle was not only one of the greatest philosophers of all time, but also the greatest fence sitter. His fundamental *modus operandi* was to search for the middle ground, not any middle ground but the *golden* middle ground, the golden mean. When it came to tackling the problem of formalised reasoning, he would have been confronted with two conflicting demands. The first task involves formalising the concept of valid reasoning. This requires formalising logic and involves the science, not of truth, but of truth *as a value*. This is involves the science of the true. To be more prosaic, the science will be a logic of *truth values*.

The second requirement is to provide a formal apparatus for explaining truth. This involves truth as a placeholder. It is only in the geometry of the placeholder that such notions, for example, as the interplay between the qualitative and quantitative can be formalised.

As we shall see, modern symbolic logic totally ignores any such ontological structure as being superfluous to logic. In that sense, the modern logicians would be correct. However, they would err if ever they claim that rationality is totally encompassed by logic. In addition to the logic, there is the requirement of an accompanying ontologic. The missing ingredient of modern formalised rationality is the ontological.

In developing his science of the categorical syllogism, Aristotle provided the first formalisation of logic. In addition, he also provided an ontological structure that was capable of providing, not only valid answers but validity explanations.

Aristotle's categorical logic consists of propositions, each made up from two terms. The syllogistic is sometimes referred to as term logic. Aristotle called the terms *limits*, which is quite a good nomenclature as two of the three qualifiers involved are the *All* and the *None*. Interpreting the syllogistic, as a "logic of limits," emphasises the metaphysical. The classical interpretation as a term logic emphasises the more mundane logical aspects.

In between limits of the *All* and the *None* is that which is neither *All* nor *None*. This translates down to the *Some*. However, even *Some*, as an undetermined quantity, can be taken as the opposite pole to the determined quantity of *None*. This is really a logic of limits. Properly formalised, these three *All*, *None*, and *Some* qualifiers are clearly not of empirical origin but formal generic constructs that can be constructed from first principles. As such, they should be expressible in terms of ontological gender as the gender construct provides the two building blocks for generically constructing literally anything. The objective of this section is to show how categorical logic relates to the gender construct and thus has an ontological and hence right side science interpretation. Straddling as it does, both left side and right side paradigms we also catalogue the left side interpretation of the syllogistic.

The basic generic structure of the syllogistic can be seen in the form of Aristotle's Square of Oppositions. This artifice is basically a semiotic square that can be constructed out of two oppositions. The left-right opposition is that between the affirmative and the negative. The front back opposition is between the universal and the particular. This leads to the semiotic square shown in Figure 36. The EAIO labelling was added to the syllogistic during Medieval times by the Scholastics.

universal

affirmative	A Every S is P	E No S is P	negative
	I Some S is P	O Some S is not P	

particular

Figure 36 The semiotic square for the four terms of Aristotle's Syllogistic logic. The square is formed from two oppositions, the negative/affirmative, and the universal/particular.

An important underlying assumption is what the modern logicians call *existential import*, which demands that the members of the S and P classes in the particular side of the square must actually exist. Thus, there must be at least one particular in each of class in the I and O propositions. From a right side perspective, such a requirement is necessary as otherwise, there would be a violation of FC: There would be an absolute dichotomy between particulars that are existents and the non-existents. They are all existents.

In modern times, these four kinds of two term propositions are often represented by Venn diagrams. Venn diagrams are more at home in the abstract world of left side science, however, we can employ them to illustrate how the gender construct fits into the syllogistic.

Our first step is to try and interpret the left-right opposition between the affirmative and the negative in terms of gender. If our polarities are correct, the left-right opposition should be an instance of a standard feminine masculine opposition. It turns out that the gendering of the terms in the four EAIO propositions corresponds the classic distribution typing. The "distributed" class corresponds to masculine typing; the "undistributed" class corresponds to feminine typing. The definition of term distribution can be found in any classical logic textbook and is rather obscure as well as tortuous to comprehend and apply. One can only commiserate with the countless students over the centuries that have had to torture themselves over this construct. We prefer the gender construct.

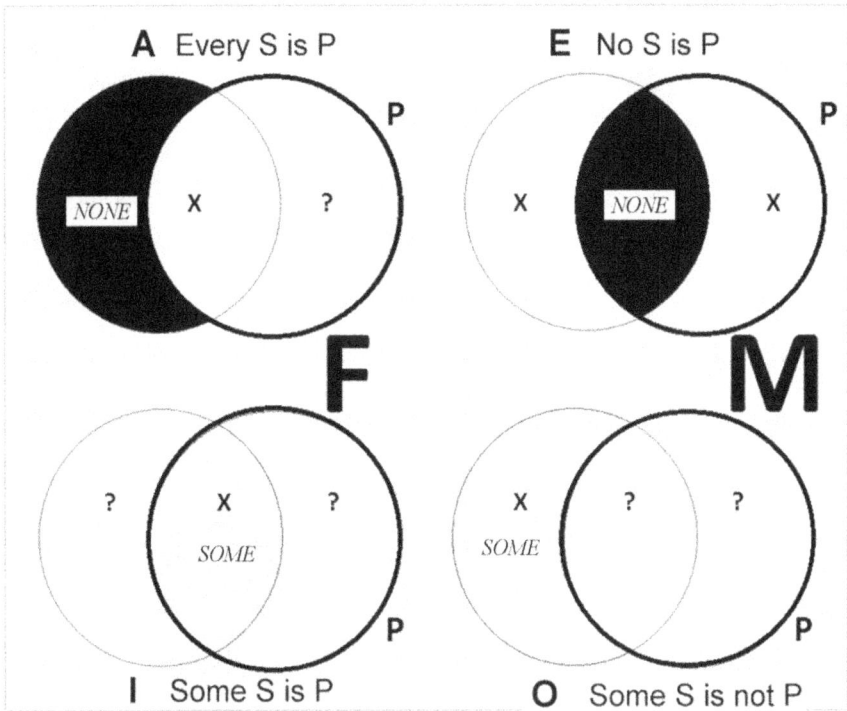

Figure 37 Venn diagrams illustrating left-right gender opposition applied to the syllogistic terms.

Figure 37 shows the Venn diagrams for the four EAIO propositions, highlighting only the P class. This illustrates the left-right opposition of the underlying semiotic square. This opposition, according to right side science must satisfy the Socratic Uncertainty Principle. One pole must be totally determined and the other pole totally undetermined. Aristotle would have said that they must be proper extreme "limits." In his case, the limits were the poles of the logical affirmative and negation. We want to interpret that opposition using the extremes of the feminine and masculine gender. Looking at the diagram, we see that the two provisionally masculine gendered P classes on the right side can be considered as being free of any determined divisions. For the P class in the E term, the intersection between S and P is null and so no P is undivided, hence masculine. For the P class in the O term, P must contain at least one member but how and in what quantity is totally undetermined. Thus, P is like the previous case, undivided and hence masculine. On the other side of the square, The P class in both the E and the I terms, definitely contain determined structure. Both of them contain at least one member shared with the class S. Relative the P classes on the right, the left side P classes each exhibit internal structure. Thus, we have an opposition between P classes on the left containing determined structure and P classes on the right not containing determined structure. This is an extreme Aristotelian limit. It is an opposition satisfying the Socratic Uncertainty Principle. It is an instance of an opposition between two opposite genders. For those that still prefer the age-old classic version, it is an opposition between undistributed P classes on the left and distributed P classes on the right.

This process can be repeated for the front back opposition as shown in Figure 38. In this case, the argument applies to the other class involved in the four propositions, the S class. In this case, the S term in the front section of the square are typed masculine. The S class in the A proposition is undivided because there is nothing to divide as it is empty. The S class in the E proposition is masculine for the same reason as for P in the previous case.

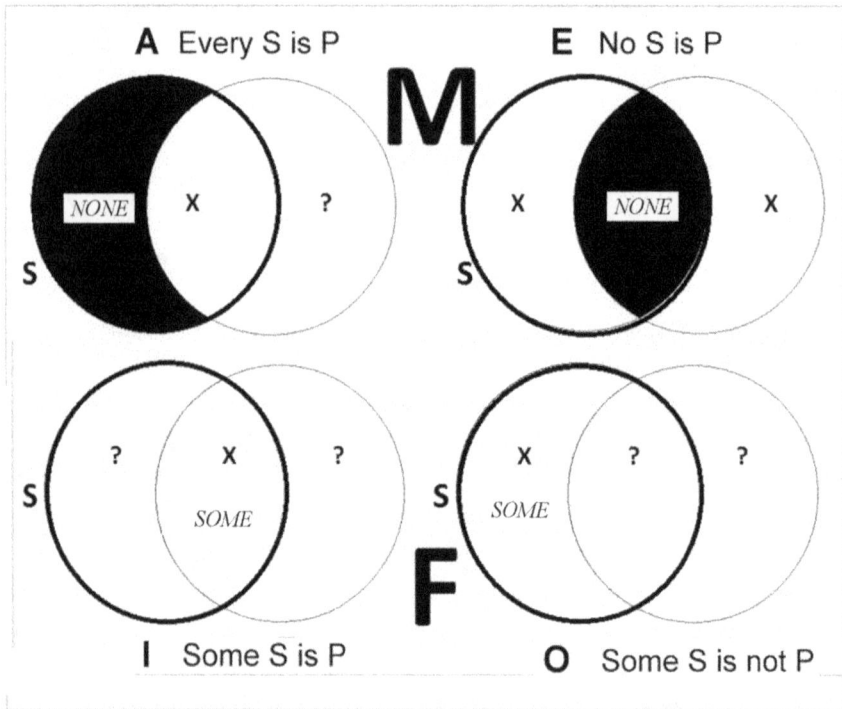

Figure 38 Venn diagrams illustrating front back gender opposition applied to the syllogistic terms.

The Oppositions

Aristotle described how the four kinds of terms could be placed in a square illustrating the various oppositions between them. He then went about characterising each kind of opposition, although the subalterns were not mentioned explicitly.

The oppositions between universal statements are contraries. Contraries have the property that both cannot be true together. One may be true and the other false. It is also possible that both can be false together. On the other hand, subcontraries involve oppositions between particulars. In this case, both cannot be false together.

Of great interest to us is an opposition at a higher-level altogether, the opposition between Aristotle's syllogistic structures and modern logic. The dramatic difference between the two approaches was clearly illustrated by George Boole in what has become the modern version of the Square of Oppositions.

Modern logic differs from the classical logic by simply replacing the universal with the general, in other words with the abstract. This can be achieved by using labels and the logic becomes *symbolic logic*. Thus, the term 'All men' is replaced by the abstract version 'All X'. The thing gets replaced by a label and introduces different semantics. The label becomes simply a placeholder and as such, like any placeholder, may be empty. The logicians explain this as relaxing the requirement of existential import. From a classical mathematics perspective, the generalisation introduced by modern logic is to allow sets to be empty. Once the reasoning becomes abstract, the logical difference between yellow centaurs and canaries evaporates. Not only have that, all the oppositions except the contradictories, also evaporated. For example, both sides of the contraries opposition 'All centaurs are yellow' and 'No centaur is yellow' are true. The contraries opposition has evaporated.

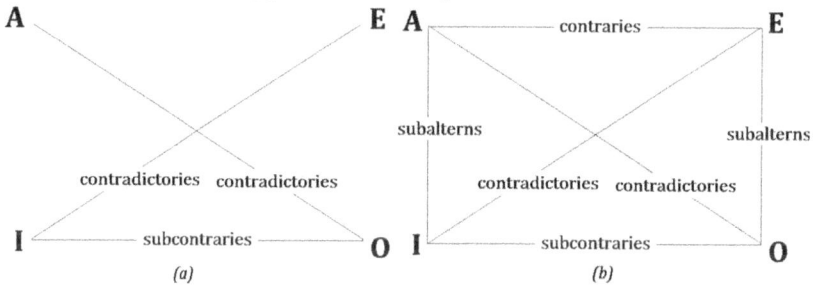

Figure 39 (a) The modern abstract logic version of the oppositions.
(b) Aristotle's square of oppositions.

Figure 39(a) shows the resulting modern logic version of the square of oppositions. The square has virtually collapsed and only the contradictories and the subcontraries survive. We have deliberately drawn the modern version on the left side relative to Aristotle's square to illustrate that this is the left side variant of logic. The other variant is Aristotle's seed for the right side version. The left side involves abstract, symbolic logic. The right side in the diagram represents Aristotle's version of elementary generic logical structure. In practice, the modern symbolic logic approach boils down to a simple bipolar nominalism where the basic opposition is between two particulars, I and O. The letters A and E act as pure label signifiers for the I and O respectively, acting as the signified. The contradictory oppositions A-O and E-I model the relationships between signifier and signified. In essence, the system becomes a simple two-letter system labelled by A and E.

Stoic Logic

Nothing remains of the work of early Stoics and there is certainly no mention anywhere in the literature of a Chrysippus Square of Oppositions. However, we do have a scanty knowledge of Chrysippus' logic in the form of his five undemonstratables. Using four of these fundamental syllogisms, it is not difficult to construct a corresponding semiotic square that fits in smoothly with the general thrust of our approach. As for the remaining syllogism of the five, it can play an even more pivotal role than those four making up the fundamental square of oppositions.

Stoic logic differs dramatically to that of Aristotle. There is no static classificatory apparatus. There are no species and no genera. This is no extension or comprehension of terms. The figures and modes of the syllogistic evaporate into thin air. To the Stoics, Aristotle's syllogistic was "useless." (Chénique, 1974) In contrast, Stoic logic is starkly oriented to the particular. As such, it incorporates one aspect that might entitle the logic to be considered a *generic* logic, logic free from abstraction. It is this kind of logic one needs to construct and deconstruct a real world, not an abstract world. However, what precisely is a generic logic? Our immediate task is to answer this question.

Generic Bases

The five undemonstratables that form the core of Chrysippus' logic were discussed in the previous chapter. Chrysippus' Logos refers to qualities that can be taken into possess by the subject. Interpreted as a simple left side logic, the qualities can be expressed as simple propositions and the subject can be ignored. An examples that comes straight from Chrysippus' is the hypothetical syllogism:

If it is day, it is light: but it is day: so it is light.

This example, pertains to the realm of logic, a left side form of rationality. The qualities involved are expressed by the propositions "it is day" and "it is light." What interests us is the application of the Chrysippus Logos to the more noble Parmenidean realm dominated by what *is*, the realm of ontology. Rather than logical reasoning we want to examine ontological reasoning. This takes us to the right side form of rationality where the subject becomes the fulcrum. Rather than abstract qualities, any specificity involved must be determined relative to the subject and perhaps even by the subject. At this point in the development, the scenario is so generically undetermined that it can be difficult to grasp the concept in hand.

To make for light reading and to avoid thinking abstractly, we abandon the quest to grasp the concept. Instead, the subject grasps a chook. This situation is easy to understand. The subject can thus declare with confidence that there is a bird in the hand. Appealing to popular wisdom we can also note that a bird in the hand is worth two in the bush. Thus, relative to this subject, there are two kinds of chook, the one in the hand and the ones in the bush. The bird in the hand is in a highly immediate and determined rapport with the subject and so will be of masculine gender and the birds in the bush will be of feminine gender. Gender, of course is always determined relative to the subject. This scenario illustrates the opposition, always relative to the subject, between what *is* and what *is not*. This kind of opposition is a limit that satisfies the Socratic Uncertainty Principle where absolute certainty on one side faces off against equal uncertainty on the other. This opposition can serve as the left-right opposition in a semiotic square. The subject involved here plays the role of the impersonal. The scenario can be repeated for the personal subject. This can be represented by a semiotic square as shown Figure 40. In order to emphasise the non-linguistic nature of right side science, the four different determinations of the semiotic square are represented iconically. Each icon consists of two circles with the each particular binary gender determination indicated by a horizontal line. Of particular interests is that the circles on the right of each icon represent the determinations due to the impersonal subject whilst the circles on the left represent the personal subject determinations. Each circle is typed by a single gender as either F or M.

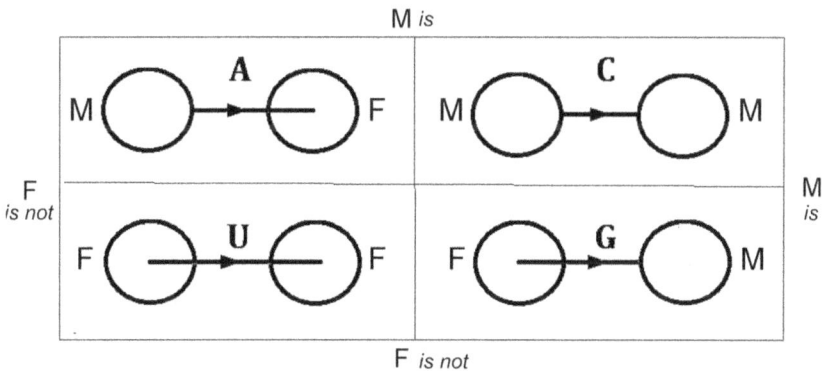

Figure 40 The Parmenidean Semiotic Square based on the opposition of what is an what is not, applied to itself. The diagram provides iconic representations of the four generic building blocks for constructing systems based on FC. These will be called generic bases and have been labelled with the four letters of the genetic code. Note that the 'is not' can read as 'has'.

The four icons shown in Figure 40 can be taken as the representations of the four binary gendered generic building blocks of any organism based on the principle of FC. These building blocks will be called *generic bases*, Like the four letter EAIO labelling scheme of classical logic, we have also chosen to label them. To throw the cat among the pigeons, we have labelled these generic bases using the same CAUG (RNA) lettering scheme as for the bases in biochemistry. The match between generic bases and biological bases and why, was determined after long deliberations by the author which will not be detailed here. As this story progresses, the match shown here will become increasingly compelling and even obvious.

The Generic Lottery

Since right side science is non-abstract it must often rely on the metaphor, allegory, and the icon in order to satisfy the thirst of intellect's intuition. One way of understanding the generic semiotic square and its two rounds of determinations is to think of it as a lottery. Consider a barrel full of lottery tickets. One of the tickets in the barrel is very special: It is the winning ticket. All the other tickets are forcibly losing tickets or destined to be thus. This is an example of an opposition where the two opposite poles satisfy the Socratic Uncertainty Principle. There is nothing more extreme than the opposition between winners and losers. In this case, the winning ticket will obviously be typed masculine relative to the losing tickets, which will all be typed feminine. The winning ticket in the lottery game can be thought of as corresponding to the impersonal subject and, of course, is typed masculine. One could say that the impersonal subject *is* the winning ticket, the key pin for the game. As to the personal subject, it corresponds to the winner of the lottery, the buyer of the winning ticket. Once again, the winner will be typed masculine.

It is now a worthwhile exercise to examine in detail the resulting generic structure of this generic lottery. The structure is articulated in terms of the relationship between player and lottery ticket. There are four such generic relationships. The reasoning, somewhat like a Naya syllogism, goes in five steps.

Step 1

The first step is the *C* type relationship corresponding to a single arrow where the source and target sides are both typed masculine as shown in Figure 41(a). The source side of the arrow corresponds to the winner of the lottery while the target side corresponds to the winning ticket. This C type structure represents the winner with his winning ticket in hand.

Step 2

The next to consider is the **A** type relationship shown in Figure 41(a). The **A** type relationship is represented by an expanding cone of arrows emanating from a masculine type entity. This entity corresponds to the player. The open end of the cone of arrows is typed F and corresponds to the all of the lottery tickets, of which one must eventually be chosen.

Step 3

The next relationship is of type **U** and consists of a sheaf of arrows with both ends open and so typed as feminine. In the diagram, the arrows are shown parallel to each other. One end of the sheaf of arrows corresponds to all the tickets in the lottery and of course, any one could win. The other end corresponds to the winning ticket. Since which ticket is the winning ticket is unknown at this stage, there will be the same number of arrow ends as there are for the other side of the sheaf. Thus, the sheaf has a winnable end and a winning end. The direction of the arrows between these two ends goes from the winner side to the winnable side. This might seem counter-intuitive but is of utmost importance. The basic idea is that if the arrows went from the winnable to the winner side of the sheaf, this would imply that winners predictable from the winnable, when in fact it is the opposite. Given a winner, one knows that there was a winnable, but given a winnable does not imply a winner. The orientation of this **U** type base must be kept in mind when we come to bolt these bases together to build our generic lottery.

Step 4

This step corresponds to the **G** type structure consisting of a cone of converging arrows. In this case, the source side is open and so is feminine and corresponds to the winning ticket. Since the draw has not yet taken place, the winning ticket is still in the barrel and cannot be distinguished from all of the loosing tickets. Thus, there will be as many arrow shafts on this open side of the sheaf as there are tickets in the barrel. The convergent end of the **G** sheaf corresponds to the particular ticket that is destined to win the game. This ticket is clearly of masculine gender.

Step 5

The first four steps of the development involved associating the four atomic bases with the four generic bases of a lottery. The final step is to bolt these bases together to provide a composite structure based on the AUG bases. This triadic AUG structure is shown in Figure 41(b) and determines the C base consisting of the winning player at the source end matching up to the winning ticket at the target end. Both are masculine. Tongue in cheek, this completes the logic of how to win the lottery, any lottery whatsoever. The concepts presented here will be formalised more clearly later. This rather allegoric example is much more profound than it might first appear. However, we will not linger here too long as this only forms part of the story. To give a taste of what is to come, we will use biochemistry terminology call any triadic structure made up of triplets of the four generic bases generic codon. As will be seen, these generic codons act as the basic building blocks for determining virtually anything whatsoever, at least generically. As for the AUG codon represented schematically in Figure 41(b), this particular will crop up time and time again at the beginning of any story. For example, those with a keen eye might be tempted to interpret the A and G cones of arrows and the sheaf of parallel arrows U with, respectively, the timelike, spacelike and lightlike arrows of the Special Theory in physics. If so, then you are on the right track. This is a remarkable story.

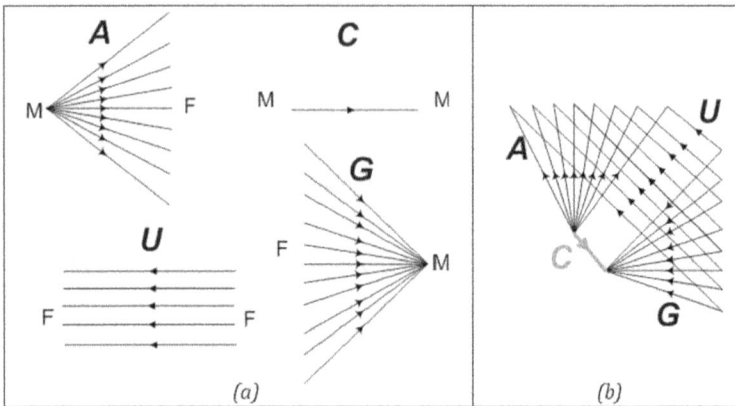

Figure 41 (a) Arrow interpretation of the four generic bases. (b) The AUG bases bolt together to determine the C base. This is the generic "start" codon

Lottery Cosmology

The lottery example above provides an allegory for the whole Cosmos or any other organism organised along the principles of FC. It sketches out the embryonic formal structure, based on the gender calculus, of a lottery game where the winner is formally distinguished from the losers. It can be interpreted as an allegory for making something out of nothing, a scenario that many people dream about when they buy their lottery ticket every Saturday. This AUG "start codon" structure articulates how one entity pair, the winner and winning ticket combination, can achieve privilege at the expense of others. As a Cosmology, the allegory breaks down as it violates FC. According to FC, there can be no entity absolutely more privileged than any other. The rigid dichotomy based on winner and losers would violate the principle. For FC, anyone is a winner, or at least anyone that exists. At the beginning, the AUG structure must be applicable to any player.

It would be foolhardy to try to erect a Cosmology based on what we so far have here; All we wish to do is to shine a torch on the kind of apparatus that can begin to articulate how to create something out of virtually nothing. The necessary science that we are advocating must be based on a generic code capable of articulating this kind of scenario. This generic scenario must be applicable to any entity that wants to enter into being. The lottery prize will be winning the ticket of being itself, the ticket *to be*. Thus, the generic lottery formalised by the AUG start codon must play a fundamental role for the start of anything that exists. It is, indeed, a real start codon.

The Chrysippus Square of Oppositions

We now intend to build a Chrysippus version of the Square of Oppositions using four of his five undemonstratables. According to our interpretation, the premise of each syllogism involves a subject that *is* or *is not* in possession of a "quality" or something or other. Leaving out the third undemonstratables, the semiotic structure of the remaining four fit into the generic base scheme considered above. The circle on the left of each base icon corresponds to the first quality and the circle on the right corresponds to the second quality. The direction of the arrows in each icon corresponds to the flow of logical inference.

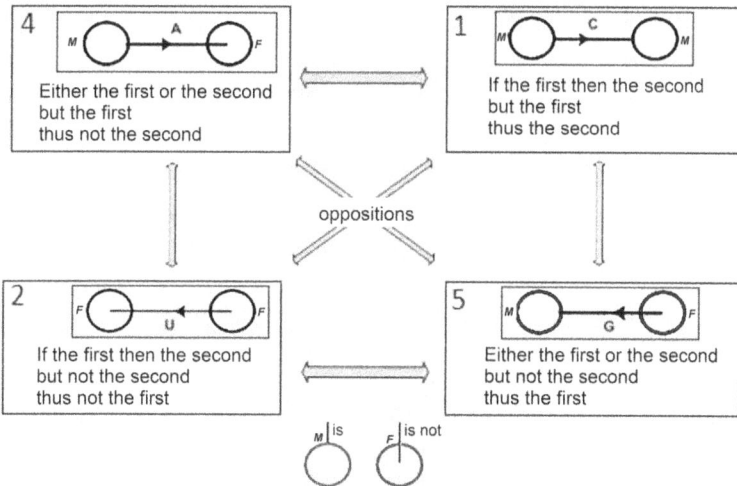

Figure 42 Four of Chrysippus' five undemonstratables can interpreted as generic bases to provide an alternative to Aristotle's Square of Oppositions.

This construction or reconstruction of the Chrysippus Square of Oppositions is a very fundamental generic structure. One facet that fascinates the author is that the direction of the flow of inference on the top of the diagram is left to right whilst the direction for the bottom two icons flows from right to left. In a previous cultural example of generic structure, the top two bases were interpreted to correspond to Indo European cultural paradigms (Christianity and Hinduism) while the bottom two corresponded to Eastern and Middle Eastern cultural paradigms. (Buddhism and Islam). We can reinterpret the four generic world religion analysis in the form of four generic cultural paradigms. Now it may be drawing a long bow, but the generic inference flows of the ontologic interpretation icons correspond to the reading order of the dominant languages: Indo-European languages read left to right while Arabic and ancient Chinese read right to left. Alternative explanations are hard to come by. In the lack of any credible alternative explanation, this observation may have some merit.

The Fifth Element

The four generic bases correspond to the binary valued generic types MM, MF, FF, and FM. We have labelled each of them by the four letters C, A, U, and G respectively. We have sketched out the historic link to the four roots of Empedocles that he identified as Fire, Air, Earth, and Water, respectively. For several thousands of years this became the accepted four-element theory of physics. Aristotle added a fifth element that he called aether to fill the apparent emptiness between Earth and the stars. The Stoics also argued for a fifth substance and called it pneuma. Pneuma played a fundamental role in their physics.

> *The pneuma is in constant motion. It is a process into itself, and from itself. The inward process produces unity and substance, the outward process dimensions and qualities. The pneuma is a disposition (hexis) in process. As a disposition, the pneuma holds the Cosmos together, and accounts for the cohesions of each individual entity. The pneuma is the cause of the entity's being qualified: For the bodies are bound together by these. [Chrysippus views on the pneuma (Reesor, 1989)]*

The coherence, the very being of an organism is sinuous with it maintaining Oneness. The mechanism for achieving and maintaining Oneness is through the establishment and maintenance of gender typing. The organism must know, without a shadow of doubt, what it has and has not, and what it is and is not. These are the key determinants of consciousness. Also, the determinations are purely relative. They are purely subject-ive. This, one must admit, is truly a beautiful, self-referring system. Somehow, the very coherence of the system is mediated by pneuma.

Beautiful indeed, but how does this pneuma work? With profound beauty, one would expect an accompanying simplicity, a profound but simple principle. Seeing that everything involved in this kind of self-organising organism is relativistic, there should be some fundamental relativistic principle at play. In the traditional sciences of our day, the only relativistic principle known is in physics. There is no known equivalent in biology. In physics, we see relativity theory expressed as demanding that the laws of physics remain invariant from one reference frame to another. Perhaps more pointedly, as shown by Zeeman, the principle of relativity is intimately bound up with the non-violation of the causality principle. It is here that one can grasp the simplicity and elegance of the theory. System coherence demands the coherence of causality. The claim of generic science is that this is not enough. A much more demanding form of relativity is required, what we call *generic relativity*.

If the work presented here is to be more than the usual exposition of inconclusive philosophical prose, then we should be able to advance an equally simple and elegant formulation concerning the essence of generic relativity, the cornerstone of the generic science we are trying to develop. Fortunately, we do not have to look far. The principle is located at Ground Zero and there is no one who knew this spot in the Cosmos better than Chrysippus, the Stoic logician par excellence. Ground Zero is the location of the Logos, the reasoning faculty of any subject whatsoever. The form of the Logos can be understood in terms of the dialectic of having and being, which leads to the semiotic square. Chrysippus provided the logical framework of the Logos semiotic square in the form of four of his five undemonstratables. We have resurrected this structure as an alternative to Aristotle's Square of Oppositions. We have named this Chrysippus' Square of Oppositions. The fit between this structure and the four undemonstrable is comfortable and reasonably self-evident. This structure effectively provides an additional logical impetus to the thrust of our argument. The four undemonstratables provide a logical dimension to the interpretation of the four-element theory and the corresponding four letters.

Absolute Incompatibility

Five undemonstrable minus four leaves one. The missing syllogism is the third undemonstrable, the incompatibility syllogism: *One can't have one quality and the other at the same time.* We now come to the fundamental tenet of generic science. It is founded on the premise that there is nothing more incompatible in this world than the masculine and the feminine. It is this incompatibility principle that holds not only the Cosmos together, but any being whatsoever that exists,

In the case of biological organisms, the concept should be relatively easy to grasp. A stumbling block might be in accepting the fact that the genetic code is more than a mere transcription language which curiously somehow accidently became a convention adopted by all living organisms since the year dot, without exception. Accidents do happen, but this accident does seem a little bigger than most. Life might be subject to evolution, but the language of life seems absolutely impervious to change. Nature seemed to have got it absolutely spot on, right from the start.

The reader may rest with that interesting accident hypothesis or move on to considering that the code may be based on a generic semantic and ontological structure. According to our take on the question, this structure is based on the dialectics of *being* and its naturally orthogonal counterpart, that of *having*. This can be formalised in terms of the gender construct and leads to a four letter code based on the four possible binary combinations of the two genders. It is generally accepted that all biological processes are coded by the genetic code, what we claim to be the *generic* code. Moreover, in multi-celled creature, the same code is repeated for each cell. We say that this code expresses a relative typing on all aspects of the organism. At the very grass roots level, the typing is in terms of complex combinations of gender typing. We claim that the organism relies on this form of organisation in order to arrive at knowledge and consciousness of itself. It is via this absolutely relativistic gender typing that the entity knows what pertains to it or what does not. This is the most elementary and most essential feature of life.

Moreover, the basic health of the organism will be placed in peril if this typing mechanism starts breaking down. The cohesion of the system demands the constant maintenance of the integrity of gender typing through the organism. The Stoic picture of a pneuma permeating every aspect of the organism is very helpful The pneuma is constantly attracting and repelling, constantly maintaining the equilibrium of the organism.

The Stoics claim that there are two primary principles working through the pneuma, the active principle, and the passive principle. This terminology is also helpful, as long as we recognise that the active and passive ultimately refer to the masculine and feminine, in a particular configuration. For example, we refer to the *feminine as active* by the mixed gender term FM. The *masculine as active* becomes MM and so on for the passive MF and FF variants.

Maintenance of the integrity of gender typing throughout the organism is paramount. Since the system is changing and reacting to its environment, this integrity must be synchronised. This brings us back to the key logical ingredient that guarantees such coherence.

The Gender Coherence Principle

The organisational coherence of an organism is regulated through gender typing. The maintenance of organisational coherence is synonymous with maintaining the integrity of gender coherence. This can best be expressed in the form of Chrysippus' third undemonstrable, the incompatibility syllogism. The premise can be restated in the form

> In no singular moment can an entity be both masculine and feminine at the same time.

We will call this the *gender coherence principle,* the fundamental organisational principle of Nature.

Note in passing that an entity can have multiple gender typing. However, it cannot have two different gender typings at the same time. This raises interesting question regarding the degeneracy of the genetic code. Take the amino acid asparagine, for example. It can be coded by the bases either AAU or AUC. In gender terms, this translates to the gender typing MFMFFF or MFMFMM. According to the gender coherence principle, such an entity has two possible "quantum" gender states. At any time, it can be functioning as either MFMFFF or MFMFMM, but not both at the same time. Remember that gender typing at any instant of time is not absolute and cannot be measured deterministically by a third party. Gender typing is relativistic and dynamic and in coherence with the organism so typed.

Note that the so-called superposition of states addressed by Quantum Mechanics disappears if they are considered to be more like relativistic gender states. Any observer that tries to deterministically measure a relativistic gender state of an organism will encounter superposition. For the organism in question, there is no superposition whatsoever. Relative to its integrity system, the gender coherence principle demands that the very opposite apply at each and any instant.

Physics Interpretation

In the Part 2 of this work, elementary particle physics will be interpreted from a generic point of view. This results in elementary entities like quarks and leptons being gender typed in terms of codons, similar to in biology. In this way, any being in nature code itself in terms of the generic code based on gender typing. This includes the Cosmos itself, as a dynamic self-organising being.

In traditional relativity theory, one can discern an elementary organisational coherence, which can be stated in a form comparable to the gender coherence principle. In this case, it becomes the *principle of causal integrity.* The principle states the dialectic of cause and effect.

The cause event is always antecedent to the effect event.

This is the most fundamental organisational principle known to traditional physics. The law must not be violated in any context (i.e., in any reference frame) and so demands a system that obeys Einstein's Special and General Theory of Relativity.

One can see that the form of Einstein's relativity has a certain resemblance to the generic form expressed, not as a causality coherence, but as gender coherence. There is also a fundamental difference; Einstein's relativity demands a coherence across time: Causes must precede effects in time. In other words, Einstein's relativity is diachronic in nature. In contrast, the generic version of relativity demands a coherence at the *same* time. Generic relativity is thus synchronic in nature, and up until now, has been totally ignored in physics.

Computer Science Interpretation

It is important to keep in mind at all times when dealing with the generic that it is not an abstract science. Generic science is capable of formalism but not as an abstraction, which is necessary dualist. Generic science is monist and non-abstract. Some effort is required to become accustomed to this totally different paradigm. Interpreting some of the concepts in a Computer Science setting can help, in this regard, unlike axiomatic abstract mathematics, Computer Science is a constructive science and naturally synthetic in nature. It also leads ta natural monism where the theory of code can be expressed in code.

Generic science is a discipline that has for its vocation the task of articulating the structure and organisational principle of any living being. The science is naturally constructionist. This raises the question of how to construct an organism based on generic science principles. Such an organism would have to be based on gender typing and be organised on the gender coherence principle. In addition, the whole system must not violate the principle of First Classness.

What we wish to do in this section is to provide a very simple example of how Computer Science, unknowingly, has already started to go down the path of Generic Science. Our example is the very computer itself, the Von Neumann computer.

Before Von Neumann, there already existed programmable calculating devices. However, they all had one thing in common. They were based on an absolute dichotomy between data and program. For example, the program might be hard wired into the device and the data fed in via paper tape. If we want to put some gender typing into the mix, we could say that the program was masculine stuff and the data feminine. With this arrangement, the gender coherence principle could be satisfied because at no time is any confusion possible between what was program stuff and what was data stuff. Data was always on the paper, and program in the machine. The only problem was that such a device violates First Classness, which is incompatible with such a blatant and absolute duality. First Classness cannot tolerate a world cut up into two, one made of paper and one made of the other stuff.

Von Neumann started the process of moving the calculating machine into the realm of a generically organised entity. He made two innovations. The first innovation was *shared memory* where there was no longer to be any absolute dichotomy between data stuff and program stuff. They were all loaded into shared memory in the same format as small chunks of information. Von Neumann was then faced with the problem of how the computer could tell the difference between program stuff and data stuff. It was here that Von Neumann decided to invoke his version of Chrysippus' incompatibility principle. The principle was that:

No chunk of information in shared memory could be both data and program at the same time.

In order to implement this principle, he came up with his second innovation. It was called the Program Counter; The Program Counter is a pointer into the shared memory of the computer. The rule was that a program instruction was the chunk of memory pointed to by the Program Counter at a particular instance in time. All the rest of the chunks were considered data. Having executed that instruction, the Program Counter would be incremented to the next memory location, and that would then be considered a program chunk and no longer potential data. In the case of a JUMP instruction, the Program Counter would be moved to some other distant place in memory and the process continues. The computer was born.

Like practically every major advance in computer science, the Von Neumann's computer was an exercise of eliminating violations in First Classness, in this case, eliminating the fixed dichotomy between data and program. Hence, forth, the distinction became relative to the dynamically changing Program Counter. What was program and what was data depended on content.

However, such a device is far from freeing itself from violation of First Classness. The Program Counter itself becomes a rigid privileged memory location, totally estranged from the run of the mill information chunk in shared memory. That is yet another dichotomy to be eliminated by generic engineering principle and so on and on it goes. Like out lottery example, the Von Neumann compute is only an allegory.

The Next Step

The next step in the development of the science involves how the four generic elements combine. As we argue at several reprises, the universal, generic structure always consists of triads of the four elements. Each triad is made up of three letters chosen from the four letter alphabet **A, U, G,** and **C** corresponding to the four generic types **MF, FF, FM** and **MM** respectively. This is covered in the appendices.

In the appendices, we look at using arrow diagrams to model the generic structure underlying the four generic letters and types. In the process we briefly look at the way traditional left side mathematics uses "arrow-theoretic" methods in the most abstract kind of abstract mathematics, notably Category Theory. Appendix A risks to please nobody. Unless you are a Category Theory theorist, this discipline usually inspires a feeling of fear and dread amongst ordinary folks like us. However, the appendix won't even please the theorist who will undoubtedly be aghast at our rendition of the topic.

What to do? Should the reader read the appendix or skip it? Our advice is to at least skim it before perhaps coming back at another time. This advice also applies to the second, Appendix B, which is our last version of our story, presented without worrying about losing readers along the way.

To skim or not to skim, that is the question. The reader is recommended to at least skim.

Chapter Summary

In this chapter, we have developed the theme that there are two diametrically opposed epistemological and ontological takes on reality. These have been termed left side and right side kinds of science. The left side take corresponds to the familiar dualist, atomist, reductionist paradigm of the traditional sciences and mathematic. These sciences can be characterised as attribute based sciences and based on abstraction. Right side science provides the solution to the Kantian problem. This requires a science that is monist and characterised by a complete absence of *a priori* determined attributes. All attributes of the monist science have to be constructed from first principles. The first principle involved was identified with FC. As for the right side alternative to abstraction, that was seen as being an approach based on the generic. The fundamental formal apparatus for the unification of the science into a monistic whole was proposed in the form of a calculus based on ontological gender. This leads to a generic algebra as the formal cornerstone of the any generic system. This chapter includes an initial interpretation of the generic gender based calculus as having the same structure and role as the genetic code for biological organisms.

The cornerstone of generic science is made up of four parts, what the ancients called the *Four Elements*. In this chapter, we have presented a rational explanation of the four elements that should satisfy a modern mind. However, in understanding how this unfolds one should remember at all times that that the kind of reasoning involved is right side reasoning and thus is not abstract reasoning as in traditional mathematics and the mathematical sciences. This is perhaps the most difficult aspect of this book; to tear readers, particularly those from Western culture, to tear them away from the narrow constraints of the left side paradigm based on abstraction. This does not inevitably lead to nihilist, mystical, or ":non-scientific" forms of reasoning as many have claimed. To the contrary, right side, generic science is formalisable as a science in its own right. The only difference is that it employs another kind of formalism to that employed by traditional left side science. Moreover, unlike its left side counterpart, formalisation is not just a superficial appendage to be studied by obscure specialists; right side science is itself a pure formalisation. The formalisation at its purest presents itself in the form of the generic code; something that the ancient alchemists, together with Newton would probably readily admit: "Tis true without lying, certain & most true." Here lays the Philosopher's Stone, the eternal code, resting as it does, on the principle of FC.

12

Stoic Structures

Although an amateur student of religion, the author does not consider himself a religious person. He wouldn't even describe himself as particularly spiritual. Above all, he is quite wary of superstition. However, he has recently experienced that rare moment of finding himself face to face with such absolute beauty and perfection that he is at a loss for words to describe it. He can only say that this is the work of the Divine or, if one prefers, Divine work. There is no other word for it.

This project started many decades ago. Over the past decade, the project has become more part-time than usual as the author has been working in industry doing other things. There is not much of a demand for generic scientists at the moment. However, there were many breaks of a few weeks or more to work on this project. Each time, the author homed in on one single problem. He was grappling with his generic approach to the genetic code as the universal code of not just the biological but of physics, Quantum Mechanics and practically anything else that comes to mind. One test of his embryonic generic science would be to explain one glaring, unexplained mystery about the genetic code, notably its degeneracy. There are sixty-four possible codon combinations, but they only code twenty amino acids and three stop codons. Why?

Left side science has come up with some explanations. For example, the genetic code has evolved to become very error-tolerant, hence the degeneracy. However, there is no evidence that the generic code ever did evolve. The contrary in fact, the code today is the same it was two billion years ago. Coded creatures evolve but the code does not. Error tolerance seems indeed to be a property of the genetic code, but that property doesn't explain anything. The author knew that a traditional, left side mathematical approach would never come up with an explanation. Traditional mathematics wouldn't work. Mathematics would see the degeneracy in terms of equivalence classes and then search for symmetries using Group Theory, for example. No one has succeeded at doing that, and never will. Left side mathematical sciences are dominated by symmetries and dualities. However, if the Code is a right side construct, all the emphasis will be on asymmetries not symmetries.

The Special Theory and the Start Codon AUG

We present here a revelation that begs belief. The author first became aware of it over a decade ago. Rather than being a Eureka moment it was more of an anguished "Oh, my God!" moment. "Nobody will ever swallow this story," he thought. Well the story didn't go away and became more pressing with the passage of time. Things were going from bad to worse. Finally, in Appendix B, we have managed to craft together a reasonably coherent technical account of what is involved. At least if the story won't go away, we are starting to get it off our back. However, you, the reader, should exercise great caution, as once bitten by this story, you may not get it off your back.

According to the author, the genetic code is based on right side science which is non-dualist and consequently, fundamentally non-symmetric. After all, right side science is founded on the gender construct, the opposition between the masculine and the feminine, the opposition between *to be* and *to have*. There is nothing more non-symmetric than that relationship.

Why then, is the genetic code degenerate? The author has finally made some progress on this question. In so doing, he glimpsed the work of the Divine. Let us briefly explain, before entering into the finer details. In what follows, because of the embryonic state of our science, we only provide a detailed explanation for three specific codings. We come up with a generic, non-problem domain specific explanation for each of the three examples. This is where we see the hand of the Divine. Each of the three codings turn out to articulate three different ways to articulate FC, that is to say, pure First Classness.

The first two codings involve the only two codons that have non-degenerate coding in biochemistry. These are the AUG and UGG codons. We do not look at biochemistry as that kind of science can only provide descriptions, not explanations. Biochemistry is a contingent, left side science. Our argument is that the observed degeneracy in the genetic code can be explained in terms of FC. FC plays no direct role in left side sciences, dominated as they are by rampant dualism, the anathema of FC, Instead we must turn to a domain where we can get some traction, in this most slippery of terrains. We turn to classical Relativity Theory and its spacetime geometry. This provides an easily understood left side version of what we have to present, Right side science itself must embrace relativity. In face that is the whole point of the paradigm, to provide a first class relativity based science. That is the route we take, all in order to explain the degeneracy of the genetic code. It is like climbing up and over Mount Everest in order to get to the shops. It mightn't be the best route but at least you can do some shopping.

This all might seem like absolute madness. What on earth has spacetime geometry and the speed of light have to do with the generic code of an earthworm or a bed bug? In order to answer this question, we provide a generic form of reality to replace the traditional abstract version. For a start, it is well known that a more generic way of expressing relativity that avoids direct reference to the invariance of the speed of light is to see it in terms of causality. Travelling faster than the speed of light violates causality. However, we go further and provide a much more generic version of relativity. In Appendix B, we present the generic version of the Special Theory of Relativity. The right side version is about as generic as any theory can possibly aspire. Instead of a theory based on the invariance of the speed of light, generic relativity demands that the theory itself must be the invariant. What this means is that the theory must be starting point invariant. Wherever and with whatever you start, you must always get the same theory. The theory must be generically universal and universally generic.

Sidenote

A novel paper by a group of psychologists (Caelli, et al., 1978) looked at the implications regarding the maximum perceivable velocity of movement of human vision. It was argued that this maximum value could be used as a constant just like the speed of light. The idea was to explain phenomenal distortions of perceived motion using the Lorentz transform of Special Relativity.

A theory that is its own invariant is enough to make the head spin. However, we don't stop there. Unlike our favourite nemesis, Richard Dawkins, we then turn to religion. After all, this kind of thinking is enough to make one turn to drink. Why stop there? We go the whole way and plunge into Dawkins' world of hopeless delusion, religion. For the rabid atheist, religions articulate views that are absolutely contrary to reason. Religious thought is based on falsehood and delusion; it is as simple as that. Our position is the opposite. Religious knowledge, particularly when shared by a large community of people across the ages, more often than not, can be based on deep generic truths. In Appendix B, we look at the most fundamental principle of Hinduism as expressed by Sankara in the Middle Ages, the principle of Non-

Duality. This principle expresses the relationship between Brahman, the cosmic soul and the individual soul ātman. These two entities must be different but indistinguishable from each other: they must be *non-dual*. This is an easier way of expressing that the object of theory must be non-dual to the subject of theory. The starting point invariant ends up as a theory of Self, where Self can be the Brahman or ātman; two different entities of different degrees of determination. It is quite remarkable that this non-duality principle provides the philosophical foundations for one of the four great World Religions.

The story becomes even more remarkable when we argue that this generic paradigm is generically articulated on one single codon of the genetic/generic code, the AUG start codon. It is fitting that this codon is the first that we encounter in our unfolding story recounted here. The AUG codon and its associated generic geometry can be understood via its manifestation in physics. It leads to a Minkowski space interpretations. It can also be studied as a particular manifestation of pure generic geometry. Further insights can be gained by studying the Vedanta philosophy of Sankara. The biological genetic manifestation of the generic code should also provide further insights. Here we see the AUG codon playing the universal role of start codon. It also encodes the amino acid methionine in all eukaryotes, but we will refrain from making any generic interpretations in this work. It is safer for us to stick with what we know. Our expertise is in generic knowledge. Armed with this knowledge, particularly once it becomes more developed, we will have an incredible tool for starting to unravel the secrets of life, biological and otherwise.

Cosmic Censorship

The term *cosmic censorship hypothesis* was coined by Roger Penrose in 1969. The concept expresses a particular form of FC. It arose out of the work he carried out with Stephen Hawking on singularities. In physics, singularities are somewhat of a mystery. To the physicist, every black hole has a singularity hidden in its interior. The Big Bang is another case of a singularity. The question to be determined is whether such singularities are observable or not. This brings in the relationship between the observer and the singularity to be observed. As we have mentioned many times, left side sciences such as physics only allow a very restricted kind of observer, the one with the "God's eye" view on the situation. The physicists formalise this observer as the "observer at infinity." Presumably, "at infinity" means the largest possible distance away. Thus, when they say whether a singularity is observable or not, they are talking about whether it is observable by an extremely distant observer, the only kind of observer allowed in classical physics.

Penrose presented two distinct forms of the cosmic censorship hypothesis, one weak and one strong. Of particular interest at this conjuncture is the weak form. The weak cosmic censorship hypothesis states that, with only one exception, all singularities in physics are hidden from the observer. The only exception is the Big Bang. This is observable by the far distant observer because this is where he is at.

One way of expressing the weak cosmic censorship hypothesis is to say that the only way that you can observe a singularity is to be in it.

Singularities will always be an enigma for traditional physics. This is due, not to the nature of the physical world, but due to the nature of the enigmatic physics observer stuck out in the middle of nowhere. From the perspective of right side science, the notion of singularity changes dramatically. It should be kept in mind that right side science is a kind of operational version of the left side science. Operational methodology is synchronic rather than diachronic, as we have already mentioned. However, our methodology is somewhat more elaborate than simply transforming from a time domain to a frequency domain perspective as in the traditional operational approach of left side mathematics. Here, we transform across paradigms.

The paradigm of right side science is founded on the singularity in its most generic form. The generic singularity par excellence is none other than the pure masculine entity, the generic subject, the generic observer, the generic point of view. In this science, the pure singularity as masculine entity dances with the pure non-singularity, the feminine and sometimes even with itself. Combined in the four MF, FF, FM and MM combinations, this couple form the sub-morphemes of the generic code with its four letters A, U, G and C respectively. These four morphemes combine together in triplets to articulate the 64 syllables of this universal generic code that codes any being that exists.

Our first major excursion into this generic code has been to study the generic geometric algebra of the AUG codon, the start codon in the genetic code. In so doing, we discovered that the underlying structure lent itself to a relatively standard interpretation of Einstein's Special Theory of Relativity. However, our AUG structure contains finer structure than is possible with the left side paradigm. Just as in standard relativity theory, the A, U, and G can be interpreted as coding a cone of timelike lines, a bundle of lightlike lines and another cone of spacelike lines respectively. There are only two singularities involved, notably the focal point of each of the two cones, as shown in Figure 43(b). The timelike singularity can be interpreted as the stationary observer, just as in traditional relativity theory. This is an observer that we can identify with. One can say that this observer is me or speaking generically, the Me. The other singularity, the spacelike focal point, is not the Me but a singularity that accompanies the Me. One could call it the It or even the At. Traditional left side physics says that the Me and the At are identical. The right side version says that this is nearly the case. The two singularities are different but, for all intents and purposes, remain indistinguishable. Nevertheless, they are different.

We noted that the Vedanta philosophy of Sankara provides a more generic expression of the principle. In this case the Me becomes the ātman and the At or It becomes the Brahman. Sankara expressed the difference and indistinguishability of the two entities as the Principle of Non-Duality, which is a nice way of putting it. Classical relativity theory assumes that the two singularities are identical. Viewed from the tight side perspective, we see an important nuance. The two singularities are different but indistinguishable. They thus appear identical. The same sort of reasoning applies to the monism that this construct helps implement. The system appears as a monism because it is indistinguishable from a monism. Sankara understood this nuance. That is why he called his system non-dualist, not monist. The non-dualist monism forms an integral part of the great Ruse that enables all of this to be stitched together. Lucky the Ruse worked, otherwise we wouldn't be here.

From the point of view of purely generic right side science, the paradigm coded by the AUG codon expresses the necessary constraint for our theory to be truly generic. Our theory must be starting point invariant. No matter where you start, you must always get the same theory. The principle of Non-Duality articulated by the semantics of the AUG codon assures that this will be the case.

Is the Weak Cosmic Censorship Hypothesis True?

A question often asked is whether the Penrose weak hypothesis true or false? From within the left side science paradigm, this question cannot be answered and so Penrose was correct to leave it as a hypothesis. However, from a right side perspective the problematic becomes more nuanced. From a right side perspective, there are singularities all over the place. In fact, anyone of us can rightly claim to be a singularity. Even a cane toad can claim to be a singularity, at least before you squish it with a cricket bat. Anything that is capable of autonomously maintaining an independent existence can claim to harbour a singularity. This includes the universe we are living in, galaxies, and so forth.

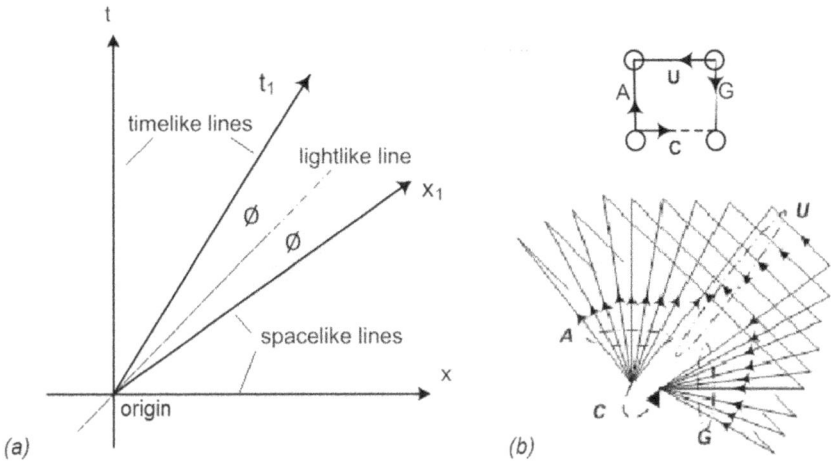

Figure 43(a) The left side science traditional illustration of the Special Theory in terms of cones of lightlike and spacelike lines. (b) A right side paradigm perspective of the Non-Duality Principle requiring that that the tips of the timelike and corresponding spacelike cones must be indistinguishable from each other. Nevertheless, they are different.

All of this indicates that the Penrose weak hypothesis is indeed true. However, there remains the niggling "What if?" question. What would happen if your Me got out of kilter with your At, or your ātman gets out of kilter with the Brahman. Clearly, this would involve a serious violation of weak Cosmic Censorship, albeit on a smaller scale. Is this life threatening? The answer to this question is quite reassuring. You will never observe such an event in your whole lifetime, so don't worry.

When the Penrose hypothesis is moved over to the scrutiny of right side scientific critique, the hypothesis remains a hypothesis but takes on a new allure. Instead of an abstract proposition, it takes on flesh and bones or some other kind of stuff. Any being in this world can be considered as an ambulating, self-justifying, Penrose hypothesis. This gives the being in question something to live for. Every waking and sleeping moment of its life, it must continually prove the Penrose Weak Censorship hypothesis to be valid. Failure to do so leads to a fate too dreadful to countenance...

Of course, this is not the only hypothesis that any self-justifying, autonomous being must continually prove to be true. For example, there is the Penrose strong version of the Cosmic Censorship hypothesis. To understand that hypothesis we must move away from the AUG codon and look at UGG, to be considered later.

In the spacetime geometry interpretation of the paradigm, the non-duality of the two singularities translates into the non-duality of the two endpoints of the dyad C, shown in Figure 43. The approach also replaces the abstract geometry of Minkowski space with the generic counterpart. This is all quite straightforward but not easy to explain.

AUG Generic Geometry

We now take a brief look at the generic geometry behind the AUG codon. The top part of Figure 43(b) shows the schematic logical diagram for the AUG codon. These schematics can be thought of as the right side equivalent of Venn Diagrams. In Appendix B, we started calling them *Chrysippus diagrams*. Each of these triadic structure is formed from triplets made up from the first, second, third and fourth of the Stoic undemonstrable. These triadic structures resemble in some ways the AUOE syllogistic structures of Aristotle's Organon. However, in accordance with the right side paradigm, all logic involved must be first order logic only. It is from triads of these first order logic structures that are constructed the elementary generic forms making up *anything*. Each elementary form consists of the generic logic, and generic geometry underlying each codon. Here, we have considered the first of them all, the start codon AUG and discovered some of its marvellous generic secrets. The AUG codon shown in the Figure can erroneously be interpreted as the geometry of s space where entities can reside. Due to the non-dualist nature of the right side paradigm, any such space-entity duality is forbidden. Instead, there are just entities with extent, that is to say, bodies. A physics interpretation for the AUG codon would see it as an elementary particle with extent. The extent would presumably correspond to the physical dimensions of its A and G components, the wavelengths so to speak.

In a later section, we will briefly look at how generic properties of entities can be derived from the logical properties of their codons. Such generic properties as "spin" and "charge" are readily suggested. These generic properties are easily interpreted in the physic manifestations of the generic coded codons, but the interpretation in the biological manifestation is far from obvious.

In the physics manifestation of this generic structure, the triads appear as kinds of quarks. It would appear that the AUG codon codes a photon and so the photon possesses a triadic like shape much like the hadrons whilst the left side science perspective can only conceive of it as point-like.

AUG Codon Summary

As we have seen, the semantic interpretations of the AUG codon are vast. We have been talking about the need higher order semantics. There is no better example of polysemic science than the semantics of the AUG codon. Just to ram the point home, we have provided the following different interpretations of the one underlying generic structure:

- AUG manifests as the start codon
  in the genetic code
- In physics AUG articulates Einstein's
  Special Theory of Relativity
- Sankara's Principle of Non-Duality
  is a generic form of the Special Theory.
- Penrose's weak Cosmic Censorship
  Hypothesis is yet another AUG viewpoint.
- Special Theory light speed invariance
  and generic starting point invariance.
- AUG and the generic geometric
  algebra interpretation based on gender.
- The generic logic represented by the
  AUG in its Chrysippus diagram.

More technical details are provided in Appendix B. The main point concerning the Special Theory is that the constraint can be replaced by a generic, non-problem oriented version based on the Non-Duality Principle, thus achieving starting point invariance of the whole theory cum organism (the same thing).

The Second Theory of Relativity and the UGG Codon

The Hindu mystic Sankara was the first to touch on the generic form of the Special Theory. In what was to become the central pillar of Vedanta philosophy, he advocated the Principle of Non-Duality. Einstein discovered the manifestation of this principle in physics with the Special Theory. We have sketched out how this generic principle can be formalised in terms of a generic geometric algebra, as covered in Appendix B. The fundamental generic principle underlying the whole scheme is that of FC. Any truly autonomous organism must satisfy, or strives to satisfy, the principle of FC. Any such FC abiding autonomous organism can be thought of as a life form. In this context, we must agree with the Stoics that the universe is organised along the same principles as us human beings, or even cane toads for that matter. This is all spelt out in the 64 letters of the genetic cum generic code, the fundamental geometric algebra of anything that aspires to exist.

Having completed our investigation of the AUG codon we move on to the next codon of interest. The only other uniquely encoding codon in the genetic code is UGG. It is here that our story becomes even more interesting. There is no trickery involved, no smoke and mirrors. There is not even any abstraction. In Appendix B, a reasonably straightforward application of the generic geometry principles developed there leads quite naturally to Einstein's Second Theory of Relativity, the General theory embracing gravitation. Although our generic science framework differs somewhat from the traditional left side approach, the terrain will be familiar to physicists, as explained in the Appendix.

In addition to the physics spacetime interpretation, there is also a theological interpretation that surprised the author. The UGG geometry leads to a totally isotropic form of spatiality. There are literally no singularities allowed. This is totally the opposite of the AUG case and its two primordial singularities, the totally undetermined *any point whatsoever* (Brahman) and the determined *any point whatsoever* (ātman), In this case the achievement of FC is that these points are essentially, if not identical, at least non-dual. There is a centre of the universe. Where is the spot where this centre is located? You are standing on it. You are experiencing FC in action. Just as you always knew, the whole world is gyrating around none other than little you, yourself.

FC can be quite comforting. We are starting to get a handle on this most slippery of all concepts.

The UGG codon involves looking at the universe in a different way to the AUG perspective. The UGG codon is uncomfortable with singularities. This second mini-paradigm avoids violation of FC brought about by some egotistical singular entity posing as superior to all others. FC can simply be achieved by abolishing all singularities! Historically, the most generic expression of this principle was due to Siddhartha Gautama's Buddhist doctrine of *anatta*, the Doctrine of Non-Self.

For a long time the author held the mistaken belief that the Doctrine of Non-Self was inferior to Sankara's Doctrine of Non-Duality. However, on closer investigation he realised that this Buddhist doctrine was on equal epistemological footing to the Non-Duality, doctrine although appearing to contradict it. Like any other major religion, Buddhism must provide knowledge of Self. Who am I? What am I? These are the stock questions that a religion must answer. Buddhism responds to these questions by declaring that the answer is that there is no answer. It goes even further and declares that the very notion of an "I" is erroneous. Any attempt to "know oneself" only results in falsehood. Whatever is perception, whatever is feeling, whatever is consciousness, past future or present, the judgement must be that "This is not Mine, this am I not, this is not my self." So declares the doctrine.

By positing that there is nothing permanent in an ever changing world and that even any conceived notion of Self is erroneous; Buddhism might be thought to lead to nihilism. However, the opposite is the case. Nihilism is where an errant Self takes control and follows every capricious whim that might come to mind. No moral effort is required. There are no rules to be violated because there are no rules. On the contrary, Buddhism has a rule, the rule of anatta, the Doctrine of Non-Self. Applying this rule demands a great deal of training and discipline. Moreover, unlike nihilism, the rule can be violated, in so doing proving that the rule is a rule. What's more, you know when it has been violated. The Doctrine of Non-Self is violated every time you allow ego to enter on stage.

One fundamental way to understand the essentials of a fundamental religion like Buddhism is to study the geometry that best illustrates the doctrinal principle in action, the geometry of the stage. In the case of Buddhism, the pertinent geometry is that articulated by the UGG codon. In what follows, we will quickly summarise this kind of geometry from a right side perspective. Essentially, the geometry is based on the requirement that the construct must result in a spatiality that satisfies FC and is totally devoid of singularities, just as Gautama would have wanted.

Generic Geometry of the UGG Codon

In the genetic code, there are only two codons that code a single amino acid, these are the codons AUG and UGG. We have already sketched out the generic structure of the start codon AUG and found that it expresses a generic form of the Special Theory of Relativity. It is now time to look at the other codon that shares this singular coding feature. We now look at UGG, initially interpreting it in its manifestation in physics. Here, there is no room for tricks. From a physics perspective, we already know that U codes a bundle of lightlike lines and each G codes a cone of spacelike lines. In the UGG codon there are no A's and hence no timelike lines. There are also no troublesome C dyads either.

Figure 44(a) shows the Chrysippus diagram for the UGG codon. Figure 44(b) shows a sketch of the generic geometry of the UGG codon. This is, as always, a three-0plus-One structure with three "imaginary" generic attributes and one real. The three imaginary attributes are made up of one U dyad and two cone-like G dyads. The real part appears to unambiguously be yet another G dyad. As always, the real part can never directly be known. It can only be known by its "appearance" in the form of three imaginary dyads. In what follows, we will avoid any discussion of the real part of the diagram and simply assume that in any context, it will appear as an infinitesimal relative to the other dyads and will be ignored here.

Figure 44 takes a single arrow chosen from the U bundle and two G cones of arrows. Each lightlike arrow from U stretches across the diameter of the circle. The arrows from each of the G cones are orthogonal to each other and appear to join up to the other U arrow across the circle diameter. However, the space involved is not Euclidean but hyperbolic and so this construction is only approximately valid near the centre of the disk. As the two orthogonal lines approach the circumference of the circle, they will appear to become parallel, although still orthogonal to each other. In the spacetime interpretation of this diagram, locations towards the circumference of the circle are moving away from the centre. The further from the centre, the faster a point in this space will be receding. At the circumference, the speed of recession will be the speed of light. At that speed, all spacelike lines appear parallel. The point z, where the two G arrows are supposed to meet, is just beyond the circumference and so they never actually meet. The same applies to each G line and the lightlike line forming the diameter. The end result is that there are no singularities. The space is purely isotropic.

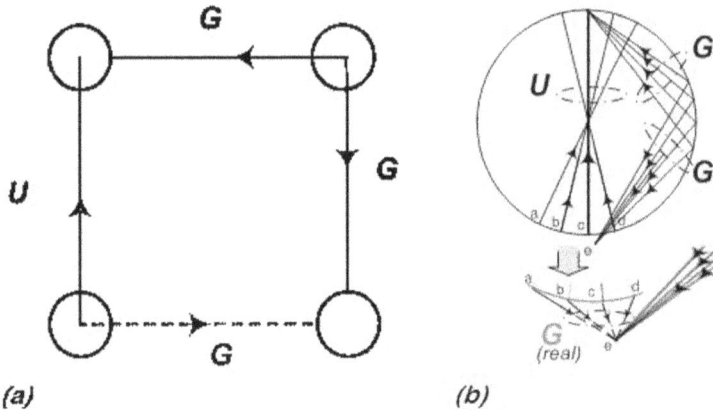

(a)   (b)

Figure 44(a) The Chrysippus diagram for the UGG codon (b) A sketch of the generic structure of the UGG codon. The real part of the Thee-plus-One structure appears to be another G base.

The Second Relativity Theory

From the General Theory of Relativity point of view in physics, this kind of space can be interpreted as a de Sitter space. The space is maximally symmetric and completely isotropic. It can be thought of as a sphere in Minkowski space, the space of the Special Theory. This space can be interpreted as the zero energy solution of Einstein's equations for General Relativity. We don't want to get caught up with the details here, but suffice to say that it has something to do with Dark Energy and the expanding universe. What interests us is purely the correspondence with the implicit generic geometry of the UGG codon.

Penrose's Strong Cosmic Censorship Hypothesis

The strong version of the Penrose Cosmic Censorship Hypothesis states that absolutely no singularities are tolerated in Nature. This hypothesis pertains to a different context to that of the weak hypothesis. In that case, we admitted a world containing timelike lines. In the context of the UGG construct, there are no timelike lines, just a picture of raw space stuff, Dark Energy perhaps. It is caught in the act of expanding at an approximately light speed, plus or minus a little bit. As shown in Figure 45(a), the only singularities in the construct are the vertices of triangles like xyz. If these triangles were in Euclidean space, they could be construed as singularities. However, the space is not Euclidean but Hyperbolic as shown Figure 45(b). In this case, the lines forming the sides of the triangle never ever meet. This disk like structure has all the characteristics of a disk minus its boundary points, a kind of open set. It is truly isotropic and so the strong Cosmic Censorship Hypothesis will be valid in this context.

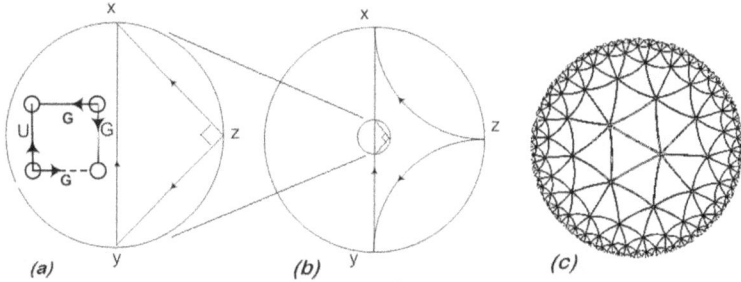

Figure 45(a) The arrow xy is from the U bundle, xz and yz are arrows from each of the two G cones. (b) The UGG codon geometry determines a two dimensional hyperbolic space geometry. (c)The geometry can be conformally mapped onto the Poincaré Disk here shown with a triangular tessellation just for affect. In the Disk, the angles in hyperbolic space are conformally maintained, but the metrics are not the same as in true hyperbolic space.

Now Einstein's equations do not demand that the Cosmic Censorship Hypothesis be valid. Thus, the UGG codon, interpreted as a relativistic constraint, appears to be a bit s more demanding, but we are straying outside our depth here.

UGG in the Genetic Code

In the physics manifestation of the generic code, the AUG and UGG codons both play the role of regulators of the Special and Second Theories of Relativity. The right side interpretation is that both are entities. According to our generic method for calculating "spin" and "charge" from their Chrysippus diagrams, both have zero generic charge and a spin of unity. This usually means that they are both bosons, that is to say, carriers of force. For AUG, there is no problem and we associate it with the photon entity. The UGG entity is different however as it has no timeline and so can't carry anything anywhere from where it is at.

In the biological manifestation of the generic code, the UGG encodes the amino acid tryptophan. It has been associated with a growth effect on rates. It is a biochemical precursor to serotonin, niacin, and auxin. In plants, auxin plays a cardinal role in coordination of many growth and behavioural processes in the plant's life cycle. Because of it is a precursor to serotonin, tryptophan is often taken as an anti-depressant in humans.

Just like all the other amino acids and biological substance, the biochemistry is usually well understood, but the systemic role of these building blocks of life remain unknown. If indeed, the fundamental organisational princ8iple of life forms is based on the principle of FC, a whole new world of science opens up before us..

The GGG Codon and Video Games

We are considering the degenerate coding of the twenty amino acids in the genetic code. We only feel confident about three of these twenty groups. We have already treated two of them, AUG and UGG. These are the only two one-to-one coding of amino acids. The third codon that we are to consider is the GGG codon. From a straight out naïve interpretation based on our generic geometric algebra, each G letter signifies a Euclidean dimension. From a left side perspective, this says, not Minkowski, not de Sitter space, but a bread and butter three-dimensional Euclidean space. From a right side perspective, these letters don't code spaces, they code elementary bodies. In the biological case, they encode amino acids. More generically, they code geometric entities.

In the biological realm, GGG codes the amino acid Glycine. Thus, if our generic approach to the code is valid, somehow Glycine must is associated with a speck of three-dimensional logico-spatial structure in some way. For the moment we lack a fundamental knowledge of how life forms organise themselves and so we are somewhat at a loss to see how this all meshes together. We can simply note a few details concerning the Glycine molecule. For what it's worth, we note that Glycine is the only protein-forming amino acid that is superimposable on its own mirror image. That looks a bit promising. It is also the smallest of all the amino acid molecules. Thirty per cent of collagen is made up of Glycine and the details keep coming, and our frustration grows. All we can say is that if life forms are organised along the principles of FC, then there should be a match here. Obviously, this is a question to be resolved. Prove us wrong, please!

We have no doubt that life forms are organised along the lines of FC and so we will not abandon ship. We assume that everything is going to plan and the precise fit between our generic code, the genetic code and the organisation principle of life will be settled down the track. The next matter is that in the biological genetic code, GGG is not the only encoding. The codons GGA, GGU, and GGC also encode Glycine. This is our first example of the so-called degeneracy of the genetic code. Why the degeneracy?

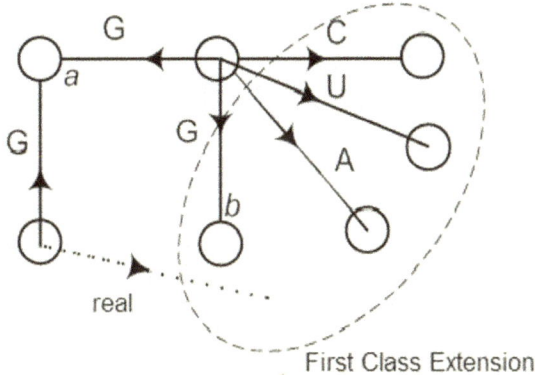

Figure 46 The Chrysippus diagram for the GGG codon and its equivalents in the genetic code. In the genetic code, the codons GGG, GGA, GGU, and GGC all code the same amino acid Glycine. Our explanation is that, on its own, the generic geometry of GGG cannot satisfy FC. By retaining the three degrees of freedom but adding more dimensions, a conformal geometry becomes possible that does satisfy FC. This FC explanation is advanced to explain all cases of degeneracy of the genetic code.

We only have the one stock answer for all such questions, First Classness. The Divine is anchored in the ultimate perfection of FC. It turns out that the generic geometrical structure encoded by GGG on its own is riddled with Second Classness and thus is unacceptable, particularly when its concerns the ultimate tool of the Divine, its code. Unlike the AUG and UGG codons, the GGG codon needs helpers.

The Divine is not alone when it comes to smelling that something is amiss when it comes to GGG. The codon GGG codes a hunk of vanilla, 3D Euclidean spatiality. As David Hestenes points out (Hestenes, 2001), the problem is that the model of Euclidean space founded on a set of basis vectors will have an origin. The origin becomes a *distinguished point* enjoying a status not shared by any other point in the space. This is a clear violation of FC if there ever was one.

In our generic version of geometry, we do not use a bunch of basic vectors, all centred on the origin. We use triadic structure like the GGG codon illustrated in Figure 46. The greatest menace to FC is the singularity. The GGG codon has two such singularities, one marked *a* and the other *b*. The *a* singularity could be interpreted as the point at the origin of the geometry and *b* as the point at infinity. Both points reside in 3D Euclidean space. Both points are distinguished points and so violate FC. In order to obtain a geometry compatible with FC, these two singularities must be removed from 3D Euclidean space, whilst retaining the same degrees of freedom. How can this be accomplished? Hestenes took the classical conformal geometry pioneered by A. Wachter (1792-1817) and reinterpreted the work in a Geometric Algebra framework, leading to Conformal Geometric Algebra (CGA). He moved the classical framework from a left side perspective towards a right side perspective.

This all sounds very technical, but can help to explain everyday things like why modern day computer programs are becoming so spectacular at handling 3D. More likely than not, the software was written in an environment supporting CGA. It also might explain what Google is up to when they drive down your street in a pick-up truck with a couple of cameras mounted on the back. How do they end up showing where you live as a 3D image? Once again, solving that problem is a piece of cake if one resorts to CGA.

The trick with CGA is to add some extra dimensions to the base space and hide the point at the origin in the extended space, whilst not increasing the degrees of freedom. It can't be any kind of origin though. It must use the same technology as in the AUG start codon, where the origin is totally generic. This requires a Minkowski space kind of geometry. The same applies to the point at infinity. This will require the hyperbolic geometry we met up with in the case of the UGG codon where the point at infinity was forever just over the horizon. In this way we can aspire to a solution compatible with FC.

We will not go into any details here. For the curious reader, we recommend reading Hestenes, overviews (Dorst, et al., 2011), (Doran, et al., 2002), and the textbook (Dorst, et al., 2007). CGA can seem counter intuitive at first and so practical experience, for example with the GA Viewer, is highly recommended. There, one can enjoy a practical experience with First Classness, rendering geometric object and manipulating them with the lettered names of geometric objects. Leibniz' Geometry without Number starts to become a reality. In a second class environment, circles and lines are separate kinds of entity. In CGA any line or circle can be defined by three points. If one of the points is the point at infinity, the circle becomes a line. Four points determine a sphere. If one point is at infinity the sphere becomes a plane.

Classical, second-class environments are based on coordinates and matrices, with numbers flying around all over the place. Nowadays, this is the kind of stuff best cut out for people that read telephone books for a hobby. Rotating a geometric object needs a different kind of transform than translating a geometric object: everything is a hassle. In CGA, where First

Classness rules just as with GA in general, there are not even any transforms, just geometric entities. Rotations of a geometric entity are accomplished via the geometric product with another geometric entity. Here things start becoming rather counter-intuitive because in order to translate an entity in Euclidean 3D, one rotates it in the 5 dimensional enveloping space. Rotations, translations, reflections, joins and intersections can also be carried out in the same manner, all by using the simple algebra of geometric multiplication. This is FC in action.

CGA evokes the image of Plato's cave. The shadows of reality are cast on the 2D Euclidean space at the back of the cave. The left side science approach to geometry, and affairs in general, is to study the shadows cast and try to make something out of it all. The right side version of geometry is to look at the whole and this requires a bigger world than the immediately apparent, as revealed in the shadows.

The technical details of how to extend Euclidean geometry to a First Class Conformal Geometric Algebra are reasonably simply. In our case, we will start with a 2D Euclidean space made up of two G dyads, in other words, two cones of spacelike arrows meeting at *a*. The usual next step is to add another G dyad and an A dyad to the geometry. There is also an implicit U dyad that comes with such a construct.

Sidenote

The traditional technique is to add the basis vector e0 and e1 where e1 squares to 1 and e0 squares to -1. These two basis vectors are then replaced by the two null basis vectors n0=e1 + e0 and n1=e1-e0 where both n0 and n1 square to zero and so are in the "optical plane," so to speak.

We are attempting to avoid any mathematics here. All we are trying to do is to patch together a story of how the A, U, and G dyads, rather than the G dyad alone, provide an extension to the spatiality that is compatible with First Classness. The end result is that the "real dyad" determined by this structure, instead of being a Euclidean blade of arrows, lives in this extended space.. This is just a sketch, as we are illustrating our philosophy with pseudo mathematics. A complete, formal solution is called for.

Please note the one thing we have neglected is the presence and role of the C dyad in Figure 46. This kind of geometric entity is totally absent in the Clifford Algebras and in GA in general. This kind of basis would "square to zero," just like the an arrow taken from the bundle of U arrows,. However, it would differ as such an arrow, treated as a vector, would be orthogonal to all arrows present. It is the *singular* arrow. This kind of dimension seems so degenerate that it is totally ignored in applied mathematics and mathematical physics. However, from the point of view of a science rigorously respectful of

FC, this dyad "dimension" must be treated as of equal status to the A, U and G. The A, U, and G dyads can be readily understood from a spacetime perspective and via familiar Clifford Algebras. Nature seems to have some use for it. What is the C dyad for?

In brief, ignoring our embarrassing ignorance of how to handle C dyads, our story works reasonably well in explaining how FC can be achieved using a standard Conformal Geometric Algebra approach to make the geometry implicit in the GGG codon compatible with FC. The approach refers to a 2D Euclidean projective space determined by the first two G's of the GGG triad extended with the usual conformal extension. There will be no FC violating "distinguished points," at least in 2D.

The First Classness Degeneracy Conjecture

We now come to our explanation for the degeneracy in the genetic code. Firstly, we claim that the genetic code is a manifestation of the generic code that we have been developing. The generic code has a wider scope than the generic as it applies universally to any being capable of autonomous existence, not just biological life forms. Thus, we argue, the underlying specificity of the genetic code is founded on a generic geometry manifested in all beings. Any observable algebraic structure manifested by the genetic code, will be explainable in generic terms, that is to say, in terms of FC. Thus:

> The FC Degeneracy Conjecture states that the observable degeneracy of the genetic code does not result from any quirky biochemistry but is due to the underlying generic geometric structure based on FC that it implements. Codons are grouped together (hence "degenerate") to form geometric wholes satisfying FC. In the cases of AUG and UGG, only one codon is necessary for a FC compatible geometry. In some case, up to six codons are necessary to form an FC compatible geometry.

In support of the conjecture we advance our explanation of the generic geometric algebra of the start codon AUG and the UGG codon. In this case, FC could be achieved without degeneracy. We then cobbled together an account of why the GGG codon needed to extended to include the GGA, and GGU codons. The role of the GGC codon was left unexplained. In the context of a total lack of any other feasible explanation, the FC degeneracy must have some merit.

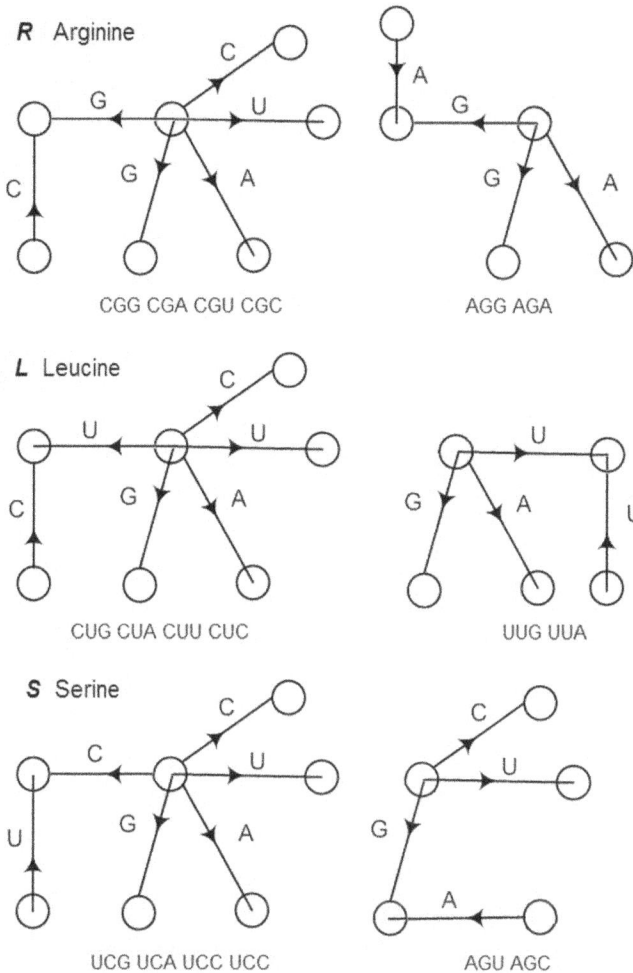

Figure 47 Showing the Chrysippus diagrams for the codons of Leucine, Serine and Arginine. Each of these amino acids is coded by six codons in a 4+2 structure.

As shown in each diagram, there is absolutely no hand-waving, or glazed eyes speculation concerning the actual geometry involved. This is all spelt out right down to the finest detail. All that remains is to discern the predicted underlying First Classness, using the AUG, UGG and GG* codons as an example of what is meant.

One thing to notice for these sextuplet codings is the paucity of A and G dyads. Each structure seems to organised around a single pair of A and G dyads. These are differently types cones. However the cones don't meet at the apex, as in the AUG codon. Most of the other dyads are C and U dyads. We have a conceptual handle on what a U dyad involves. From the space-time interpretation, U represents a bundle of lightlike lines. Considered as a geometric basis vector, it squares to zero, It is orthogonal to itself. However, when it comes to the C dyad we must admit that our geometry is far from maturity. It is the only "real" number, as A, U and G are all "imaginary." Like U, as a basis vector it also squares to zero, or something close to it when treated from a Non-Standard Analysis perspective. The kind of orthogonality relationship involved with this tricky C dyad is that the orthogonality is universal. The C dyad is orthogonal to *everything* in its presence.

The standard enumeration of all the geometries based on the quadratic form involves the signature approach of the Clifford Algebras. The signature of a geometry is defined as the signature (p,q,r) where p is the number of basis elements that square to 1, q the number that square to -1, and r the number that square to 0. In Space Time Algebra (STA) these three kinds of basis vectors generate spacelike, timelike and lightlike lines respectively. Left out of the mix is the basis element corresponding to singular lines. A singular line is not only orthogonal to itself but to anything present.

In order that the enumeration of geometries be complete, the singular lines associated with our C dyad must be included. Thus, the signature of a geometry should be written as (p,q,r,s) where s is the number of basis elements that not only have an inner product with themselves which is zero (hence square to zero) but the inner product with anything else must also be zero.

Of course the classification and enumeration of possible geometries based on a sequence of scalars such as a signature is not an adequate mechanism for the task. The author knows this as he has it from the most reliable and authoritative of sources, the Divine. The Divine has come up with a more elaborate system for enumeration geometries using the generic code. However, one cannot help but just stop and stare in awe at its solution, filled with wonder. What we are after is a Divine Geometry. From a more technical point of view, such a geometry will be First Class. It will be a First Class Geometry.

The Most Mysterious Dimension: Gaining Traction

The story presented here essentially involves the generic geometry of life and its code. The story sounds quite grandiose but please don't blame this author, he is just a reporter. We have presented many parts of the puzzle, but there is one important ingredient not properly accounted for. The troublesome missing ingredient has dogged the author for over a decade and even well into writing this book. The missing ingredient is our understanding of the geometry behind the mysterious C base, the MM dyad.

We have quite a good handle on the bases A, U and G founded on the MF, FF, and FM respectively. We interpreted the geometry behind three letter is the very terms of spacetime geometry and looked at the more generic ways of interpreting this kind of geometry. The "spacetime" term can be quite misleading as we have shown how computer graphics and computer video games use this "spacetime" geometry to great effect. Video software can be more simply developed in Conformal Geometric Algebra (CGA) by patching on a two dimensional "sopacetime" geometry, a two dimensional Minkowski space, to ordinary Euclidean geometry. In this way our video games are simpler to develop and provides a greater reign for creativity for the developer. Thus, the biologist and biochemist need not balk at the notion that the geometry of the biological world might be soaked through and through with so called "spacetime" geometry. Spacetime geometry is only the physics manifestation of generic geometry and so the basic mathematics (or anti-mathematics) is the same.

Thus, we can talk about cones of timelike and spacelike lines and it is clear what we mean. The same applies to lightlike lines, In no way are we inferring that this kind of geometry has anything to do with the invariance of the speed of light. As we have already discussed here and in the Appendix, there are much more generic forms of invariance at play that in traditional spacetime geometry and relativity theory.

Our story has bumbled along quite happily through relatively familiar territory, at least as far as the geometry goes. Then we turn to the troublesome C dyad. This is not a cone of spacelike or timelike lines. Neither is it a bundle of lightlike lines. What then is it? The C dyad adds an extra dimension to the mix, but what is it for? Referring to Goldblatt's text, we know that our A, U, and G entities, interpreted as geometric entities, can be understood in terms of their orthogonality properties. For example, A squares to -1, G to +1 and U squares to 0. We know that C also square to zero, or something close to it. Thus, like the U base, the C base is orthogonal to itself. Moreover it is orthogonal to everything else that is around. Goldblatt calls them singular lines. From Goldblatt's abstract mathematic perspective, this construct is too degenerate to be of much interest. His text, which is notable for its many examples, has no examples featuring singular lines.

This C type entity has been a real headache for us, as witnessed in our constant referring to this problem across the pages of this book. Finally, as a consequence of looking at our CGA treatment of the GGx code, our intuition has made a break through. Instead of asking what use God, or the Divine would make of this extra apparently completely degenerate dimension, we asked a more mundane question. What use would this Parmenidean dimension be to Computer Graphics gurus? Are they missing something really useful? What is missing in their CGA toolkit? In order to answer this tricky question, we proceed as follows.

To begin with, we ask why they use CGA at all? Why not use just ordinary, unadorned, vanilla, Euclidean space technology, just like all those Unenlightened Ones that still manage to eke out a living in the industry? The answer is that they want to emulate the Divine and practice First Classness. With First Classness, everything becomes so homogenous. Life becomes so much easier. There are no special cases for this and for that and so on. Instead of using traditional clay tablet technology, they can express themselves in a high level geometric language. However, by looking at the GGx codon in the genetic code, it would appear that they could do better. This codon indicates that the Divine is also keen on the CGA solution in this context. However, in addition, the Divine has thrown in our troublesome Parmenidean dimension for some reason or other, and only God knows why. Instead of just GGA, GGU, and GGG coding the same amino acid, the Divine chucks in GGC as well. Knowing the Divine as we do, there can only be one reason. The reason must be First Classness. In what way then does the C base ensue FC and how can this be exploited by the Computer Graphics community, and mankind in general for that matter?

To answer this question, we must find a blatant case of Second Classness within the conventional CGA geometry currently in use. Looking into this, we actually discover a glaring case of Second Classness in Parmenides central doctrine, the doctrine that the C base is supposed to implement. The poem of Parmenides stated that:

That what *is* exists. That what *is not*, does not exist.

What *is* corresponds to the real, what *is not* corresponds to the imaginary. This is the central opposition in Parmenides and there it is for all to see. It should have been glaringly obvious. The opposition of Parmenides violates FC! This is because the paradigm declares an absolute dichotomy between what *is* and what *is not*, between what exists and what exists not.

Figure 48 Illustrating the Parmenidean dimension. "Either it is or it is not." One proposition is true and the other false. Which one? The truth involves having both propositions present in the same picture.

Now Parmenides wrote his poem over four thousand six hundred years ago and so it has attracted a fair share of commentary across the ages. Thus, if we have anything to add to the debate it will have to be somewhat novel. What we intend to show is how to implement a geometric construct so that Parmenides opposition between what *is* and what *is not* will not violate FC. In other words, the absolute dichotomy between the real and the imaginary becomes relativised. The trick to solving the problem is the Parmenidean dimension itself.

The Parmenidean dimension can be thought of as the *IS* dimension. Talking geometrically, consider a geometric entity, a line say, in a GGx space. How can one say that it *is*? We can say that it *is*, if and only if, it has a non-zero component in the *IS* dimension, in other words, it has a non-zero component in C. We will illustrate what we mean with the famous vase image, as shown in Figure 48. This image, taken as a whole, contains a Figure and a Ground. Figure is defined as that what *is*, the real part of the image. Ground is defines as that *what is not,* the imaginary part. We will assume that we are employing the advanced image technology of CFA made even more advanced by the addition of one *IS* dimension. This is the spatiality determined by Nature's four GGx codons. We place the vase image into this "five dimensional" space and ask the question: what is Figure and what is Ground? The Figure is the part of the whole that has a non-zero component along Parmenides' *IS* dimension. Looking at the image, the reader can see what we mean.

First Classness is illustrated in this image as there is no rigid absolute dichotomy between Figure and Ground, but nevertheless Figure and Ground do form a dichotomy, a dynamic dichotomy that depends on the disposition or location even of the subject. The disposition or location of the subject can be formalised in terms of Parmenidean *IS* dimension. The IS dimension allows the real and the imaginary part of a whole to be simultaneously present. This is really FC in action. Parmenides is let off the hook.

From what we have said here, we do not expect a rush of Computer Games developers adding in an IS dimension to their current CGA visual graphics technology. The Parmenidean dimension is much too subtle for that. For the time being we will simply use this illustrative example as an allegory for what really is at stake here. However, we are at last getting some traction on this most allusive aspect of generic geometry.

Cosmic Shorthand

There are many explanations for the degeneracy of the genetic code. One of the author's favourites used by many online encyclopaedias is the explanation:

Degeneracy results because there are more codons than encodable amino acids.

The genetic code is degenerate because it is an example of degeneracy. There are also wobble and coding reliability arguments. We won't get embroiled in the debate at that level. Our argument is based on FC. We argue that generic code and its biological manifestation as the genetic code is uniquely based on the principle of FC. The shape of things, the structural geometry of any being is subject to the constraints of FC. We have illustrated this concept for the cases of the AUG, UGG and GG* codon codings. The implicit generic geometry behind these constructs provides a mini-snapshot of FC in action. Our FC degeneracy conjecture postulates that this is applicable across all the codings.

From our generic geometry perspective, we argue a one single codon like UUG, for example, actually uniquely codes one part of the generic geometry implicit in the amino acid Leucine.

How FC is implemented in detail does not concern us here. What is interesting is that, from our perspective, the genetic code can be considered as a kind of Cosmic Shorthand. For example, consider with the single codon UUG. We argue that there is no degeneracy in the coding whatsoever. This codon uniquely codes one component of the implicit generic geometry of the amino acid Leucine. The codon UCA uniquely encodes another component of the very same geometry. The same amino acid is involved in both cases, hence the so-called degeneracy. It is like having different coloured jigsaw puzzles. You only need one piece from the puzzle in order to know from which jigsaw it came from. Thus, the organisation of life forms has a very powerful inherent problem solving ability. Even at the very elementary level of the codon, one can see Nature's Ruse was of maintaining compatibility with FC. For the last few billion years, and even beyond, all the various life forms possess this most universal of all forms of knowledge. Each and every life form knows this fact without ever being taught. It owes its life to it.

In brief, we argue that the genetic code is not a degenerate labelling scheme based on first order semantics. Rather embraces a very precise non-degenerate mechanism based on second order semantics. Our conjecture is that several different codons code the same amino acid because each codon codes different subparts or "subspaces" of the same extended generic geometric object. The extended geometric entity is like the extension of the GGG codon. On its own, the GGG codon has privileged points and so violates FC. The extension of dimensions, but not of degrees of freedom, leads to the multiplicity of codons for the same geometric unit; four codons in the case of the GGG codon. In the generic code, these "generic units" correspond to elementary amino acids, the building blocks of proteins.

Only one geometric aspect of the entity, the amino acid building black, is indicated in each case. If one has a rigorous knowledge of FC, one should be able to impute the rest of the structure. Even within the most minute brainless microbe, this is occurring all the time, without any thinking. This is truly a form of Cosmic Shorthand.

13

The Good Model

Introduction

There are two scientific takes on reality, the left side and the right side take. The traditional sciences of today, including mathematics, are left side sciences. Our challenge is to develop right side science. In this section, we investigate the right side approach to the physics of matter. True to the left side paradigm, the conventional approach to studying matter is empirical. The principle empirical methodology of particle physics is to try to observe the interaction of elementary particles in motion and develop abstract theory to model the behaviour. The approach is fundamentally diachronic. All left side sciences are fundamentally diachronic. This means that these sciences study a flowing, focused, localised reality. The flow might be a flow in time, the usual case for physics. However, diachronic can equally involve other kinds of flow such as a suite of deductions as in mathematics. The diachronic paradigm sees a localised reality in movement.

The right side paradigm represents the other side of the equation, the synchronic. The synchronic take on reality is not a reality where matter comes to rest and movement ceases. The right side take on reality is a different take, not a different reality. Instead of the science playing the role of the impartial spectator of the speeding bullet, the reality becomes that of the bullet; seeing the world from its point of view. The entity, together with knowledge of the essence of the entity, must be in synchronisation. As we have seen in Appendix B, a simple ruse is applied at this point in order that the generic and the individual entity are non-dual. From there, the voyage of discovery is not empirical but based on reason as dictated by the principle of a perfect, autonomous, no strings attached, world. In other words, the reasoning is dictated by the principle of First Classness, as explained elsewhere in this book, from many angles.

The diachronic approach is like operational methods in applied mathematics, except that it operates across paradigms rather than just within the same paradigm. As an example to explain operational methods in general, take the sound from a guitar made by striking a chord of notes. The sound can be analysed diachronically as the variation in sound pressure as a function of time. This is said to be looking at the sound in the time domain. The synchronic approach involves looking at the sound in the frequency domain as a collection of frequencies, the *spectrum* of the sound wave. A more right side approach would be to endeavour to discover the correct ensemble of notes that make up harmonious wholes. This is precisely what the mathematician Pythagoras (2580 - 2500 B.C.) is credited with accomplishing. His approach was partly empirical. Pythagoras played notes on a single stringed instrument to find out the mathematics of which collection of notes went together. The criterion for choosing the correct combinations was founded on an informal notion of First Classness based on what note combinations pleased the ear. The harmonic series developed by Pythagoras was eventually perfected by Michelagnolo Galilei, father of the famous Galileo Galilei. It forms the foundation of modern Western music today.

Pythagoras was the first to come up with a science of the Good in music. For the philosophers of ancient Greece, it was natural to demand that the Good play a central role in science. For the Stoics, the whole universe was fundamentally based on the Good. Leibniz thought the same. In our work reported here, we are attempting to formalise the nation of the Good in terms of First Classness. Like Pythagoras, we aspire to discover what structures go together to form a harmonious chord to the very discriminating ear of Nature.

Any system based on the Good is a system based on First Classness. The Good and First Classness are synonymous. In this chapter, we will occasionally adopt the more terse term *the Good*, where we mean First Classness. The two terms will be used synonymously. Our generic entity of study is the Good System. Such a system will be capable of autonomous existence without any strings attached. The role played by the Good is that such a system must be organised in such a way that autonomous existence is maintained. In other words, the system must be equipped with the minimal apparatus to know what is Good for itself. This means that the system must be equipped with the capability to discriminate between what it **is** and what **is not**. It must harbour a notion of Self, in some way. The system is modelled on Nature, thus it must organise itself, not only for its own Good, but always be compatible with the Good of Nature. These two notions, the Good of the personal Self and the

Good of the impersonal Self, are built in to any Good system. As we see elsewhere in this book, there are two oppositions. There is the opposition of the determined entity (personal Self) with what it is not. There is also the opposition of the undetermined entity (impersonal Self) with what it is not. The basic semiotics of the generic entity is articulated by opposing these two oppositions together.

In Appendix B, we demystify the notion of two oppositions being applied to each other. We develop the basis of a generic geometry that can express these notions in an easy to understand geometric form. The notion of opposing one opposition to the other gives way to the more tractable notion that the two oppositions are "multiplied" together. The multiplication operation is a more generic version of the *geometric product* pioneered by Grassmann. The end result is a quadratic form, a kind of semiotic binomial. This leads to a whole range of geometric spaces along the lines of the Clifford Algebras, but with extra structure. These elementary structures are all Good structures and are the generic building block for any Good system, any system organised around the principle of First Classness.

The Elementary Building Blocks

As we have seen, the basic algebra of a Good system is based on gender. The elementary building blocks are formed from binary combinations of the two gender types, the M and the F typing. These are the four generic bases. These bases can be intuitively interpreted geometrically by sheaves of arrows as shown in Figure 49. Three of these sheaves are familiar to spacetime geometry. The MF cone of arrows can be interpreted as a cone of timelike lines. The FM corresponds to spacelike lines and the FF sheaf corresponds to light like lines.

The MM dyad has been the enigma in this research project. In the beginning of this book, we named it the Parmenidean dimension. It is a geometric dimension ignored in mathematics and physics. Goldblatt called this kind of line a singular line. It was characterised by being orthogonal to all other lines present, including itself. What kind of geometry does this lead to? This dimension does not feature in the Clifford Algebras, which is supposed to enumerate all of the possible geometries based on the quadratic form.

In our previous diagrammatic representations, we represented the MM Parmenidean dimension by a single arrow whereas all the other bases were represented by sheaves of arrows. Finally, in our discussion of Visual Illusions and Figure Ground reversal, we realised the true role of this dimension, and the MM arrow. It is just as Parmenides originally saw it. There are two sides of reality. That what *is* and that what *is not*. Only what *is* corresponds to the real. In our framework, all of these determinations are relative to the organism in question. In the example of Figure Ground reversal, what is corresponds to the

Figure. Of course, this is relative to the subject. With Figure Ground reversal, what *is* and what *is not*, flips from one to another in the flash of an eye, the flash of the subject's eye, of course. The right side paradigm demands that the subject is present at all times. This applies to anything, including neutrinos and whatever. Thus, on this side of the equation, the flip flopping Parmenidean dimension starts to make sense. More than that, it is crucial in holding the whole system together.

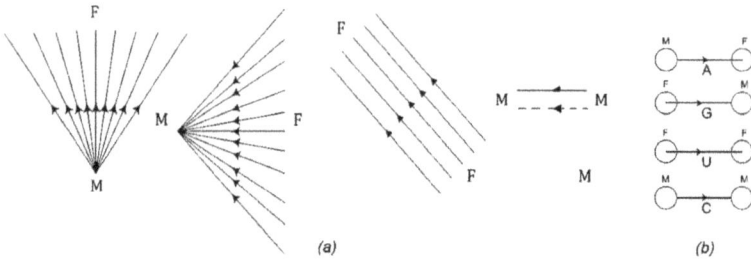

Figure 49 The four bases MF, FF, FM, and MM, as sheaves of arrows (b) The Chrysippus diagram for each of the bases. Each diagram corresponds to one of the Stoic "undemonstratable" syllogisms. Only four of the five undemonstratables have Chrysippus diagrams.

Consequent to all of this, we have upgraded our diagram in Figure 49(a), so that the MM dyad is represented by a sheaf of two arrows, one shown as a full arrow and the other dotted. The full arrow indicates what *is*, the dotted what *is not*, relative to the subject. This MM dyad, this Parmenidean dimension, enjoys an equal status with the other three bases. Each base encapsulates its own degree of freedom. The MM has its own degree of freedom, the freedom to determine what *is* and what *is not*, relative to the organism. Using our Figure Ground reversal, we see that what *is* can flip with what *is not* and vice versa. In the final count, everything is relative. Moreover, this flip-flopping may be necessary for the maintenance of FC.

Figure 49(b) shows what we now call the Chrysippus diagrams for each of the four bases. This summarises the fact that the four bases have a fundamental logical interpretation. Each base corresponds to one of the undemonstrable syllogisms that make up the Stoic system. The Stoics claimed that physics and logic, together with ethics (knowledge of what is Good), formed an integrated system. They certainly seem to be right on that score. There is no tighter scientific or philosophical system than this one. Moreover, the whole system lends itself to becoming a formalised science.

The Generic Conjecture

Before getting down to details, we take a look at the big picture. The traditional left side sciences are bottom up disciplines. A complementary top down science is proposed on the assumption that reality is fundamentally based on FC. These two sides of science are diametrically opposed and do not mix, even though they may cooperate, collaborate and even consolidate. They are like two poles of a Kantian antinomy expressing an irresolvable opposition: abstract, empirical, reductionist dualism on the left side is opposed to the non-dualist, monist, generic on the right. The generic, which is completely at home with oppositions, can understand the abstraction side but abstraction can never articulate the full essence of the generic, the underlying essence of a reality founded on FC.

Just like the patient suffering from a right hemisphere stroke, left side science only ever sees a half-reality and is totally unconscious that the other side even exists. In the case of the patient, this "hemi-neglect" can be so severe that the patient forgets to even wash or shave the side it doesn't own, only eats food on the side of the plate that it owns, and so on. This is a serious condition. For right side science, both sides of reality must be simultaneously present and a whole body comprehension results, even in the absence of any left side collaboration. Right side science might not be as agile as its left side counterpart, but it does aspire to seize the whole picture.

Based upon FC, the principle of the Good, right side science constructs a four-letter alphabet built from binary gender typing, leading to three letter morphemes and finally a code for articulating right side knowledge. The generic conjecture is that this generic code reveals underlying fundamental structure of the generic code. Secondly, the generic code articulates the generic version of geometry based on structure built from the degrees of freedom implicit in the masculine Exclusion Principle and the feminine Inclusion Principle, the most fundamental of all oppositions. This is an articulation of generic form to which any FC reality is bound.

Left side sciences like Particle Physics and Quantum Mechanics have reached a reasonable degree of maturity and the author has chosen this area to provide practical illustrations of how generic structure is organised. It is argued that the basic feature of the generic code is that, as a language, it is totally devoid of abstraction. This leads to an extreme nominalism. This can be seen empirically in the case of the genetic code with the direct correspondence between a codon and the amino acid it codes. In the case of the generic code, the correspondence is between each of the 64 codons and the corresponding generic geometric object. In what follows, a concrete match is established between these generic structure and corresponding sub-atomic entities known, and many as yet unknown, to Particle Physics.

Our claim that a biological type code such as the generic code developed here could not only explain biological organisation but also organisation of the inanimate, obviously opens the door to extreme criticism and even possible outright ridicule. Whether what follows will stand up to such critical onslaughts is left to time and for others to judge. In the following sections, the generic code is interpreted in the context of left side spacetime geometry in an endeavour to describe the right side counterpart. The level of rigour is low. This is just a quick jaunt into the science of matter from a generic point of view. The bulk of the following text was written in the early part of this project, but the basic presentation will probably be sufficient to convey the essential idea. It should be kept in mind that the approach is operational. Operational methods bring enormous implication to otherwise complex problems. We thus "do a Heaviside" and coalesce everything down to a synchronic form. This is what Heaviside did with differential equations. He replaced all the differentials with the corresponding powers of a complex number p, and solved for p. Very simple, and it worked. It needed Laplace to come along and make it nice and rigorous, but the essentials survived.

In the following discussion, we borrow notions from left side science spacetime in order to get tractable and easily understood semantic into the discussion. The intention is to illustrate that the generic code can lead to an alternative to the Standard Model for the foundations of physics.

The Standard Model

The left side version of the physics of matter has reached a high degree of maturity. The science is extraordinarily developed and detailed. The conventional and widely accepted view is summed in the Standard Model of particle physics. It is hardly within our capability to attempt to rival such a corpus of knowledge. Indeed, this is not the role of right side science. Reasoning based on the right side paradigm has a strong creative streak. It offers an alternative the blind stabs in the dark that often characterise left side scientific empirical studies. Based on the principle of FC, the research leads to glimpses of the world as it *should* be, rather than what it might be. As knowledge of generic structure develops beyond the naïve fragmentary details sketched out in this book, the science can aspire to declare what *must* be.

In the pages that follow, we take the gamble of sketching out generic right side take concerning the elementary constituents of matter. The first problem that we encounter is that such a venture risks to be branded as atomist, which is anathema to the right side doctrine we espouse. To prove our right side credentials we must fastidiously bow to the dictates of the Code. Instead of searching for the fundamental constituents of matter, as does the left side paradigm, we must claim to know already the answer in advance. Like the biological case, the 64 codons of the Code enumerate the different micro-holistic views of the organism. Each "particle" will appear in a synchronic, non-dualist representation. Thus, we know the answer in advance, albeit in a rather embryonic form. The next step is to match the answer to the particularity of the reality in question, notable that of physical matter. This demands a massive effort, and requires a dialectical opposing of knowledge acquired empirically with that acquired from generic reasoning. What follows is our first sketch, warts and all. In this context, the author should be thought of as an ancient Stoic philosopher who has wondered onto the scene and has been asked to give his impression of the Standard Model. Many people would say that such a time traveller would have nothing to say of any value. Time will tell.

The Non-Standard Model

The orthodox view of traditional particle physics is summed up in the Standard Model. According to the model, the most elementary particles are all point-like. There are three types of elementary particle, quarks, leptons and bosons.

Quarks always come together, three at a time, in triads of quarks. For example, composite particles like the proton are made up of a triad of quarks. The same applies to the neutron. Protons and neutrons belong to the family of baryons. All baryons are made from a triad of quarks. In the Standard Model, in the case of a meson, a single quark can pair with an anti-quark. The family of leptons include electron and neutrinos, all of which are point-like. Finally, we consider the bosons. The bosons are particles that carry force. Here we find the photon and the gluon. There are other bosons but that are composites.

The Standard Model is the result of empirical science. Here we present a sketch of the alternative based on the principle of FC. The approach is generic and applies to the elementary constituents

The Four Forces and Bases

From a generic point of view, the four fundamental forces in physics would match up to the four bases A, U, G and C. As a first iteration, the following match suggests itself.

The U Base and Gravity Force

The U base contains no singularities. Thus, its interaction with other bases will be independent of any singularities present. Since gravity effects objects independent of composition, U is the natural candidate for being the key player in the gravitation stakes.

The A Base and Electro-Magnetic Force

The A base encompasses one singularity and maps to the Electro-Magnetic Force. Rule: Only elementary particles that include the A codon will experience the Electro-Magnetic Force.

The G Base and Weak-Nuclear Force

The G base articulates the Weak-Nuclear Force. The most significant aspect of this force has to do with "change of flavour." As seen in our proposed Chrysippus diagrams for quarks, such a change of flavour involve a change in relative position and number of the G bases present. Unlike the A base, which has an expansionist cone of arrows, the G base is contractionist. There is also a very short-range fifth force associated with "weak hypercharge." As a first iteration, this force is also associated with the presence of the G base.

Rule: Only elementary particles with a G base in their codon will experience the Weak-Nuclear Force and the weak hypercharge force.

The C Base

From our generic analysis, the C base acts as some kind ontological dimension with two ontological states, an *is* state and an *is not* state. In the context of particle physics, the position of a base in a codon is determined by its colour according to the convention, Red for first, Green for second, Blue for third. In the case of a C base, its only specificity is its colour. The colour that either *is* or *is not*. Thus, there are six colour r,g,b, and the "is not" anti-colours r̲, g̲, b̲. The Strong Nuclear Force that binds quarks together is regulated by colour. We interpret this that the C base plays the key role in this organisation.

Generic Spin and Charge

As a first iteration, and without any explanation, we calculate the Spin of a base from its Chrysippus diagram. The generic spin is related to the number of degrees of freedom of each base, as shown below. We also introduce a generic notion of Charge. At the very early stage of the science, the author is more intent on painting a picture. The painting becomes more important that the paint.

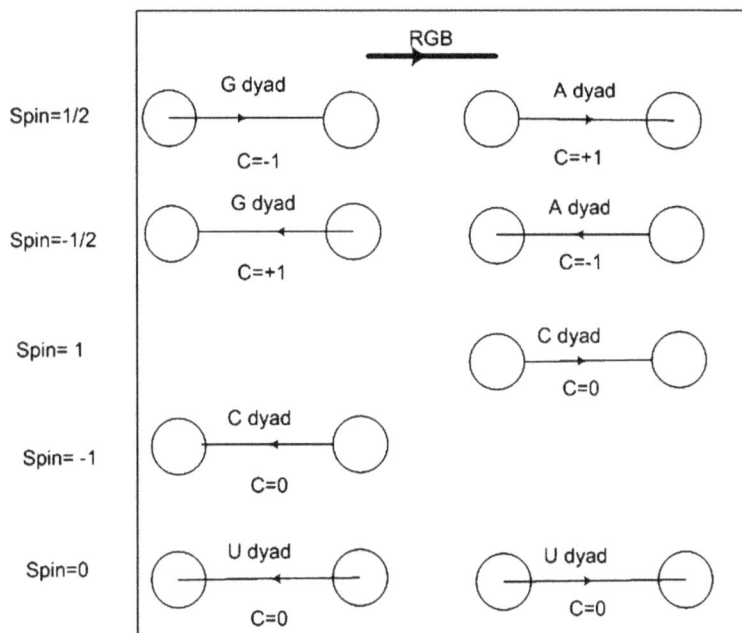

Figure 50 Diagram showing "Spin" for the four base dyads. The "Charge" is also shown, under the assumption that there is an A base in the codon.

We concentrate on the implicit logical, or perhaps we should say, the ontological structure of each of the bases. We use the Chrysippus diagram for calculating some important generic numbers. These are generic "quantum numbers" so to speak. The calculation is should be quite mechanical. The difficult part is the interpretation. For that, we throw all caution to the wind.

Rules For Calculation of Charge

For the first iteration, we state the rules for charge calculation. The charge for a base will be as shown in Figure 50, but under the condition that there is an A base present in the codon containing the base. In the case where there is no A base, the charge of a base will be effectively zero. The lack of an A base implies a lack of localised determination; no localisation, no "strong" charge. If the G base is present then the assumption is that there will be also weak hypercharge, following the same rules as "strong" charge. With this first iteration of assumptions, it is now possible to find reasonable fits between the codons of the generic code and the entries in the Standard Model.

The spin is a signed quantity, being positive in the direction of the dyad arrow and negative for the other direction. Composites can be constructed by triads of dyads. The spin for the composite is calculated by the signed sum of the three bases. Charge is also a signed entity. Intuitively the charge is a measure of divergence of the "flux" of represented by the arrows. The dyads

MM and FF have no divergence and so have zero charge. Positive divergence should indicate positive charge and negative divergence should indicate negative charge. We use the word *should* because we have to take into the consideration the sign convention adopted by conventional left side physics. Unfortunately, physics has adopted the reverse convention. After some head scratching, we have decided to go with the current convention. Thus, a positive divergence indicates a negative charge and vice versa. In the right side paradigm, one does not need a convention for charge, as it is a pure divergence calculation. Note that there is no notion of fractional charge, as in the Standard Model, as we cannot find a generic justification for the construct at this stage. If we are in error, then we have to travel with it.

As a passing comment, the notion of the arrows representing some kind of flux goes well with the Stoic notion of the fifth element, the *pneuma*. In this respect, all physical phenomena can be explained by the *pneuma*. This is the unifying force of all the four forces.

AUG and the Generic Photon

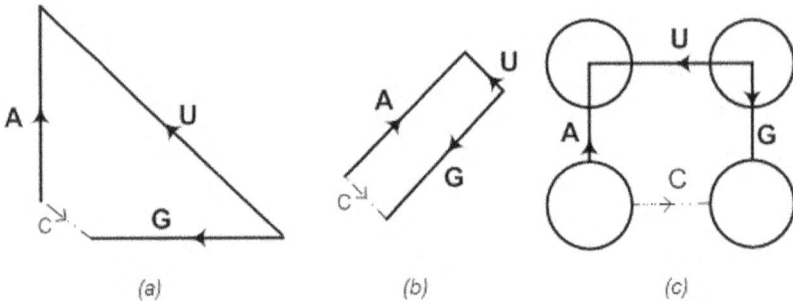

Figure 51(a) The synchronic geometric structure of generic photon from the point of view of the photon. (b) The geometry of the photon from the point of view of third party. (c) The Chrysippus diagram for the generic photon. Spin of 1 and the charge 0 can be calculated from this diagram.

The first generic body we investigate is the one coded by the start codon AUG. According to an elementary estimate, the Spin of AUG will be ½ + 0 + ½, which is a Spin of 1. The elementary charge comes to zero. This, and other factors, leads us to the right side science interpretation of this entity, not as an important particle participating in the principle of relativity, but rather as the implementation of that principle – as a body. Consequently, we will make the audacious and possibly outlandish move and refer to the elementary body coded by AUG as *the generic photon*.

Complementarity

From the perspective of the left side scientific paradigm, photons appear as quanta exhibiting both the particle and wave like behaviour. In Quantum

Mechanics, all elementary particles exhibit this wave-particle duality. It is our view that Copenhagen View of Quantum mechanics corresponds to the most correct orthodoxy of the left side paradigm. According to this view, the particle-wave duality is an aspect of Complementarity where a phenomenon can be viewed in one way or another, but not both simultaneously. Both perspectives cannot be present at the same time. Even though this dogma might not please everyone, one should not attempt to tinker with it. Instead, to get another take that might be more intuitively pleasing one should change gear and move over to the other paradigm, the right side paradigm. Paradigms should be kept pure and never tinkered with.

Crossing over to the right side paradigm, we find that all perspectives must be present in the same moment. Note that the particle, viewed from a left side perspective, is point-like. All of the elementary particles of particle physics are considered to be point-like, being totally devoid of extent and interior structure. The left side paradigm doctrine has no choice but to declare that its final aim is to demonstrate that "everything" is constructed from these elementary point-like particles. Quantum Mechanics saves the day somewhat by claiming that the particle has another side to it. All particles exhibit wave-particle duality.

The right side paradigm has its own take on Complementarity and the wave-particle duality. As far as Complementarity is concerned, that principle is encoded right down at the core of the paradigm. For the right side paradigm, the fundamental Complementarity is between the two most elementary entities of them all, the pure feminine and the pure masculine entities. It is the masculine that expresses the pure singularity side of the equation, as the masculine is devoid of extent. However, it does have presence. The feminine on the other hand enjoys extent, but is devoid of singular presence. In order for something to enter into the realm of real existence, it must marry extent with presence. According to the right side paradigm, an entity capable of autonomous existence will be organised along the lines of gender typing. The four basic building blocks of right side science are those with the binary typings Mf, FF, FM, and MM. The ontological, generic geometric structure of these four building blocks is articulated in four of the Stoic undemonstratables.

The undemonstratable left out of the mix is third, the incompatibility syllogism. Interpreted from a gender perspective, this syllogism declares that gender typing must be absolutely black and white. Gender can be masculine or feminine but not both at the same time. This is the most fundamental organisational law of right side science. The systemic coherence of an organism organised along gender lines is guaranteed by this principle. It is here that we see the right side version of the Complementarity Principle. As we have endeavoured to show in this book and its Appendices, any organism organised by gender does not lead to duality, but to non-duality. On the right side, the Complementarity Principle becomes an essential aspect of any organism aspiring to enter the realm of the Good, that is to say, the realm of the living.

A lot of ink has been spilt on the particle-wave duality question. What are elementary entities really like? Are they particle-like or wave-like? Perhaps they are both. The answers that have been given to these questions depend more on the underlying epistemological assumptions than what matter "really is." A key tenet of this book is that many dualities and apparent symmetries, real or hypothesised, can ultimately be explained generically. What something fundamentally looks like from a left side perspective and from a right side perspective, should be capable of being worked out by reason alone. The way something appears to be depends on the particular epistemological take operational at that moment. This applies right across the board, not just for the physics of matter.

The Generic Gluon

In particle physics, the gluon is an exchange particle responsible for the strong force that binds quarks together. The residual force also binds the neutrons and protons in together. The gluon is seen as a point-like particle with the following physical attributes:

1. Mass 0
2. Charge 0
3. Spin 0

In addition, it can carry "colour charge," of which there are 8 possible states. Because the elementary particles of the Standard Model lack any interior structure, the whole notion of the role of "colour" in the physics becomes very opaque indeed.

Viewed from the right side paradigm, the situation becomes much clearer. Instead of point-like elementary particles, the right side version presents the elementary particles as RGB triads. This "colouring" was introduced to label the first, second and third dyads making up the triad. The non-sequential structure of the triad was established due to the requirements of FC. Thus, the RGB triad provides the template for all of the 64 codons making up the generic code. Each dyad in the triad is binary gender typed, making up the 64 possible codon combinations. This RGB convention, together with the gender typing convention establishes the overall "polarity convention" for the organism. In the case of particle physics, the organism is the universe, as a vibrant, self-organising organism.

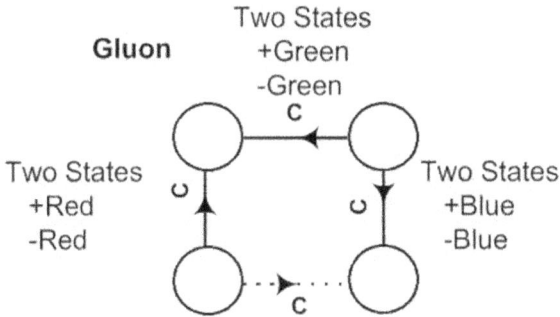

Figure 52 Proposed Chrysippus diagram for the gluon represented here by a CCC codon. Composed uniquely of C bases, the generic geometry of the gluon is totally lacking in spacetime type structure. The spin and charge of a C base is zero and thus, this will also apply to the gluon.

The best candidate for a gluon in the generic code is the enigmatic CCC codon. It is made up of three Parmenidean dyads! Figure 52 shows the proposed Chrysippus diagram for the gluon represented by the CCC codon. Composed uniquely of C bases, the generic geometry of the gluon is totally lacking in determined spacetime kind of geometric structure.

In our explorations in Appendix B, we finally got some traction on understanding this most enigmatic geometric construction. Each C dyad has two states related to some kind of Figure Ground reversal. This is a generic property of C dyads, not an empirically determined property and certainly not specific to physics. The CCC codon can be thought of as a kind of raw RGB codon. All codons are based on the RGB triadic structure, where each of the "colour positions" are gender typed. The typing of a C base is MM, that is to say, doubly masculine. The conventional geometry of such a structure is the singular line, a line of zero metric and orthogonal to everything. On the face of it, the geometry is absolutely degenerate. However, as an ontological dimension, the construct becomes very interesting. Instead of each base enjoying one of the RGB colours, the C dimension introduces another nuance. Instead of a "Red" C base simply declaring "this Red *is*," an additional ontological state is catered for, the "this Red *is not*," state.

Avoiding getting bogged down here, we will have to suffice with a general overview of the role of the gluon in the "management of colour." In the Standard Model, the gluon is a boson acting as a carrier of "colour charge." From the point of view of generic science, the gluon manages colour via "ontological switching."

Since each base of the CCC codon has two colour states, one positive and the other negative, the codon will have eight states. Here we see the difference between the left and right side perspectives. From the right side perspective, the states are directly observable. This is because the observer is in a synchronic relationship with the observed. Both are present. In other words, the observer subject is none other than the organism itself, as subject. The observer subject could be the gluon itself, or some composite organism. From a left side perspective, the observer is a foreign third party and so not in a synchronous relationship with the gluon. The non-synchronous observer can abstractly reason that the state of the gluon will be a superposition of the eight states, which is the limit of knowledge obtainable in that circumstance. The situation changes when the third party makes a measurement, in which case the superposition collapses and the non-synchronous observer enters into a synchronous relationship with the organism, of which the gluon is a constituent. The observer becomes a part of the system, and as such modifies it. Direct measurement of the gluon entity being is out of reach. This is not the same synchronous relationship that the gluon has with itself. Thus, calculations of what constitutes a gluon state will be probabilistic. The guiding principle will be that the pobabalistic states should all be equiprobable, an aspect of FC. Left side theorists have come up with several solutions that satisfy these constraints. One has to be very clever to work this all out from the left side perspective. Fortunately, gluons are not very clever at all and so viewing things from the gluon point of view can be much less challenging.

It is interesting to reflect on the Stoic notion of the active and passive principle. In this respect, the C base, which has MM gender typing, stands out as active. The C base enjoys an ontological function, arbitrating between what *is* and what *is not*. By contrast, the A, U, and G bases seem more passive. They play a role that is more familiar to us as spacetime geometric constructs. The CCC, although rather lacking in substance plays a most active role in the state of ontological affairs.

Generic Quarks

In left side physics, quarks are elementary point-like particles with attributes. Some of the discerning attributes are the quantum number of Spin equal to ½ and non-negative charge. Right side science deals exclusively in bodies and so without much effort, the most obvious fit of elementary base bodies with the quarks is shown in Figure 53. The fit was established by matching spin, charges, and relative mass where the latter was estimated as follows. The lightest bodies correspond to the open-ended part of the triad being typed as MM, the next lightest being MF or FM and the heaviest being FF where the inclusion principle of the feminine should provide the highest relative mass.

These six bodies are our candidates for matching to the Standard Model quarks.

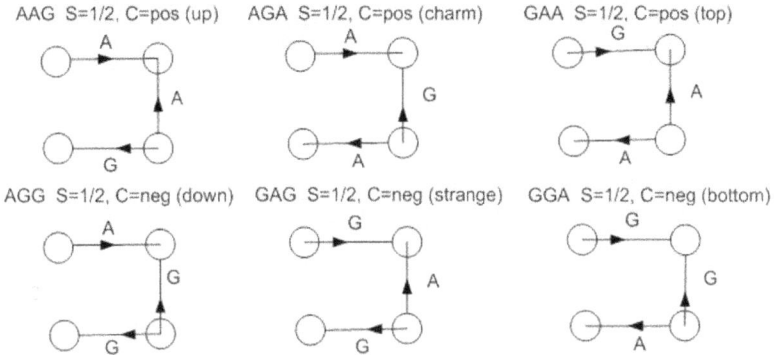

Figure 53 The Chrysippus diagrams for the 6 quarks of the Standard Model. The correspondence has been established by matching generic spin, charge and estimates of relative masses to the Standard Model versions. In the left side paradigm positive charge, translate

Generic Neutron and Proton

Figure 54 shows our proposed Chrysippus diagrams for the electron and neutrino. The electron has an A base and so would experience the Electro-Magnetic force, but not the Weak Nuclear Force. The neutrino has a G base and so experiences the Weak Nuclear Force (and the weak hypercharge interaction). The C bases have zero spin, so both particles have the spin of the A and G base respectively, which is a spin of ½. The single A base of the electron gives it a charge of 1. The neutrino has no A base present and so no charge, only weak hypercharge.

Shown in the diagram, are the "first generation" electron and neutrino. In the Standard Model, the elementary particles come in so-called *generations*. There are three generations. In the case of the quarks, the first generation quarks are the *up* and *down* quarks, The second generation has the charm and bottom quarks. The top and bottom quarks make up the third generation. As shown in Figure 53, each quark is made up, either of two A bases and one G, or two G bases and one A. Our right side perspective provides a simple explanation for the generations. There are three ways of arranging the bases. If AAG codes the up quark, then AGA and GAA codes the other two "generation." The same applies to the down quark. Some scientists wonder if there are more than three generations. From the perspective of the right side paradigm, it appears that there are only three combinations of the constituent bases possible and hence only three "generations."

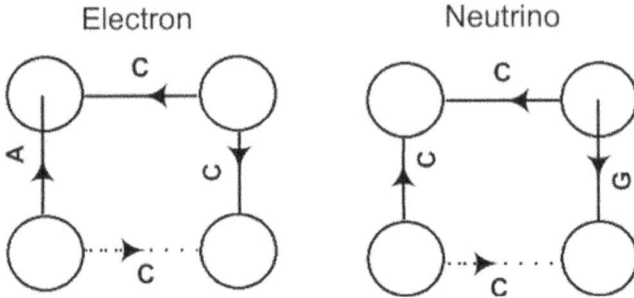

Figure 54 Proposed Chrysippus diagrams for the electron and neutrino. The electron has an A base and so would experience the Electro-Magnetic force, but not the Weak Nuclear Force.

Applying the same reasoning to the neutrino, we have CCG for the first generation, and CGC and GCC for the other generations, in an order to be determined. The same applies to the electron with ACC for the first generation and ACA and CAC for the other generations. The only fly in the ointment is that the A base in CAC codon. Applying our *ad hoc* rules for calculating charge, would indicate that the charge of CAC is of the opposite sign as for ACC and CCA. The *ad hoc* charge rule would have to be modified to take this into account. A more mature version of the right side paradigm should be able to provide replace the *ad hoc* rule with a properly generic version.

The generic code for biological matter manifests as the genetic code. The basic building blocks of life, the amino acids, are coded by the four letters of the genetic code. On the assumption that our generic conjecture is valid, a similar situation must prevail for the generic coding of physical matter. The elementary constituents of physical reality are coded by the four letters of the generic code. So far we have proposed generic code mappings for quarks, thee photon, gluon and the three generations of electron and neutrino. Figure 55 shows two non-elementary composite entities constructed from quarks, the neutron and the proton.

In addition to the elementary particles already considered, there are obviously many more indicated by the generic code. It would thus appear that there should exist many other kinds of elementary entities that are totally unknown to present day physics. Many of these unknown entities will have no A or G bases and so will be undetectable by any direct means. Nevertheless, if this right side paradigm based science holds up, it will serve as an intellectual spotlight for bringing into view, what has previously been invisible. The science will then take on one of its natural vocations. It can play the role of, what the philosophers call, the Logic of Discovery.

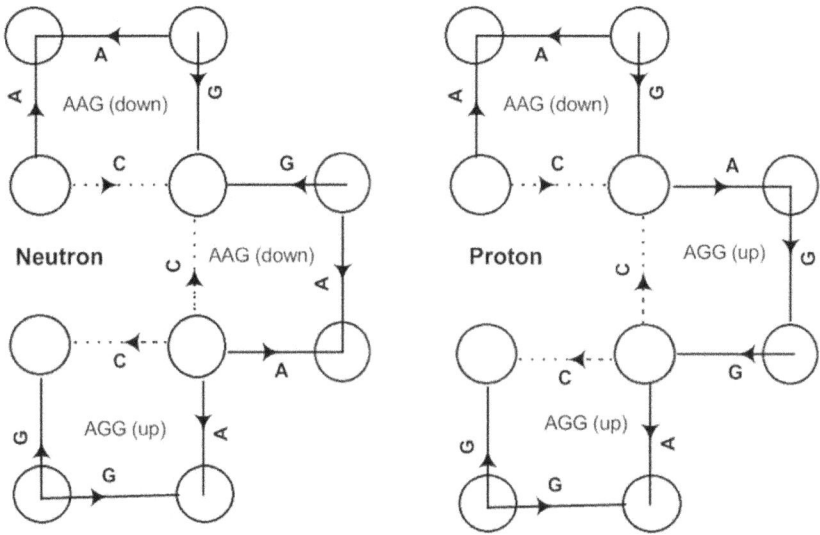

Figure 55 Proposed Chrysippus diagrams for the Neutron and the Proton, each constructed from a triad of quarks. Each quark is a triad of base elements. First Classness would probably be achieved and maintained by a complex structure of CCC gluons.

14

Epilogue

The Dawkins Ignorance Hypothesis

Dawkins' polemic, *The God Delusion*, is an excellent example of left side reasoning and so it is not surprising that he presents a worldview that is totally at odds with the main thrust of our project. His polemic has been widely contested on many fronts but one of his core assumptions seems to have escaped criticism. His assumption can be paraphrased as the declaration:

You can't get knowledge out of ignorance. Everybody knows that!

From there he goes on to claim that God is nothing more than a place-holder for ignorance. He claims that we get nowhere by labelling our ignorance God. Once again, we could add the implied comment, "…and everybody knows that."

In his polemic, he is committed to the street language of the rant that has become so prevalent over recent times. This populist discursive style does not lend itself to considering the more measured and considered aspects of the problematic. However, someone of Dawkins' education and statue would have come across Kant. In the Critique Kant addressed precisely Dawkins' question:

How can you get knowledge starting without any a priori knowledge or experience whatsoever?

In other words, how can you get knowledge out of ignorance? For Kant, metaphysics was the science that was supposed to find answers to this question. Many philosophers argue that such a science is impossible. You simple cannot get knowledge out of ignorance. Towards the end, Kant might even have ended up with this view. If these philosophers are right then any notion of metaphysics or God must indeed be quite vacuous.

The key part of the riddle is that everyone knows that you can't get knowledge out of ignorance. The ignorance at first glance becomes, on closer inspection, a kind of knowledge, particularly if everyone knows it. We find ourselves right in the middle of Socrates' Confession of Ignorance. It is this very paradigm that we use to construct the most awe-inspiring knowledge of all. However, to Dawkins this kind of awe is "silly," declaring

I'd take the awe of understanding over the awe of ignorance any day.

Hawkins claims that you can't get anywhere by labelling ignorance God. He particularly riles to the God label, but presumably his claim goes for any label. You won't get anywhere by labelling ignorance, full stop. Dawkins is providing a good service for us. He is providing a formal declaration about ignorance and knowledge. We will call it the Dawkins Ignorance Hypothesis. It simply states:

It is impossible to obtain knowledge by labelling ignorance.

This statement is meant to be obvious to the audience that Dawkins is addressing, namely the mythical god fearing, banjo plucking, pig farmers of the Appalachian mountains. However, on closer inspection we see that not only is Dawkins a vehement hater of the Gods, he is also a hater of mathematics. For instance, the most foundational notion of elementary algebra is to obtain knowledge by precisely "labelling ignorance." Let x be the unknown in the equation. The value of x is unknown. Solve for x. Presto, knowledge from labelling ignorance.

Now it might seem that we are just being disingenuous here. In fact, we are deadly serious because, in our work, we use this vary technique of labelling ignorance of the most profound kind, in order to reverse engineer the Code. This is exactly the same Code that Dawkins has spent so much of his professional energy on analysing from the bottom up. Working from the bottom up, Dawkins plunged into the algorithmic genetic entrails of Nature. He emerges and declares that Nature is organised by nothing but algorithmic genetic entrails. He solemnly reports that he found no trace of God.

Our approach is top down and so is less messy. From pure ignorance, we can obtain the algebra of knowledge, just as Kant called out for in the Critique. In order to accomplish this task, we have to negate the Dawkins Ignorance Hypothesis. One must label ignorance in order to access this higher knowledge, the Code of the gods.

As we have already discussed, there are two ways of labelling ignorance, one is with the feminine F and one is with the masculine M. F labels the ignorance of the wildcard and M labels this ignorance as being the only singular bit of knowledge capable of dragging us out of this morass. Armed with these two ignorance labels, the four letters of the generic cum genetic code can be constructed. This gender calculus takes us back to the time of Empedocles and his theory of the four roots or letters. The Stoics interpreted them as the four elements in their physics. For us, this leads to a generic algebra that can be interpreted as the reverse engineering of the genetic code. This genetic code, we call the generic code as it codes more than just the biological. It can code virtually anything. We find this literally awesome. Knowledge can be obtained out of ignorance. Also, whenever and wherever there is an M in the equation, there is the possibility of interpreting it as the finger of God.

The enigmatic key to it all is the Socratic confession of ignorance.

All I know with absolute certainty is that I know nothing with absolute certainty.

Who would think that this provides the ontological building block for the science of the generic? Nature is a little wilier than indicated by the Dawkins Ignorance Hypothesis.

The Two Savants

For light relief.

In the Appendices, we start developing the notion that left side sciences are based on second order logic and first order semantics. On the other hand we explain that right side science is based on only first order logic and but second order semantics. For light relief we make a little diversion.

Our dilemma is that it is difficult to convey an understanding of right side s scientific reasoning using a left side writing style. It is like asking a rustic to describe life in the big city. Our immediate task is to explain what is meant by higher order semantics versus the vanilla first order semantics of traditional left side reasoning. We will attempt to illustrate the difference using the allegory of the *idiot savant* and his lesser-known accomplice. Besides, it is time for some light relief.

In our attempt to find an easily grasped understanding of the difference between first order semantics and second order semantics, we can look at the difference between the *Idiot Savant* and the *Savant Idiot*. The first kind of *savant* is a left side dominant thinker and the latter is right side. Thus, there are two generic kinds of brilliant idiot in this world. It is the left side thinking *Idiot Savant* who consistently applies first order semantics to understand his world. For example, in order to understand the city he is passing through he will read the local telephone books. His favourite occupation is counting cards in casino blackjack. On the other hand, the *Savant Idiot* applies second order semantics to each situation, dialectically musing over this and that. He is very philosophical. He is truly wise and resembles in many ways, the perfect Stoic sage. He is never wrong. His only weakness is that he cannot do up his own bootlaces or even fill out a tax return by himself. In fact, he is totally inept at most things practical. This is his idiot side.

Somehow these two *savants* seem to get along together, or at least they used to.

The Shape of Knowledge

In any ground, breaking work there is a polemical streak and this work is no exception. Our presentation has raged across the axis of traditional left side science and our proposed right side science. The arena for this epic tussle has been the nature and structure of scientific knowledge. What we have failed to do is clarify exactly what we mean by scientific knowledge. We have argued that there is another kind of scientific knowledge than left side conventional science. This was the right side science. Now we must ask the question as to whether there are any other kinds of knowledge, knowledge that escapes the scientific tag. Embarrassingly, there is another axis of knowledge that is dramatically distinct from the scientific. In this book, we have ignored this other axis, an axis of equal importance as the left-right science axis.

To begin with, our topic here is universal knowledge, knowledge that includes scientific knowledge as a special case. Universal knowledge is composed of two, and only two, fundamental ingredients. These ingredients are semantics and logic, in the large sense. In order to understand the universal shape of knowledge, one must understand that there are two orders of logic and two orders of semantics: both have a first and a second order form. The notion of first and second order logic is a quite familiar to present day logic, and has been even formalised from an axiomatic perspective. However, this is not the case for semantics. Even the prevailing notion of semantics is hazy, let alone any notion of first and second order semantics. We must rectify that situation as we proceed.

Attempting to explain the shape of knowledge can take up reams and reams of pages and still not get anywhere very fast. Our best recourse is simply to illustrate the shape of knowledge with the semantic square. As usual, everything finds its place and we get the fit shown in Figure 56.

The Four Kinds of Knowledge

MF	MM
2 order logic-1 order semantics	2 order logic-1 order logic
FF	FM
2 order semantics-1 order semantics	2 order semantics-1 order logic

Figure 56 Semiotic square of the four kinds of knowledge.

From the diagram, we identify logic with the masculine gender and semantics with the feminine. As we have seen, the pure feminine F typed entity
has extent but no presence. On the other hand, the pure masculine M typed
entity enjoys presence (it *is* presence) but has no extent. Logic plays in the
masculine register and so becomes the logic of presence. The masculine
becomes the ultimate determiner of what *is* and what *is* not. Semantic plays in
the feminine register involving the interplay of that which can claim some
substantiality. The minimal requirement for substantiality is extent. The pure
feminine has extent but no presence.

Now we come to the question of scientific knowledge. We start with a
rough-hewn definition of science in the context of the semiotic square in
Figure 56. The parts of the square that qualify as science are the boxes where
both M and F are present. M provides the logic and F provides stuff with
extent. Science thus fits into the slot of being the "logic of stuff." This would
be the MF version corresponding to traditional left side science. Reversing the
order, we get the FM science corresponding to the "stuff of logic." This
corresponds to the right side science that we have been developing. One could
surreptitiously slide in the comment that the MF science studies dead stuff
whilst the FM science studies living stuff.

Note that we are using the semiotic square in Figure 56 as our compass in
our attempt to avoid Kant's curse of the "fine spun argument." Keep in mind
that one does not need to write a doctoral dissertation each time one consults a
compass.

Thus, it appears that the main dialectical opposition of this book has raged
along the MF-FM diagonal of the semiotic square. There we find two kinds of
science, the traditional left side and the right side science that we are promot-
ing. The semiotic square nicely characterises these two kinds of science. Left
side science is based on the paradigm of second order logic and first order
semantics. In other words, all the traditional sciences, including mathematics,
are based on abstraction provided by a second order logic and a shallow first
order semantics. On the other hand, there is the generic, and universal
oriented right side paradigm, which is totally devoid of abstraction and its
higher order generalisations. To remove abstraction from the pudding, the

paradigm only allows first order logic. Where it shines, is that it can handle non-trivial second order semantics.

It appears that if you want a science with non-trivial semantics, you have to throw away abstraction and its higher order logic. *Vice versa*, if you want the generalisation power of abstraction, you have to throw away higher order semantics and use the rather trivial default version based on first order semantics.

Just before going on to explain what is meant by these different orders of logic and semantics, we cast one more glance at the semiotic square of knowledge. Apparently, scientific knowledge works along the MF-FM diagonal of the square. This leaves two other kinds of knowledge left out of our science equation. The other diagonal consists of FF and MM knowledge. This is a topic that we will have to come back to later. For the moment, whatever this kind of knowledge may involve, we will refrain from characterising as "non-scientific." Rather, we will call these potential sciences, the *subtle sciences*. Such sciences have a distinctly Eastern flavour. We shall briefly discuss these subtle sciences later.

Our immediate task is to clarify what is meant by first and second order logic and semantics. We start with first order logic. From a traditional left side perspective, first order logic comes down to the logic of propositions, the *propositional calculus*. The propositional calculus involves well-formed formula, called propositions. Each proposition has a truth-value of either **true** or **false** and is made up logical conjunctions, disjunctions, and negations. In brief, propositions are mathematical logical expressions made up of abstract symbols combined with AND, OR and NOT primitives.

Second order logic is an extension of the propositional calculus and is called the *predicate calculus*, In addition to the propositional calculus structure; the predicate calculus allows the abstract symbols to be treated as variables with values ranging over sets. Each variable x can take on a range of value restricted to a particular set of values A. This simple construct provides the necessary equipment for abstract logical reasoning. The reasoning is formalised by the addition of two logical primitives called universal and existential operators. The so-called *universal operator* is used to mean that a predicate is valid "for all x," The *existential operator* means that "there exists an x" for the logical expression to be true.

Second order logic is the basic construct that enables abstract reasoning. An essential characteristic of abstract reasoning is that the objects of reason are not required to exist. Whether something exists or does not exist may be true or false, depending on the assumptions. The reasoning is based on generalisations and, despite the "universal quantifier" terminology, has no concept of universals. The universals belong to right side science.

Sidenote

Generalisations and general laws apply to everything in a closed world. For example, the Second Law of Thermodynamics is a general law and so applies to everything in a closed system. The law states that everything in in the confines of such a system drifts to a state of maximum entropy, that is to say, to a state of thermic death. One can say that general laws apply to everything but not everywhere. They only apply to within the closed system. The general law is only valid in the confines of a sealed bottle. The bottle may be made of glass and contain a mixture of gases. Glass is a favourite material for making closed systems for the left side sciences. Another favourite material for building a closed system is axioms. Axioms make very fine watertight bottles and ensure that everything enclosed within is sure to be headed towards thermic death.

On the other hand, universal laws apply everywhere but not to everything. Instead of applying to the closed system, the universal law applies to the open system, the system that, instead of living in a bottle, lives within itself. Instead of drifting to thermic death, the universal system will tend to proliferate and diversify, producing life.

In addition to second order logic, left side science must have recourse to a first order semantics. We associate logic with the masculine and its punctual nature. We associate semantics with the feminine and its extensive, non-punctual nature. Semantic expresses itself in the form of oppositions between contraries. In the case of first order semantics, there is only a single opposition involved. On the other hand, with second order semantics there are two oppositions, often involving one opposition applied to itself. The traditional left side sciences, including mathematics, only use first order semantics.

Traditional left side mathematics only uses semantics of the first order. Mathematics constructs its semantics from the fundamental opposition between a collection on one side of the opposition, and the objects making up the collection, on the other. This leads to Set Theory, the general expression of first order semantics for practically all of mathematics. The Set is on one side of the opposition and the Elements of the set are on the other. The elements of the set provide some kind of primitive notion of extent. For example, the set of points making up an interval of the real line is such an example of extent. It is quite remarkable that this is the only semantics that axiomatic mathematics really needs. Feed it Set Theory and off it goes. No fuss.

It is equally remarkable that all of the traditional sciences of our day are based on first order semantics and second order logic. These sciences operate under the heading of the neon light, flashing MF typing of knowledge. It would appear that this kind of knowledge is favoured by those ethnicities that belong to the Christian tradition, cultures with an MF disposition.

We now turn to knowledge of the FM type. According to our analysis, FM type knowledge is based on second order semantics and first order logic. From a linguistic-cultural point of view, we have associated the FM disposition the Islamic tradition supported by the Semitic language Arabic. However, from a philosophical point of view we are lead to the Stoics. The Stoicism was the "least Greek" of the ancient Greek philosophies. Moreover, all the early founders such as Zeno, Cleanthes and Chrysippus, were all of Semitic origin. Thus, the cultural typing of Stoicism might justifiably be classed as coming from a Semitic background. Be that as it may, we type the knowledge speciality of the Stoics as being of type FM.

The Stoics had their brand of first order logic and they consistently expressed an aversion to employing second order logic and its attendant preoccupation with abstraction. The Stoics only reasoned in particulars arguing that generalisations do not exist. Socrates can exist but Man and mortals do not. There is no such thing as Man. There is no such thing as mortals. Abstract generalisations do not exist. They rejected the species and genus of Aristotle saying that they had no need for them. In modern mathematical terms, they rejected sets. All of modern mathematics is based on sets in the form of Set Theory. Without Set Theory, there can be no traditional mathematics. If a Stoic were alive today, he would still reject Set Theory. The Stoic has no need for such abstractions. The Stoic is content with the logic of Chrysippus, which faithfully avoids anything but the particular. After all, only particulars can exist and that is what concerns the Stoic.

Of course, traditional mathematics goes the other way and reasons over the elements of an abstract set of objects, the set of green apples, the set of prime numbers, for example. First order logic avoids such abstract thinking and only talks about qualities relating to the existence of a particular entity. In their purest form, the qualities involved have nothing to do with the greenness of apples or even the primeness of a number. The qualities are the generic qualities of the generic entities. What matters is whether one has or possesses the quality or not. "If you have the first and the second quality ..." is the premise of Chrysippus' first of the five undemonstrables. The logic does not say what the quality *is*, but rather *whether it **is*** or ***is not***. Relative to you, the quality ***is*** if and only if you happen to have it possession at the time. This is an ontological logic. Despite avoiding abstraction, the first order logic reasoning of the Stoics becomes surprisingly profound, as explored in the appendices.

We now turn to semantics. Before moving on to second order semantics, we take another look at semantics of the first order. Firstly, who uses first order semantics? We know that modern mathematics uses first order semantics and only first order semantics. We notice that this statement did not make the reader suddenly sit bolt upright, which is the reaction we wanted. In fact, the reader's eyes seemed to have even started to glaze over. In search of a more engaging means of explanation, we come back to earth where people and things actually exist, and not just in the imagination.

We remark that if one looks around us hard enough, one will surely discover an acquaintance, a relative even, who only uses first order semantics in their everyday life. Such people are easy to spot. Moreover, not all of them are mathematicians. The key giveaway is that the person concerned is totally incapable of putting themself in someone else's shoes. For example, such a person is incapable of putting themself in *your* shoes. In order to accomplish such a feat, one needs second order semantics. In brief, first order semantics implies a total lack of empathy.

The inability to put yourself in someone else's shoes leads to the worldview that you are the centre of the universe. This is an inevitable consequence of a first order semantics view of the world. The most famous exponent of this worldview was Ptolemy, of the first century AD. Ptolemy was a gifted mathematician that wrote on many scientific topics. The most famous was his geocentric model of the world based on a set of nested spheres. This incredibly complicated system held sway for over a thousand years until finally replaced by the much simpler heliocentric model.

One wonders whether there are any extremely over complex Ptolemaic scientific abominations around in modern times. One does not need much prodding to come up with a likely candidate, String Theory. Perhaps we should express our admiration for the String Theorists. Their achievements are even more laudable when you realise that they have accomplished so much, and only using first order semantics.

The above explanation of first order semantics is probably as clear as mud. Perhaps we will have to turn back to mathematics itself to bring some sort of rigour to bear on the question. We must turn to the empathy free zone of modern mathematics.

Without going into details, we can say that the kind of mathematical geometry possible with first order semantics is rather trivial compared to the geometry possible with higher order semantics. This is very important as we rely on mathematicians to describe to us the shape of the universe we live in.

However, no mathematicians or mathematical physicists to our knowledge have ever pointed out the fine print in their deliberations. They simply inform us that, as a consequence of applying their mathematical theories, it turns out that the world is shaped in this or that particular way. Nowhere in the description is the caveat that, by the way, the expressed views herein have all been based on first order semantics and only on first order semantics. Sadly, there are no labelling laws for modern mathematical products. This must change.

So what kind of geometry do you get when you only use first order semantics? The answer is surprisingly simple. Some mathematicians even boast about how simple it is. They see it as a triumph of applying abstraction. To begin with, they claim that all spaces are n dimensional. Mathematicians cannot stop themselves from generalising. The letter n is a very general number. That way you cover all bases and so it is hard to be wrong. Then comes the decisive factor. All the various mainstream versions of space mathematics have exactly the same geometry! Technically, they all have the same *affine* geometry. This is truly remarkable. Lines behave like lines and points behave like points in all these vastly different mathematical spaces. The only difference from one mathematical version of spatiality to another is the distance between points. Mathematicians handle this detail by ascribing a different metric artifice, called a metric tensor, to each space. In this way, for example, an ordinary Euclidean space can become Minkowski spacetime geometry by simply swapping the metric tensor.

Practically all these mainstream mathematical spaces are special cases of a Hilbert space, and so the construct goes back to David Hilbert. A ferocious critic of Hilbert was the great Henri Poincaré. Curiously, as an aside, Poincaré was ambidextrous. We could certainly say that about his mathematics too, but he was both genuinely left and right handed with the pen and, it appears, also with the mind. The ambidextrous Poincaré goes head to head against the (presumably) right handed, left paradigm dominant Hilbert: it is a nice image albeit without any grand significance..

Anyhow, history has it that the abstract axiomatic geometry of Hilbert eventually prevailed over the objections of Poincaré. However, the battle is not over. Armed with the realisation that the Hilbert kind of geometry is only based on first order semantics and that there is our second order semantic alternative, the picture may indeed rapidly change. However, this next time round, there will be no conqueror nor conquered. The only thing to settle will be as to which side of the semantic equation is the Master and under what circumstances.

In brief then, mathematics relying on first order semantics results in a very simple, abstract kind of geometry. Simplicity is always an admirable quality when it comes to scientific explanations; according to Ockham's razor the simpler the better, however, the simple always runs the risk of falling into the abyss of being simplistic. Ptolemy's thesis that the earth was the centre of the universe was also simple, but looks at the headaches that gave him, and all the poor astronomers that followed him for a thousand years. Modern day String Theory theorists utilise the simplicity of a geometry based on first order semantics and seem to get the same kind of headaches. We are not qualified to criticise the details of their work. However, looked at from afar, it might be that things could be simplified by a paradigm shift or two.

In Appendix B, we look at geometry based on second order semantics. In the process, we are lead to alternative interpretations of imaginary numbers, the basis for any fundamental geometry. In fact, we are lead to back to our starting point. The imaginary numbers interpreted as MF, FF, and FM typed entities! These same typed entities can be interpreted from a spacetime geometric perspective as cones and heaves of lightlike, timelike and spacelike arrows. In addition, we investigated the enigmatic MM typed entity and intuitively started to understand it as a flip-flopping Figure-Ground, "*is*" and "*is not*" kind of geometrical dimension.

The Subtle Sciences

In this book, we have long argued that there are two kinds of science, sciences of the traditional kind and a science of a new kind. The main thrust of our effort has been to provide a tractable way of understanding the new kind of science. Each of these two kinds of science deal with the relationship between subject and object in reality. In the traditional science, the subject, suitably dressed in a white dustcoat, is synonymous with the "View from Nowhere." The corresponding objects are those observable by the spectator "located nowhere." This leads to the MF type of knowledge based on second order logic and first order semantic structures. All traditional sciences, including axiomatic mathematics, occupy this epistemological position.

The other kind of science is our right side science, which leads to FM typed knowledge, knowledge characterised by first order logic and second order semantics. With this kind of knowledge, the subject is no longer "nowhere" but a participant in the scene. The subject is always present. The main thrust of this book is to understand and demystify this kind of knowledge. We bundle this right side science FM type knowledge together with the left side MF knowledge. We then place them under the title of *scientific knowledge*, or to be more precise, *gross scientific knowledge*.

The reason for this extra precision is that there is another family of knowledge of an entirely different ilk, knowledge of the FF type and knowledge of the MM type. We will bundle these two forms of knowledge together under the title of *subtle scientific knowledge*. These are the *two subtle sciences*, one left side and one right side.

First, there is the FF type of knowledge that also deals with second order semantics. However, in this case there is a complete absence of the masculine, that is to say, a complete absence of logic. By an absence of logic, we mean an absence of punctual reasoning. Logic is replaced by semantics resulting in a paradigm involving an application of first order semantics to second order semantics. The presence of logic in one of the fundamental knowledge paradigms is signalled by the presence of an alphabet-based language. Where the logic aspect of the paradigm is not present, it is replaced with semantics. In this FF paradigm, case there will be no alphabet-based language but rather a language based on signifiers which have intrinsic meaning, in other words, a language based on ideograms. The FF paradigm is unique in this respect.

FF type knowledge is nevertheless a left side knowledge, as is the traditional MF type knowledge of the Western sciences. Western science is based on the rational knowledge of entities based on their attributes. This knowledge is expressed abstractly through second order logic and involves abstract generalisations about reality. In the case of FF type knowledge, there is no abstraction, there are no attributes, and there is no M. This is certainly a subtle kind of knowledge. We postulate that the cultures with a natural disposition and genius for this kind of knowledge are those vehiculed by ideogram based languages like Chinese and Japanese. It is remarkable that the extremely ancient Chinese book, the *I Ching, the book of transformations*, came up with the fundamental template for the generic code, right at the dawn of civilisation. The *I Ching* is based on the 64 combinations of gender typed hexagrams. This is precisely the structure that we have unearthed in our research into the generic code of reality.

The disposition of Chinese culture is in the opposite direction to any metaphysical or philosophical speculation. Science will present as a rule based procedure such as Acupuncture, rather than any pretence to scientific theory. In a similar way, the philosophy of Confucius presents as a rule based doctrine of ethics rather than philosophy in the Western tradition.

Sidenote

In the opening chapters of this book, we mention the "rich and poor dads" of Kiyosaki. The initial reason for discussing Kiyosaki and his two fathers was to provocatively illustrate the limitations of abstract thinking. The implication was that Kiyosaki's "poor dad" was university educated into abstract thinking. In this respect, the poor but well educated dad would have a disposition to thinking in MF terms, that is to say, in terms of abstract reasoning and shallow first order semantics.

An interesting question is to characterise the natural reasoning disposition of Kiyosaki's less educated rich dad. One might be tempted to think that he would have been a "right side" dominant thinker. On careful reflection though, the author now believes that he would still be left side dominant, but with the abstract logical reasoning replaced by the second order semantics. In other words, he would have a disposition towards FF reasoning. We leave readers to make their own judgment on this fascinating topic.

The other kind of subtle knowledge is of the MM type and lies on the right side and so has a monist vocation. This kind of knowledge is the natural partner for FM type knowledge. Here we see second order reasoning applied to itself. Rather than the second order reasoning ranging over the elements of sets, the reasoning ranges over the attributes of a singular generic entity, Since the attributes of the generic entity must themselves be generic there is no generalisation involved and hence no abstraction. Instead of generalisation, we find universalisation. The generic attributes of the generic entity are universal attributes applicable to any entity (satisfying FC).

The domain of MM style knowledge is that of our anti-mathematics, the natural right side version of abstract left side mathematics. On the left side, mathematics is bundled into the same MF paradigm as the traditional sciences and suffers from the same fictionalism tendency and tentativeness. On the hand, right side mathematics, our anti-mathematics, enjoys its own unique paradigm. Unlike its left side counterpart, right side MM type mathematics is untainted by any cohabitation with entities that may have been polluted with the subjectivity of a third party. There is no mathematician behind the scenes choosing which axioms to put into the mix. Any subjects involved in the mix are already present in MM type knowledge.

From our analysis, what we have been calling anti-mathematics can be known as Generic Geometric Algebra. GGA emerges out of traditional Geometric Algebra. GA articulates an ensemble of geometries based on a triadic signature. From our perspective, GA is based on a three-letter alphabet. After making the necessary renovation, GGA emerges with an geometry of triads based on a four-letter alphabet, This leads to the generic code, the reverse engineered version of the genetic code. It comes equipped with its own generic semantics in the form of its geometric generic forms.

Such knowledge is very challenging. Knowledge of this type involves the interplay of first order and second order logic and is notable for its complete absence of phenomenal world semantics. It involves the application of the singular to the singular. It is a subtle science of entities without bodies as there is no explicit Feminine, no possibility of physical extent. This is transcendental knowledge in its extreme. Here we find logic applied to logic.

Platonic Forms and Neoplatonism

Plato introduced the idea that above the perceived world of change there existed another more fundamental reality. The higher reality was made up of abstract, ideal forms. Fundamental knowledge was not knowledge of the perceived world but required knowing the ideal Forms. Such a philosophy is known as Platonism. Because of its inherent dualism, Platonism must be regarded as a left side philosophy.

Plotinus, author of the Enneads, attempted to eliminate Platonic dualism by developing a philosophy founded on the science of the One. In so doing, Plotinus is credited as the founder of Neoplatonism. The Forms of Plotinus are no longer the abstract Forms of Plato. Rather than mere generalisations, the Forms become universals. They become generic Forms.

Plotinus made very little progress regarding any specific knowledge of the Forms. It is the Hindu culture that has a natural disposition to this kind of knowledge. It is perhaps not surprising that Vedic and Hindu mystics engaged in the quest for such knowledge often ended up taking a complete disregard to their bodies, either that, or taking over total control. The ultimate aim seems to become one of the Ms in the MM configuration. Plotinus was also noted to dislike "being in his body." It would appear that the monism of the Hindu thinkers as well as the Neoplatonism of Plotinus involve a non-dualist monism in the mind only.

Sparseness

In this book and its appendixes, we have sketched out the foundations of a new kind of geometry based on the right side scientific paradigm. When talking about the shape of knowledge, we must also talk about the shape of geometry.

Left side geometry, like Hilbert space for example, is notable for its lack of shape. Arguing from the left side perspective, this is seen as a good thing. Space does not have shape; only the objects, which inhabit the space, have shape. Thus, the shapelessness of space is a necessity for left side science. Consequently, one finds that the dimension of a traditional space is the same as the number of degrees of freedom. In the final analysis, the traditional left side geometry leads to a space made up of a formless blob consisting of infinitude of points. Such a space is uniformly dense in structure. One could say that the space is devoid of sparseness. The only sparseness is the apparent void between any objects that might inhabit the space. However, even that sparseness is illusory as objects are all considered ensembles of points. In other words, such point-like ensembles do not take up any space at all, as each point is lacking in extent.

Right side geometry takes another tack. To begin with, there is no dichotomy between space and the things that inhabit space. Space and the thing are the same stuff. One space per thing. In this context, space must have shape as the thing has shape. As shown in this book, the thing is made up of a conglomerate of spaces, as spelt out in the Code that describes it. Each space is a particular type of Three-in-One structure corresponding to a codon of the Code. However, each of these micro cum macro spaces can invoke a higher dimensionality. We have mentioned this to explain the so-called degeneracy in the genetic code. If our hypothesis is valid, each elementary space can be coded by different codons from the Code, all according to the observed degeneracy of the code. Thus, even at the micro level, the dimension and degrees of freedom are far from being the same thing. Such space, where the number of degrees of freedom is less than the number of dimensions, can be said to be sparse. Sparseness in form is a prime requirement for shape in form. According to our analysis, any entity whatsoever that exists, is an amalgam of such sparse structures. The specificity in terms of gender typing of these constituents articulates the shape of each organism in question. In addition, to note, there is no absolute dividing line between biological forms and the material forms of physics. They are all constructed from the same generic principles.

Strife

To Karl Marx, change presents as the history of class struggle. According to Empedocles, change was the outcome of the incessant struggle between the forces of Love and Strife. Love unites the elements together to become all things. On the other hand, Strife brings about the dissolution of the one back into the many. The elements become unmingled, naturally attracted to their like. Division and morselisation intensifies.

Whatever it may be, the class struggle has been settled and it now appears Strife is everywhere and gaining the upper hand every day. Voracious corporate CEO's, bankers, and money merchants are no longer constrained by social consciousness. Free to gouge, they turn on their cringing critics, accusing them of a politics of envy. Apparently, Marx's class struggle has become a fight between Greed and Envy. All of this takes place in the fractured milieu of rampant, unrestrained, deregulated, Strife.

However, our principle concern is not the Strife that is rampant in our society, but the Strife that is ravaging unchallenged in our educational institutions and particularly in the sciences and mathematics. The Humanities are the historic critics of science and things at large but have disintegrated into enclaves of irrelevance. As like attracts like, scholars run to shelter to form their own hermetic communities on their chosen island of increasingly fractured speciality. Some still nostalgically ruminate over long lost causes. Others attempt to find inspiration in nihilistic Post-Modernist worldviews. For the rest, there may be meagre pickings at the bottom of the barrel. There must be at least one unsaid word yet to say about someone who once said something somewhat interesting about something or other. Surely.

Meanwhile, blinded by the success of technological revolutions, the sciences march on victorious as each island of specialisation breaks up into even more islands of specialisation, creating the greatest oceanic world of knowledge ever known. A consensus develops that all is well. This is all there is. This half-world is the world. There is no other view but this one. This is it.

We pause for a moment to offer our thoughts of condolence to our brilliant and best young minds that are at this very moment being ushered into the half world of present day science, there to be trained to think with half a brain. Cloistered from the distractions of the real world, these young minds are being moulded to provide an army of intellectual technicians trained in the karate of abstract thought. This is the era of the abstraction technician. Abstraction brings together like with like. In the process, the abstraction becomes increasingly stripped of specificity. The supreme abstraction is the vacuous, the vacuosity of Everything. As like attracts like, this is surely leading to Strife.

Strife plays a major role in the traditional left side sciences. For example, axiomatic mathematics has for its very vocation, the creation of Strife. This is because such mathematics is fundamentally based on abstract generalisations, the lumping together of like with like. The advancement of mathematics thus follows a similar trajectory to the all the other left side sciences. It explodes into a myriad of every increasingly specialised islets of specialisation.

In response to this section on Empedocles' notion of Strife, it is tempting to write a complementary section on Love in all of its cosmic, physics, and even emotional manifestations. The pitch might be that the right side scientific paradigm sees the world coming together as a unified whole, under the influence of Empedocles' notion of Love whilst the left side paradigm emphasises a more strife torn scenario. However, in other parts of the book, we have already analysed the opposition between these two paradigms from many other, more tangible, angles. Thus, we resist the temptation and write a section on Evolution and Emergence instead.

Evolutionism

From a left side science perspective, the Theory of Evolution, enriched with modern genetics, is probably the best natural fit to the paradigm. As a theory, it is devoid of ontology. Evolution theory has no explanation or description regarding the transition from inanimate matter to animate biological organisms. It has no theory of how life started. However, Evolution theory has a teleological dimension, albeit very simplistic. Teleology is the science of final causes. The final cause for a biological organism is its purpose and goal in life. For Evolution Theory, the purpose of an organism would probably be summed up as simple survival and the propagation of its own genes. This is all that life is about, according to the Evolutionist.

It is our opinion that the best philosophical foundation for the Theory of Evolution is Epicureanism. The whole world is totally deterministic and built out of elementary atoms of some sort. The only exception to determinism is that of the occasional unpredictable, random Epicurean cum Heisenberg Swerve in the motion of the atoms. This philosophy, suitably modernised can provide a left side philosophical account of Evolution. It even provides an account for how life started; atomic matter micro-swerved from the inanimate to the animate to produce the first microscopic prototypes of life. These microorganisms continued micro swerving via genetic mutations. Guided by the Natural Selection, they ended up as us. Richard Dawkins seems to agree with this explanation and has advocated the setting up of massive computer simulations of that might be able to emulate the random transition from the inanimate to the animate. As a modern Epicurean, he wants to use modern techniques to demonstrate that the gods have no role in how life came to populate the world.

In our view, the Dawkins cum Epicurean view of reality is probably the most authentic way to present the left side science paradigm version of a deterministic world, which somehow has enough random slack to allow for change and the eventual evolution of biological organisms such as ourselves. Epicureanism even provides a tractable teleology applicable to all levels of life, including the modern day man. As we have already discussed, the Epicurean purpose is to live a happy, tranquil life free from fear. In particular, as Epicure went to great pains to point out, man can live free from the fear of the gods. Richard Dawkins also admits the existence of gods, as long as they are harmless, ineffective, and totally meaningless to the common man. He has little objection to the pantheistic gods of Spinoza, Leibniz, and Einstein, for example.

Where Dawkins seems to differ from the ancient Epicureans is his rather agitated militantism. One could say that he is an agitated Epicurean. This might seem somewhat of an oxymoron. However, the ancient Epicureans have been known to become agitated. For example, Cassius who was involved in the brutal assignation of Caesar was an Epicurean.

A true Epicurean should live without fear of anything, even the gods. Dawkins almost fits into the Epicurean mould but he allows one fear to remain. The fear of god is replaced by the fear of the god-fearing. It is these god-fearing people of America, above all, that send shivers down the spine. For example, these people can even threaten government funding of science projects. This is enough to strike fear into any scientist, atheist or not.

Dawkins is a forthright exponent of the Theory of Evolution and Natural Selection as an alternative to religion, answering all the questions. Is he serious? His version of Darwinism replaces religion. In attacking religion, he has provided us with a service by bringing into relief, the underlying moral and ethical values of Evolution Theory. We argue that the underlying morals and ethics of Evolution leads to something like Epicureanism or some other pragmatic equivalent. This applies to the underlying ethics of all of the left side sciences for that matter.

The Ruse and Emergence

Life and the very existence of anything depends on a Ruse. From the left side perspective of the Evolutionists, there is no Ruse but merely chance. Things come into being by chance. From the right side perspective, things come into being by Ruse. To understand the Ruse, one must understand the obstacle to anything happening at all in the Cosmos and even the Cosmos itself happening.

The fundamental principle of Nature is that of perfection. Perfection is its principle inviolate. As the Stoics claimed, the World is perfect. This is the obstacle for anything to happen, the obstacle for anything to come into being. How can anything happen and come into being without violating Nature's pure and eternal perfection? In this book we have moved towards a formalisation of what we mean by perfection by introducing the notion of First Classness. Nature must not violate FC. The FC requirement effectively states that the **real** Nature exists in the eternal present. Nature is not dictated by anything in the past as such things do not exist. The same applies to the future. This total lack of *a priori* and *a posterior* determinations, provided us with our one and only means to obtain scientific traction on this most slippery of all creatures, the creature determined only by the pure Virtue of its perfection.

This lead us into developing generic structures based on an extremely relativistic kind of typing system, that of ontological gender. All entities making up this creature are typed using this binary masculine-feminine algebra. Gender expresses the ultimate chicken and egg story as there is no way of proving which comes first, the masculine or the feminine. The chicken and egg story continued when we saw how the four binary types MF, FF, FM, and MM could be constructed from the masculine and feminine. These four types can be present all in the same synchronic structure and so, once again, there is no ordering involved. These four binary typed entities became the four letters A, U, G, and C of the generic algebra for coding creatures that implement the Ruse. The organism of the Ruse must have a form that keeps the binary types free from order. The solution is based on the semiotic square where the front lobes, from left to right, are typed Mf and MM, and the back parts FF and FM. There is nothing absolute about this particular convention. Other configurations are supported but change of the convention is not. This generic semiotic four-part structure provides the fundamental placeholder for the Ruse. It can be thought of as primordial Mind or primordial Body. No distinction between the two exist at this stage in the development. The principle of FC demands that no absolute distinction ever exists between the two. However, relative distinctions are allowed. All distinctions of the Ruse are relative, relative to the organism itself. The next stage in this chicken and egg story, was the Ruse of the triad. How one can order arrows without really ordering arrows? This problem was resolved (partially) by the RGB triad of arrows. This RGB triad was self-labelling. Even the mirror image of the triad wouldn't change the self-labelling. Without self-labelling generic structure, the Ruse would be doomed to failure. The triad can be labelled with RGB "colours" without ambiguity. Armed with the self-coloured triad, each of the three arrows can be typed with one of the four letters from the generic alphabet leading to the 64 possible codings. The Ruse is articulated in this FC non-violating generic algebra. We have sketched out in the appendix how the semantics of the Code can be interpreted geometrically. To be kept in mind at all times is that the Ruse must forever follow chicken and egg logic, that is to say, the logic of gender.

The crucial part of the Ruse is the gender typed RGB triadic structure. Each triad is part of a Three-plus-One structure of an elementary Whole where the three part is the triadic part. Here, we can see the very essence of the Ruse. The triadic part of the Whole, corresponds to the "imaginary" part. The three imaginary bases determine the "real" part of the Whole. Any creature exploiting this Ruse of Life will be constituted from few ranging to a myriad of such Wholes, each with one real part and three imaginary parts. The imaginary parts of the Whole can be interpreted as attributes of the Whole.

Armed with such attributes, the organism enters into the world of appearances. The real part of the Whole is not an attribute even though determined by attributes. Thus, the real does not enter into the changing realm of appearances. It rests in the world of the **real**, It is only this part of the organism that really exists. In this book we discover that the mathematical terms of real and imaginary number help provide the necessary formalisation of real-imaginary concept, once cast into a right side format.

We have sketched out many of the essential structures of the Ruse. This leads to the emergence of a world of incredible complexity. However, the underlying principle is of great simplicity. In perfection, nothing changes whilst the Ruse leads to everything *appearing* to change without violating the perfection. However, the Ruse has its limitations as, in the final analysis, every creature of the Ruse must return to from where it came. Every excursion into the world of appearances is but a momentary flutter in the big scheme of things. However, the flutter is only a flutter in appearance. From a right side perspective, every flutter is *necessary.* Corrective action or correcting action, the flutter of life is necessary for that Nature be what it is. As the Stoics maintained, one must live according to Nature.

Emergence and Democracy

The right side scientific paradigm is articulated in terms of oppositions. The algebra of oppositions is the generic code based on the four letters. The object and subject of right side science is anything-whatsoever-that-exists. This can lead to some tricky metaphysics. To avoid the metaphysics, we can think of our embryonic prototype of anything-whatsoever-that-exists in a more tangible form. This can be useful even if the analogy is not perfect. Thus, think of our of anything-whatsoever-that-exists entity as a democracy. The democracy exists, has extent, and participates in the world of appearances. It also has a real component that, except for special circumstances, never changes. The real component of a democracy is spelt out in a mass of genetic coding called the Constitution and Legislation crafted in accordance to regulations therein We can make the very simplifying assumption that the core values of democracy are a fundamental expression of First Classness. In this case, every citizen has a say in the running of affairs as upheld in The Constitution, legislated by Parliament, and enforced by the Judiciary.

A fundamental aspect of any democracy is that, wherever you look, you see a democracy. There is the local Bowls Club, professional guilds, local authorities, the Tea Club, and so on: the list is endless. All in turn, have their own constitutions inspired by The Constitution. This applies to the citizen, who will be organised along the same lines. If he were patriotic enough, the citizen would have even purged himself of his own biological DNA, replacing it with genetic material more correctly aligned to the technology of The Constitution. For the sake of argument, we assume that all citizens of the democracies considered here are patriotic.

We will now look at the dialectic of political change. How do democracies change? We propose two theories, one based on Evolution and one based on a different principle that we will call Emergence.

Now it turns out that there are two kinds of democracy, English and French. English democracy is the oldest and most venerable. However, English democracy is an embarrassment for us, as it does not even have a proper constitution. English democracy changes by evolving. Thus, it is an example of Evolution. This is probably why English academics, being good patriots, are great proponent of Evolution Theory.

On the other hand, the French have no notion of what gradual change even means. To the French, nothing changes, at least not gradually. If there is to be change, it must be a qualitative leap from the prevailing version of First Class perfection to another more appropriate form. This inevitably requires a new Constitution and a new Republic. They are now into their fifth. The transition is swift and sometimes a bit messy. Nevertheless, it is in this way that we see the Emergence of the modern civilisation, French style.

Using the discursive artifice of democracy, we can vividly illustrate the difference between Evolution and Emergence without being too encumbered by details and things that we fear the most – facts. We start with one of the greatest epistemological divides on the planet, the English Channel. On the left side we have change by Evolution. On the right side, we have the revolution of the great qualitative leaps. Adopting a Hegelian tome, we call it Emergence. In retrospect, the end result is almost identical. Nevertheless as we have said so often, it does go to show that there are always two takes on reality.

In this book, we have no intention of trying to develop the Emergence concept as the right side opposition to Evolution. That would take us way out of scope. Others such as Arthur Koestler (Koestler, 1978) have explored Evolution alternatives to explain the qualitative leaps that occur in Nature. A good example is the reptilian brain architecture that we humans share with reptiles and birds. The reptilian brain has not changed qualitatively for several hundred million years. The next qualitative leap was the addition of a second brain, the limbic brain, wrapped around the already emerged reptilian. The limbic brain emerged with the first mammals. Humans share this brain with all mammals including dogs, cats, and horses. The next layer of brain was the neocortex that emerged with advent of the first primates. From the higher primates, came the development of language capability with the emergence of man.

The Diachronic and the Synchronic

As we have previously discussed, the left side reasoning of the traditional sciences studies diachronic structure. Left side reality is one of motion and change. Paradoxically, knowledge of this changing diachronic world, presents itself as its opposite. Knowledge of the diachronic becomes synchronic in nature. The world of rapid change appears as single snapshot in right side science. In applied left side mathematics, the rapid changing magnitude ends up as a function of time $f(t)$, something that can be drawn as a single graph on a sheet of graph paper, forever fixed in time. The diachronic can thus be represented synchronically. On the other hand, right side science has for its concern, the unchanging invariance of Self from one moment to another, anchored in the unchanging *now*. However, knowledge of this synchronic world is merely represented as an undetermined, dotted arrow. The resulting synchronic arrow theoretic representation involves a non-concurrent, representation. In order for nothing to change, a diachronic reality is necessary. The dialectic between the synchronic and the diachronic adds extra flavour to a most flavoursome world.

A prime example of left side scientific analysis is the black box with an input and an output, with or without feedback. Stimulus and response, cause and effect, axiom and theorem, are the oppositions that define the traditional sciences. Evolution becomes the natural diachronic explanation of the opposition between the past and the present. The present evolves from the past. Simple.

The right side paradigm is characterised by synchronic structure where all factors are present in the same moment, here and now. The paradigm has an essentially "operational" flavour to it. A prime example of right side synchronic structure is the genome of a biological organism. All the genetic structure is present at the same time. Explaining change in terms of synchronic structure involves qualitative leaps. The start point in the change will involve a certain form of First Classness and the point at the end of the transition will represent another plateau of First Classness. Change goes from one hiatus to another. The rationale for the change will be the response to a violation of First Classness. Resolving serious violations of FC cannot be achieved by incremental change. A qualitative leap may be necessary. When you are out on a limb, sometimes you have to jump.

Qualitative leaping from one diachronic structure to the next might be a characteristic of Emergence. In fact, such a process is not that dissimilar to an Evolution account, just less incremental. For Emergence to present itself as a radically different explanation, it must provide answers to questions that Evolution Theory cannot explain. There are two brutally blunt questions that

Richard Dawkins and philosopher Daniel Dennett cannot answer from their Evolutionist paradigm. The questions are:

1. What is the origin of life?
2. What is the purpose of life?

Dawkins knows the stock answers to these questions, but rejects them both. The traditional, historic answers are:

1. The origin of life is God.
2. The purpose of life is to serve God.

Dennett, being a professional exponent of left side philosophy, would also reject these answers but probably in a more intricate way. He would probably even attack the questions and turn them inside out. But even a Buddhist would do that. Be that as it may, it remains that over the ages these two simple questions and the two simple answers have long been a part of the prevailing wisdom, particularly in the West. Is this all a delusion?

In our work reported here, we have not embraced God into our science. However, unlike left side sciences, we have definitively not rejected the construct either. As Hegel pointed out, any notion of God worth its salt, will be pantheist. Other than that requirement, the God question is left open. In the final count, one's conception of God is much the same as one's conception of oneself, and *vice versa*.

A key thrust of this book has been to understand the Ruse necessary for *anything* to come into being. There is sufficient material in this book and its appendices for the start of a tractable new kind of science based on the algebra of the Code and its generic geometry semantics. Armed with this wisdom, readers will be in the possession of a most powerful Organon. He or she can even play God.

Principia Mathematica

This book is essentially a series of evolving workbooks and diaries drawn together over a three-year period of intensive effort. It draws upon partial insights previously amassed over more than three decades. Two appendices addressing the more technical aspects of the work complement the main text. At the outset, the author had no idea that the project would turn out to be so vast, so elaborate, let alone so successful. We now come to the point where we must sum up our achievements reported here.

As a first try, we can claim to have solved the Kantian problem. Is it possible to develop knowledge free of any *a priori* conditioning? The answer is in the affirmative and can be gleaned from reading this work. We can also claim formulating the framework for resolving the great challenge sketched out by

Leibniz. The Leibniz task was to develop a geometric algebra based on a simple set of letters that could describe "natural things" including "things like plants and animals." Here we have sketched out the essence of such a universal algebra as a fundamental extension of traditional Geometric Algebra, as refined and popularised by David Hestenes in modern times.

Underlying all of the work reported here has been the Code. This algebraic Code envisaged by Leibniz appears to be none other than the generic version of the genetic code. This is the most fundamental and undoubtedly, the most outrageous claim we make. The repercussions and ramifications will be interesting to observe as it unfolds.

In writing a book like this, one starts with philosophical friends and philosophical enemies. A principle philosophical enemy for the author has been Bertrand Russell, a chief advocate of atomism, dualism, and Analytic Philosophy. However, it is interesting to note how closely the core thrust of our work runs parallel with that of Russell, and in particular, with the *Principia Mathematica* he wrote with Whitehead. Like the two hemispheres of the biological brain, there certainly are two different takes on reality. From the epistemological brain perspective, there is no doubt that Russell takes the left side and the author takes the right side perspective.

It is fascinating to observe the following parallels between the Russell perspective and that developed by the author.

Logically Perfect Language

First, there is Russell's notion of a "logically perfect language." This language would be practically devoid of semantics and be purely logical in nature. *Principia Mathematica* was Whitehead and Russell's account of such a language. The language nicely fits into the left side scientific paradigm as being based on second order logic and first order semantics. Parallel to Russell's logically perfect language we pose our "perfectly generic language," claiming that it is a reverse engineering of the genetic code. This right side language is based on the work of Chrysippus involving Stoic logic. This perfectly generic language is based on first order logic with second order semantics.

Theory of Types

A second parallel is with Russell's Theory of Types. In order to overcome paradoxes implicit in Set Theory, Russell was obliged to make a clear-cut distinction between the elements of a set and the set containing the elements. The set and its elements were considered to be of distinctively different types. Thus, the set of all subsets could not be considered as a set as it was of a different type to a set. This simple typing construct enabled Whitehead and Russel to avoid the pitfalls and paradoxes of their predecessors. Russel's

Theory of Types leads to an infinite hierarchy of types consisting of elements, sets, sets of sets, and so on.

In our case, we had to develop an alternative version of the Theory of Types, the right side version. In our case, there is no infinite hierarchy but simply two primordial types based on ontological gender. Gender typing is totally relativistic and provides the basis algebra for right side science and its geometry. It is with this ontological gender construct where our modern approach meshes into the wisdom of antiquity. Gender has long been a basic principle underlying such ancient concepts as the Theory of the Four Elements. Our task has to grab this ancient idea and drag it into the full light of day as the foundational pedestal of a new unifying science.

If our analysis is just, it is this kind of typing that dominates Nature, both in the biological and non-biological domains. The actual semantic logic of the gender calculus is implicitly spelt out in the five undemonstratable syllogisms of Stoic logic as developed in the appendices. As far as avoiding conflicts is concerned, that is covered by Chrysippus' third incompatibility syllogism. Just as Whitehead and Russel would not allow an entity to be both a set and an element at the same time, the third Stoic syllogism effectively outlaws an entity being both feminine and masculine at the same time. In other words, an entity cannot be an attribute and have that attribute at the same time.

The respect for gender coherence is the key requirement for systemic integrity and coherence of the overall organism and applies to *any* organism. From the two genders, arise the four elementary binary typed "letters" that date back to Empedocles. In this way, the gender algebra eventually provides the letters for constructing the "geometry without number," envisaged by Leibniz.

It should be kept in mind that gender typing is not absolute. Gender typing is relative to the organism in question. Viewed by a third party, the typing "collapses" and merely appears as a fuzzy superposition. The Stoic third syllogism only applies relative to the organism. Gender typing cannot be perceived by a third party.

The author recommends that anyone wishing to correct and extend our formulation of right side science should keep in mind this parallel between the *Principia Mathematica* of Whitehead and Russel. Right side science, in one of its more noble scientific dimensions, should be considered as a right side version of the *Principia Mathematica*. It should also be kept in mind that the right side paradigm is somewhat more practical than the left side *Principia Mathematica*, to put it mildly. No one will ever discern Russell's Theory of Types in Nature. If our analysis is correct, the core organisational principle of

Nature is based on the relativistic typing system described in the book, a typing system based uniquely on the binary construct of ontological gender.

Final Words

This book started off with a sense of foreboding and angst. We set ourselves the apparently impossible task of tackling the long abandoned Kantian problem, the problem of developing a science that was not reliant on any *a priori* knowledge whatsoever. In order to solve the Kantian problem, a new scientific paradigm is necessary. Using the two hemispheres of the biological brain as metaphor, present day sciences were cast in the mould of a left side paradigm. The Kantian problem became that of developing an alternative, but complementary right side paradigm. A fundamental characteristic of the right side paradigm is that it provides a science of consciousness, the foundations of mathematics, and even a scientific explanation of religion. These domains have long been beyond the reach of traditional science.

The traditional left side sciences are characterised by an adherence to anatomist, dualism. In contrast, the right side paradigm must be provide a non-dualist monism. In this most enigmatic of all scientific endeavours, the key to gaining traction involves developing a new kind of geometry. Left side geometry is easily characterised mathematically as a dualism between a mathematical space and entities contained in the space. Right side geometry must be monist. This means that there can be no dualism between space and things contained in space. Spaces and things are all bodies made of the same stuff. From the perspective of the right side paradigm, matter and space are just constituents of bodies. Bodies do not occupy space, they are space, and *vice versa*.

One key difference between the traditional left side sciences and the right side paradigm is that between the diachronic and the synchronic. The diachronic left side paradigm describes things drifting in time of a disinter4ested observer. The synchronic right side paradigm is based on the Parmenidean eternal present. What is real is *now*, everything else is "imaginary." What is now, is relative to the organism in question, regardless of scale. Ultimately, the right side synchronic paradigm becomes expressible in a Code that never changes and that can code the unchanging aspect of the organism in question. In this work we have sketched out how this universal Code can be reverse engineered from first principles. The only assumed restriction of the system is that there be no restrictions on the system. The requirement of being free from constraint turns out to be the most draconian constraint imaginable.

Sidenote

Like an adolescent that leaves home, there may be an initial burst of enthusiasm at the thought of unrestrained liberty. However, the last words on the doorstep, "You are on your own now," may came back to haunt many a troublesome night's sleep. Life is not as easy as it seems. Even staying alive can be quite a challenge at times. It is so much easier for the non-existent, particularly for the dead. They have so few expectations to live up to. Left side science exploit this facility. Such sciences specialise in the dead side of the equation, the study of life on the slab. On the other hand, right side science must deal with the very essence and science of the living. From the right side perspective, even the universe is based on the same life principles as for organic life forms.

Our analysis maps the Code to the structure of the genetic code. The genetic code starts to take on a new allure. Instead of being a mere transcription language for spelling out the biochemical building blocks of life, the Code maps out the fundamental spatio-temporal organisation of the organism as a whole. We claim that the Code applies not just to organic beings but to even the inorganic world of particle physics. This is why we call it the *generic* code. We have sketched out some examples and propose a right side alternative to the Standard Model.

We claim that the deep structure of the genetic cum generic code can be explained in terms of a generic algebra based on gender. The semantic of the code can be explained in terms of a new kind of geometry evolved from traditional geometric algebra. The underlying logic of the system is provided by the Stoic logic of Chrysippus. That is our "in a nutshell" summary of generic science, the right side, unifying science.

There are many ramifications of such a unifying and fundamental science based on the non-dualist right side paradigm. The nineteenth and twentieth centuries were dominated by exploiting the sciences based on the left side paradigm. It is now the turn of the worldview emanating from the unifying, foundationalist, constructionist, right side paradigm. It is the right side paradigm that may indeed dominate the twenty-first century before us Hopefully, this book will provide some much needed impetus.

15

Appendix A

Science Without Number

There are two kinds of scientific knowledge. The first kind corresponds to the sciences where knowledge is totally dependent upon *a priori* considerations such as experience, measurements, axiomatic structures, hypotheses, or even hunches and opinions. We call these sciences, left side sciences. All traditional sciences, the rigorous and the not so rigorous, are left side sciences, including axiomatic mathematics. The other kind of knowledge, what we call right side science, is based on a radically different paradigm, as it must be free of any *a priori* considerations whatsoever. Kant referred to this right side science as *metaphysics*. Aristotle called it the First Philosophy. Kant made impassioned calls for that this science should see the light of day. In the *Critique,* he explores the problem. Later in *Prolegomena to any Future Metaphysics* his frustration becomes palpable:

> *If it be a science, how comes it that it cannot, like other sciences, obtain universal and permanent recognition? If not, how can it maintain its pretensions, and keep the human mind in suspense with hopes, never ceasing, yet never fulfilled? Whether then we demonstrate our knowledge or our ignorance in this field, we must come once for all to a definite conclusion respecting the nature of this so-called science, which cannot possibly remain on its present footing. It seems almost ridiculous, while every other science is continually advancing, that in this, which pretends to be Wisdom incarnate, for whose oracle every*

one inquires, we should constantly move round the same spot, without gaining a single step. And so its followers having melted away, we do not find men confident of their ability to shine in other sciences venturing their reputation here, where everybody, however ignorant in other matters, may deliver a final verdict, as in this domain there is as yet no standard weight and measure to distinguish sound knowledge from shallow talk. (Kant, 1783)

Further on, as he looks into the question as to whether metaphysics is at all possible, he laments:

There is no single book to which you can point as you do to Euclid, and say: This is Metaphysics; here you may find the noblest objects of this science, the knowledge of a highest Being, and of a future existence, proved from principles of pure reason.

Kant called for the development of metaphysics as a science; However, in the process of exploring all of the obstacles to be surmounted, he made the problem appear so formidable that the enterprise seems impossible. Thus, many have concluded that Kant made his place in history as signalling the death of metaphysics. Such a conclusion ignores his impassioned pleas for such a science. It is time to answer his plea. It is time that we resolve this problem. We will call it the *Kantian problem*, the problem of developing a science free of any *a priori* determinations. This is the problem addressed here. In our terms, it becomes the problem of developing right side science, the science based on the diametrically opposed epistemological paradigm of the left side science. No *a priori* experience, no empirical measurements, no axioms, none of these are allowed. Life was not meant to be easy.

In a previous chapter, we discerned an essential difference between the two kinds of science: Left side science is reliant on attributes whilst right side science is not. Right side science must be a science *sans attributs*. This points the way towards a tractable strategy for resolving the Kantian problem. The problem becomes that of formalising a science that constructs its own attributes from scratch using first principles alone. This left the sticky problem of determining what constitutes the first principles. This problem was resolved by identifying the fundamental principle underpinning right side science as being that of First Classness (FC).

The principle of FC is nothing more than another way of stating the Kantian problem itself, the problem we intend to solve. Any formal system that requires *a priori* constructs in order to arrive at knowledge is a second-class system. Such a system violates FC. Only the system that is free of any *a priori* constructs is the kind that satisfies FC, The Kantian problem becomes that of developing knowledge of a system where the only constraint is that of the non-violation of FC. In so saying, we convert Kant's statement of the problem into a statement of the embryonic solution, all in the one breath. Solving the problem starts to become tractable.

Once a problem of this nature starts becoming tractable, one begins to form intuitive glimpses of what the solution might entail. Furthermore, since a problem of this grandeur is so central and important to all the sciences, it is highly unlikely that others had not already developed such intuitions. In this respect, one doesn't have to look too far. Just predating Kant, we find the usual suspects, Newton and Leibnitz. The lives of both these men were dominated by the obsession to discover the universal algebra of the Cosmos. Newton took the more hands-on approach, searching for signs of the cosmic algebra in alchemy. This was not to be a science of dead matter. Newton believed that all matter was permeated by the principle of life. William R. Newman summarises Newton's manuscript *On Nature* as

> *Newton's attempt to provide a synopsis of his early alchemical reading, and to come up with what is, essentially, a "theory of everything," namely a physical theory that unifies and accounts for all known natural phenomena. (Newman)*

On the science of life aspect, Newman explains:

> *As Newton puts it, there exists in nature a niter-like spirit that is 'the ferment of fire and all vegetables.' This spirit, in other words, is a principle of combustion and of life itself," It seems that Newton's alchemy research was a quest for some kind biochemistry of the Cosmos.*

Newton was after the universal code for this cosmic biochemistry.

Leibniz took a different tack and framed his intuitions in geometric and algebraic terms. His was a quest for a generic geometry. His statement of the problem was very precise: Devise a geometric calculus devoid of numbers. He called the discipline *Analysis Situs*. For Leibniz, the world was perfect and this perfection implies that it must fundamentally be simple. Perfection implies simplicity. This was not to be a bottom up science, but rather a top down science. Such a science cannot be based on abstract generalisations. Abstraction and generalisation involve a bottom up form of reasoning where, from experience and measurement abstract generalisations are obtained by inductive and deductive reasoning. The science that Leibniz sought must go the other way and be top down.

Leibnitz' principle of a world fundamentally perfect, we have expressed in terms of the principle of First Classness. Leibnitz' requirement of simplicity, we have formalised in terms of the principle of ontological gender. Consequently, the shape of the FC world, or anything that exists in this world for that matter, can be typed with only two types and can be determined in terms of a four-letter alphabet formed from these two types. There is nothing more perfect, nor simpler than that. That is the path we take and in this chapter we push forward to our solution of Leibniz's numberless generic geometry problem. A generic geometry based purely on the generic algebra of a four-letter alphabet.

To celebrate the bicentenary of Leibniz's birth in Leipzig, a mathematical competition based on his famous problem was organised by a local learned society. Encouraged by Mobius, Herman Grassmann was the winner and only entrant in the competition. His presentation described the foundations of his geometric calculus devoid of coordinates and metric properties. This marked the beginning of a new approach to geometry. The new discipline would eventually be called Geometric Algebra (GA). Hamilton followed with his geometric universalisation of complex numbers. Clifford integrated both the algebras of Grassmann and Hamilton to produce the algebras that bear his name. The geometrisation of algebra started by Grassmann then fell into obscurity and was eclipsed by what we call the left side version of geometry. The left side paradigm interpretation of Grassmann's directed number concept was developed by Gibbs and independently by Heaviside. This leads to the coordinate intensive linear algebra still dominant today. Whitehead, following on from Grassmann and Hamilton, developed Universal Algebra, a science of algebraic structures based on operations rather than relations, devoid of ordering and quantification over sets.

Over recent times there has been some revival of interest in GA. David Hestenes has advanced the geometric algebra approach to provide new insight into a wide range of physical topics from classical mechanics and electromagnetism, to Quantum Mechanics and **gauge** theory. He claims that the GA approach provides a universal language for all physics, a claim endorsed by physicists at Cambridge. However, despite the impressive achievements of Hestenes and the Cambridge group, GA still seems to rest outside the mainstream of the physical sciences.

In brief, all we have to do in order to solve the Kantian problem is to solve the problem posed by Leibniz before Kant's time, *viz.* develop a science based on a geometric algebra without numbers. This geometric algebra must explain the generic form of things, free from any *a priori* determinations. From Leibniz's viewpoint, these generic forms will be his monads, each "without windows." Each monad provides a view of the totality as whole. Leibnitz's perspective was fundamentally monist and so a right side take on reality. Grassmann provided a partial solution to the Leibniz problem. The discipline that he initiated has developed into GA and is technically quite advanced.

However, as a satisfactory solution to the Leibniz and Kant problems, it has not advanced. This is the task undertaken in this work. In resolving this most fundamental of all the fundamental problems, the solution will join up with the vision of Newton. We will end up with his *Chymistry* of the Cosmos, as we shall very well see.

Right Side Science is Abstraction Free

The dichotomy between left and right side kinds of sciences follows along the classic fault lines separating an atomistic worldview from that of a non-dualist monistic worldview. The familiar perspective of atomism is characterised by an inevitable dualism and a simple labelling technology free of any internal semantics. Most importantly, because of a fundamentally dualistic epistemology, such sciences are all based on a rigid dichotomy between theory and the object of theory. In other words, these left side sciences are all based on abstraction. It would be reasonably safe to say that, at the present time, there is no science worthy of the name that dares to promote a clear and definitive alternative to abstraction. The whole purpose of this book is to provide such an alternative. The science that is free of abstraction is right side science. The alternative to the abstraction approach is the generic. Generic methodology leaves no place for abstraction, as it must answer to the dictates of monism. Monism cannot tolerate the long accepted left side science practice of separating theory from its object.

Contemplating a science free of abstraction can be challenging. If there is no abstraction allowed, the immediate question arises: Where is the theory? The answer to this question is simple, but may appear perplexing. Just as for the traditional left side science, right side science does encompass theory. The only difference is that the theory has not been abstracted. Here the concept of abstraction is like the dental form of abstraction. Left side science attempts to abstract theory like a dentist abstracts a tooth. Like the dentist, the left side science practitioner wants his theory in hand. He attempts the abstraction whilst trying to avoid inflicting collateral damage on the writhing patient. In right side science, the tooth is left in the patient. No theory is ever abstracted. Instead, theory stuff remains an essential ingredient intertwined within the system.

In biological systems, the presence of this theory stuff can be directly observed in the form of the genetic code for the organism in question. In the case of multi-cell organisms, an identical instance of the same coded theory stuff can be found in each and every cell of the organism. As is now known, the code for this theory is based on a four-letter alphabet. The basic claim made in this work is fourfold:

1. *The longstanding classic Kantian problem can be resolved leading to a new kind of formalised science.*

   Other than initial qualitative attempts by Fichte and Hegel, no progress has been made in resolving this problem.

2. *This science leads to reverse engineering a generic algebra based on a four-letter alphabet.*

   As discussed earlier, the four-element theory of antiquity provides an insight to this kind of science.

3. *The genetic code coding all biological organisms is an instance of this generic algebra.*

   Traditional science takes a biochemical interpretation of the genetic code as merely a transcription language for coding amino acids. This simple transcription code is assumed devoid of any deeper semantics.

4. *In physics, the basic constituents of the universe itself, as an organism, can also be understood in terms of this generic algebra.*

   The limited form of relativity theory of modern physics must be replaced with a more fundamental, generic form of relativity. The natural algebra for expressing generic relativity is a generic code where the semantics is determined in a totally relativistic fashion as demanded by FC.

Summarising Ontological Gender

The object of this generic science is *any entity whatsoever*. Since nothing *a priori* is known about such an entity, all knowledge that can be gleaned from it will be of a generic nature. Thus, a generic science embraces a generic entity to produce generic knowledge. The generic entity is beholden to one and only one principle, that of the draconian demands of a monistic FC. The generic entity is prisoner of its own Oneness. The principle of FC declares that there must be no absolute determined constraint whatsoever on the generic system all must be relative - all must be relativistic. There is no one behind the scene pulling the strings. This freedom from any constraint whatsoever is the most draconian constraint in the universe. With total freedom comes an equally enormous responsibility. This creature is responsible for its Self.

The first step was to examine the generic entity from the point of view of characterising it by any distinctive attribute it may possess. Answering this question was simple as the generic entity is totally devoid of any determined attribute whatsoever. This leads to the conclusion that the only specificity possessed by this entity is that it is devoid of specificity. This articulates the nature of its attribute. This entity is highly specific by distinguishing itself from any other by possessing a totally unqualified attribute. Thus, this entity *has* an attribute, albeit totally undetermined. At this point, there is a risk that this embryonic system might violate FC. There is the danger of an absolute dichotomy forming between the entity and its attribute. FC demands that the attribute of the entity must be First Class: It must be an entity in its own right.

Thus, the first movement of the argument leads not to one entity but two. One entity has an attribute, the second entity is that attribute. In chapter 9, the difference between these two primordial entities was formalised as a difference in gender. The entity, which has the attribute, was said to be of feminine gender; the entity that is the attribute is said to be of masculine gender. There are two different entities and only one attribute between them: One *has* it and one *is* it. The two entities are different but obviously indistinguishable. This marks the beginning of the dialectic of *having* and *being,* the core reasoning process of right side science, the science of the generic. Armed with the gender construct, the limitations of a total lack of traditional attributes can be overcome. Who needs attributes when we can make our own!

The Basic Building Blocks

From a generic perspective, gender is a relative typing mechanism, where the types are determined relative to each other, independent of anything else. From a third party perspective, gender might be thought of as a kind of quantum state. In this respect, the simplest generic entity has two quantum states, one feminine, and one masculine. To the disinterested third party, the gender states will appear to be in superposition and indistinguishable from each other. The indistinguishability is inevitable as it is impossible to compare two entities that only differ by pure ontological gender, one having an attribute and the other being that attribute. However, when the third party actually starts to become interested and enters into the fray, the apparent fuzziness of superposition collapses and the gendered entities reveal themselves as what appears to be collapsed "quantum states." However, there are not two entity-states but four. In the previous scenario, there was one subject (masculine) and one object (feminine). The subject involved was the impersonal subject devoid of any specificity whatsoever. In the new scenario, there is an additional subject, the personal subject who is none other than the third party that has entered into the fray and has become a part of the action. Like the impersonal subject, the personal subject as singularity pure is always gendered masculine. The outcome of the intervention of the third party, *any* third party, is that there are now four different entities making up the whole. The four entities will correspond to the binary gender types MF, FF, FM, and MM as shown in Figure 57.

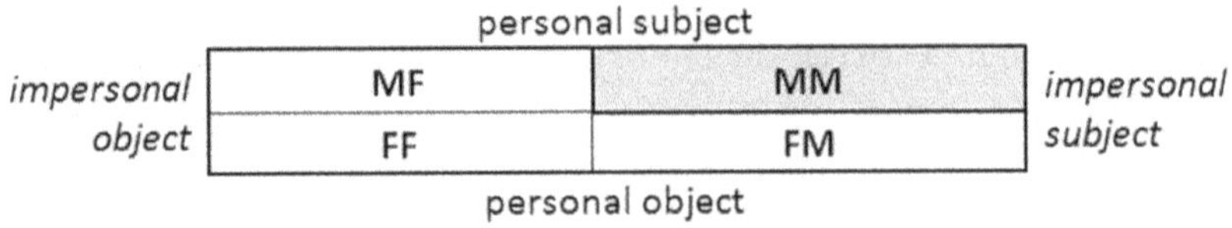

	personal subject		
impersonal object	MF	MM	impersonal subject
	FF	FM	
	personal object		

Figure 57 The four binary gender types determined by the impersonal and personal subjects

The moral of this little story is that while everything might seem very fuzzy for the outsider, once an integral part of the whole, all entities fall into place. There is coherence but this can only be seen relative to the system itself. All of this systemic coherence becomes possible thanks to the technology of the gender typing mechanism.

Just as FC demands that each of the pure genders correspond to entities in their own right, the same must apply to the four binary types. The ancients interpreted these entities as the four generic substances making up matter. To interpret this absolutely relativistic typing system in such a literal way can rapidly lead to many misunderstandings. Patience is required here, as the apparatus so far still has no mechanism for individualization of entities.

In chapter 9, we provocatively added a modern taste of things to come by associating the four binary typings with the four letters A, U, G, and C respectively of the genetic code. We interpret the code as the generic code, not for just coding the biological but coding any existent whatsoever.

Some Familiar Principles Revisited

To our knowledge, the gender construct, as presented here, will not be found in any modern scientific or philosophical literature. The construct provides new ways to interpret familiar principles known to physics. We have already looked at the concept of superposition, from a gender perspective. In chapter 9, we looked at the fundamental opposition between the two genders as a way of formalising Socrates confession of ignorance. The construct is analogous to Heisenberg's Uncertainty Principle and is stated as follows:

The Socratic Uncertainty Principle

Socrates is attributed as saying that all he knows with absolute certainty is that he knows nothing with absolute certainty. The gender construct posits the feminine entity as the bearer of the attribute of absolute ignorance. In order not to violate FC, this attribute must itself be considered as an entity in its own right. As such, it can be thought of as ultimate singular statement of certitude. In Heisenberg's Uncertainty Principle, one can know with high certainty the momentum of a particle. In this case, the position of the particle will be equally uncertain, or *vice versa*. This reciprocal nature of certainty and uncertainty typing is expressed in its most fundamental and generic form in the generic construct of gender. There, it becomes Socratic Uncertainty Principle, the elementary building block of all of right side science.

Another similar principle in Quantum Mechanics is the Pauli Exclusion Principle. Here we present the gender form of this type of construct.

The Gender Exclusion Principle

The four fundamental binary typed entities can be interpreted in a more technical way. Instead of interpreting the subject and object as either impersonal or personal, they can be interpreted as placeholder and value respectively. The impersonal subject becomes the subject as a placeholder of value. The personal subject becomes the subject as value. Such a technical interpretation can demystify the construct and also provide traction for applying the concept.

Of course, from the generic perspective there are two kinds of placeholder, feminine and masculine. The feminine placeholder is totally catholic and places no specificity requirements on content. Whilst the feminine is always an expression of the pure ignorance side of the Socratic confession, the masculine demands absolute certainty without in any way polluting the generic with any *ad hoc* specificity. How to be specific without being specific: that is the question. This can only be achieved by opposing the "anything" mantra of the feminine with the "something" mantra. The feminine can contain anything. The masculine must contain something. This can be formalised by defining the masculine as the Exclusion Principle, something analogous to Pauli's Exclusion Principle in Quantum Mechanics. The Exclusion Principle states that the placeholder can only contain one single entity in any instant; such is the nature of the single tasking masculine placeholder. The feminine can contain any number of entities in any instant: it multi-tasks so to speak. We start to see here that the masculine and feminine in this context, or *as* this context, are primordial expressions of the One and the Many. The masculine, as principle, can be thought of as the Exclusion Principle while the feminine principle can be thought of as the Inclusion Principle. It is the interplay of these two principles that dictate the basic structure of any reality that satisfies the principle of FC.

From the above it can be seen that an entity can be typed in two ways, either as a value or by the placeholder of value. In both cases, the type is either M or F. This double articulation of gender naturally leads to four possibilities. Each of these possibilities can be described symbolically by doublets of letters from the two-letter gender alphabet where the first element of the doublet comes from the articulation based on placeholder and the second element corresponds to the articulation based on value. Thus, the mixed gender construct FM announces a feminine placeholder containing a masculine value and hence subject to the Inclusion Principle. In other words, it may not be alone there. The MF construct is an articulation of a masculine placeholder subject to the Exclusion Principle containing a feminine value. We are thus lead from two M and F placeholders and two M and F typed values to four different mixed type entities that can be known as MF, FF, FM, and MM. We will refer to them as the four bases. The four bases are the elementary building blocks of the science. The four bases have been formed by entity synthesising with its own context. For simplicity and to give some indication of where this rather dry argument is going, we will denote each of the bases MF, FF, FM and MM by the single letters A, U, G and C respectively. The aim is to develop a generic language in terms of these letters that can code the generic structure of anything whatsoever, as long as it exists or subsists in a world which satisfies FC.

subject as value

object as placeholder	MF	MM	*subject as placeholder*
	FF	FM	

object as value

Figure 58 Generic entities can be placeholders and/or values.

Representation of Generic Structure

Formalising Choice

The process of formalising knowledge in the left side science paradigm is a relatively straightforward affair. The basic technology is already in place: It is called mathematics. Mathematics provides the tool for formalising the traditional left side science knowledge. When it comes to formalising right side science, one immediately comes up against a brick wall. None of the mathematics works. The obstacle is the draconian constraint of FC. FC must not be violated. The problem is that traditional axiomatic mathematics violates FC right down to its very core.

However, all is not lost. Because axiomatic mathematics is a formal system, it can be exploited to formalise the obstacle to formalising an FC compliant system. Mathematic formalises the way the problem must *not* be tackled. Axiomatic mathematics formalises the wrong way to go, that is to say, the wrong way to tackle the Kantian problem. Having a formal statement of the obstacle to progress, all we have to do is to find the way around the obstacle. If we cannot do it with mathematics as it stands, we will need something else.

Looking down at the very foundation of mathematics, we come to Set Theory, the elementary mathematic of collections. Without a formal notion of collections of things, there can be no formal mathematics. There are many axiomatic systems that claim to formalise Set Theory. Each system has a different set of axioms, but all systems contain one pivotal axiom, the *Axiom of Choice*. Faced with a Set of elements, which may even be infinitely denumerable, how can you distinguish one element from the other? How do you choose? The Axiom of Choice imposes sufficient structure on the system to solve the problem. Equivalent to the Axiom of Choice is Zorn's Lemma, which is easier to understand. The lemma effectively states that the elements of any set can be uniquely labelled with real numbers. Thus, using real numbers as labels, there always exists a unique labelling of elements such that one element can be distinguished from the other.

The very reliance on an axiom, any axiom, violates FC, as no such *a priori* constructs are permissible in a First Class system. What is of interest with the Axiom of Choice is that it situates the way that mathematics resolves the distinguishing problem. Firstly, it has to resort to a construct at the axiom level. Secondly, it is equivalent to using an *ad hoc* labelling technology, a characteristic of all left side sciences. The Axiom of Choice, and its fundamental Zorn lemma, thus articulates quite clearly the way not to proceed: don't use labels.

Structure

Structure is in the mind of the beholder. For the left side sciences the beholder is the impersonal subject providing the much sought after 'mind independent' point of view. This primary opposition between the impersonal subject and its object is ignored by left side science and replaced with an opposition of its own making, that of the rigid dichotomy between abstract theory and its object. In left side mathematics, the primary dichotomy becomes that between a set of axioms and a world of deductively explorable mathematical objects so predetermined, either explicitly or implicitly.

For right side science, the mind of the beholder is of primordial importance and is always present. Not only is the impersonal subject present, but also the personal. There are many ways of interpreting these two kinds of subject. As mentioned previously, the subject as placeholder and the subject as value is one possibility. A more mathematical flavour might be to call them the "covariant" and "contravariant" subjects, but one must be on guard not to slip into abstraction ways of thought. Both these two kinds of subject are simultaneously present in any whole considered by right side science. The science of wholes is the speciality of the right side of the epistemological brain. What matters is the generic subject formed by a highly primitive, primordial, Clifford-Grassmann style "geometric product" of these two subjects (together with their respective worlds). The end result is a the generic subject in the form of a "quaternion" kind of Three-plus-One structure, a semiotic square which can be more formally understood in terms of the ontological gender typing construct. In the right side science paradigm, this artifice occupies centre stage at all times. One could even say that it is centre stage.

One way of understanding the generic subject is to realise that it suffers from an incurable disease. The disease is called monism. Patients suffering from monism exhibit the pathological symptoms of being totally incapable of distinguishing the difference between the real world and their conception of it. Both appear to be the one and the same thing. Curiously, when not on hallucinogenic drugs or suffering from a deep schizophrenic episode, most human subjects also seem to exhibit these symptoms.

Right side science not only must articulate the basic architecture of the generic subject but also of the generic objects. There are four elementary types applicable to the generic object; four *bases* distinguished one from the other by binary gender typing. The typing of bases is determined relative to each other and ultimately compatible with the polarity conventions established by the subject, the ultimate arbitrator of type. These four bases can be represented by four binary gendered typed arrows. The problem now is to establish how these arrows can be combined to form elementary structures, without violating FC.

From a left side science perspective, if a right side science were at all possible it would present as some kind of meta science, metaphysics, or meta mathematics equipped with its own metalanguage. Such a science is not possible under the ambit of a left side paradigm inevitably plagued by its atomistic and dualistic worldview. However, even though fundamentally incompatible with FC, some accommodations can be made to achieve a kind of Partial First Classness (PFC). The resulting science will not be a true metaphysics but at least pass as a poor man's cousin.

The Sad Case of Mereology

One such accommodation is the rather obscure quasi-mathematical discipline called mereology, a left side attempt at a science of wholes and parts. Mereology is an exercise in mathematical logic. It achieves PFC by removing the rigid set theoretic dichotomy between sets and the elements that they contain. This is achieved by ignoring any explicit reference to the elements of a set and only considering containment relations between sets. The sets of Mereology do not contain elements they contain other sets. This is an example of FC as everything is a set and so the rigid duality between set and element is avoided. Contained sets are parts of the containing set. Different axiomatic schemes are set up to formalise this kind of structure where wholes contain parts and PFC is achieved by both parts and wholes being sets.

Mereology is of interest because it is essentially an attempt to formalise *has-a* relations between entities. Such structure finds echoes in the class inheritance structure of Object Oriented computer programming systems, for example. There are also echoes with our initial development of right side science where the *has-a* relation is paramount. However, right side science grants equal prominence to *is-a* relations. In fact, the basic building block involved the gender construct where the feminine ontological gender corresponds to the *has-a* relation and the corresponding masculine gender to the *is-a* relation. The core of right side science, with its ontological vocation, consists of the dialectic of the *has-a* and *is-a* relationship. In mereology the *has-a* relation is axiomatised in terms of some kind of partially ordered structure such as set inclusion. As for any ontological *is-a* structure, *is-a* relations are hard wired into the mereology axioms and hence killed stone dead with the first blow. Being a left side science mereology does not entertain any kind of *is-a* versus *has-a* dialectic. This means that the prospects of mereology enjoying a life of any significance are not very rosy: it can't even get to *be*.

A. N. Whitehead, in his philosophical quest for a holistic rationalist science, extended mereology concepts to geometry and achieved a geometric PFC (Whitehead, 1919). In this case, the rigid dichotomy between geometric objects with extension and geometric objects with no extension (points) was avoided to produce a *pointless geometry*. A pointless geometry is a right side kind of geometry.

Sidenote:

The Conformal Geometric Algebra (CFA) developed by Hestenes from the Clifford algebras, is an example of the pointless geometry for which Whitehead was probably searching. This kind of geoemetry is an expression of FC, as discussed elsewhere in this book.

However, the geometry was caste in a left side, abstract, dualistic, atomist framework. In the final count, the system inevitably violates FC on practically every other front. Nevertheless, mereology is worth mentioning here as it expresses many of the aspirations of right side science even though it fundamentally lacks the necessary equipment to deliver the goods. In passing, we note that the mereology-based paper "Steps Toward a Constructive Nominalism" (Goodman, et al., 1947) as worthy of consideration. In espousing constructionism and nominalism, the paper articulates important hallmarks of right side science. In addition, the authors start the paper with the doctrinal declaration:

We do not believe in abstract entities. No one supposes that abstract entities—classes, relations, properties, etc.— exist in spacetime; but we mean more than this. We renounce them altogether.

This rejection of abstraction is yet another fundamental tenet of right side science. However, declared within the confines of left side abstract axiomatic technology, this anti-abstract belief becomes a bit of an oxymoron. It is like the Christmas turkey that struts into the kitchen and valiantly declares to the Chef that it does not believe that turkeys are food.

From our perspective, mereology is interesting more for its aspirations than any achievements. As a discipline, mereology is so weak and vacuous that mathematicians do not even consider it to be a part of mathematics. It is a non-scientific oddity. Languishing in the domain where the left side disciplines hold sway, there is little hope for mereology. Mathematics has no need for this toothless terror that wants to formalise its structures without having any structure itself worth talking about. Axiomatic mathematics is quite capable of accomplishing the formalisation task without calling for assistance from a lacklustre bystander. The very essence of mathematics can be formalised from within mathematics itself. Formalisation demands an abstract theory of abstract mathematics, and that naturally leads to Category Theory, the meta-mathematics of mathematics.

It is with Category Theory that we can find a formal specification of the kind of structure that is anathema to our right side science. In order to advance, we do not look for a weak-kneed assistant. We look for a powerful enemy, an epistemological opposite, an epistemological foe that we can respect. We will use Category Theory as a formal negative indication of what we are up against in trying to resolve the Kantian problem.

Category Theory Structure Violates FC

Category Theory provides abstract representations of mathematical struc-ture in terms of a collection of objects and a collection of arrows or morphisms between the objects. The specificity of mathematical structure is represented by the arrows and in no way by any explicit internal structure of the objects. The approach is thus structuralist in nature. Representation of the most elementary mathematical structure starts with placing two arrows end to end. This represents the *composition* of two arrows. Composition of arrows must satisfy two axioms, identity, and associativity relying on the structures illustrated in Figure 59. Both of these structures violate FC.

Figure 59 (b) represents the composition of two arrows f and g to deter-mine a third arrow h thus satisfying associativity. The configuration violates FC because the arrow g is in an absolute ordering relationship with the arrow f. In a system satisfying FC, no entity can be absolutely before or after any other. Thus, even the two arrows shown in Figure 59 (c) violate FC. Thus, not only is associativity prohibited but also any kind of composition is off limits. We could call this disallowing of any absolute ordering relationships, the

Parmenidean condition. For FC, the only thing *that is* must be immediate, not anterior, nor posterior.

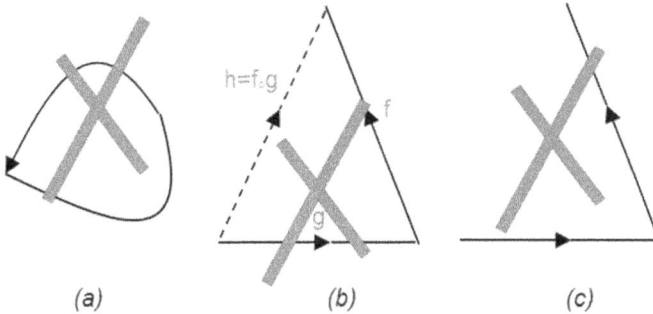

$$h = f \circ g$$

(a)   (b)   (c)

Figure 59 Even the most elementary structure necessary for a mathematical category violates FC.

A mathematical category requires the notion of composition identity to be defined for each object. This requires arrows that close back on themselves to form a loop as shown in Figure 59(a). This structure also violates FC as it infers than the same entity can be different to itself. We will call it the Heraclitus principle expressed by the saying that "You can't put your foot into the same river twice." It is a special case of the Parmenidean *condition*. This prohibition is a subtle one but suffice to say that it can be represented by a prohibition on circular arrows.

Without getting into messy details, it suffices to say that the formal axiomatic mathematical Category definition abstractly states the minimal structural characteristics that a system must possess in order to qualify as mathematics. What interests us is not mathematics, but its opposite, *anti-mathematics*. We informally define anti-mathematics, as being everything that mathematics is not. At the abstract pinnacle of mathematics, we find Category Theory. The anti-mathematical counterpart will be the Anti-Category. The only thing in common between the Category and the Anti-Category will be that they both exploit an arrow theoretic methodology.in one way or another, but definitely not the same way.

The Anti-Category and the Kantian Conditions

The conditions on the Anti-Category can be summed up as:

1. Unlike the Category, the Anti-Category cannot be abstract. This can be achieved by prohibiting dualistic structures, the essence of abstraction.

2. Unlike the Category, the Anti-Category cannot tolerate a duality between a collection of objects and a collection of arrows. For the Anti-Category not to violate FC, the mantra is that all entities are arrows. In this way, any entity will possess extent. From an ontological

point of view, we reiterate the Stoic mantra that only bodies exists. Point-like entities do not exist.

3. Unlike the Category, there can be no identity, no associativity, and not even composition of arrows.

4. There can be no axioms as any such predetermining structure violates FC.

We will call these conditions, the *Kantian conditions* for determining a formal structure that is totally devoid of any predetermining considerations. Realise an apparatus that satisfies the Kantian conditions and one has resolved the Kantian problem. In other words, one would have provided a formal basis for right side science, the monistic counterpart of the dualistic left side sciences. Not easy, but it can be done.

The axiomatic formalisation of mathematical categories is quite precise. Taking these conditions in the negative provides draconian requirements on the right side counterpart to the Category, the Anti-Category. Briefly, arrows determining anti-categories cannot form loops or be concatenated end to end. This leaves plenty of slack for finding a solution to the riddle. At least the Kantian problem is starting to look tractable.

Arrow Theoretic Methodology

Category Theory is based on an arrow theoretic methodology. It expresses its fundamentals in terms of arrow diagrams. Our task is to develop the right side counterpart of the Category in terms of the Anti-Category. If we can achieve this objective then we will have made a breakthrough is resolving the Kantian problem, the fundamental thrust of this book. Thus, we claim that, in addition to the known left side arrow theoretic methodology of Category Theory, there must be a complementary right side arrow theoretic methodology. Our task is to bring this right side version of arrow theoretic methodology into the light of day. In the process, we will see that the traditional left side version specialises uniquely in the syntactical aspects of structure and is virtually devoid of fundamental semantic considerations. On the right side of the equation, we will demonstrate that right side arrow theoretic structures are virtually syntax free, concentrating uniquely on semantics. One could say that traditional left side abstract approach to semantics leads to a syntax only account: Abstract semantics distils down to syntactical expressions. On the other hand, the right side paradigm approach to semantics leads to generic, non-abstract semantics. This kind of semantics is ultimately expressed in the gender calculus in the form of a syntax free generic code, a code capable of coding any semantics whatsoever that is compatible with FC.

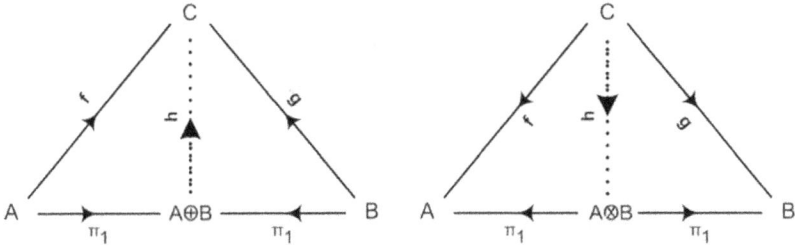

Figure 60 The arrow diagrams for the categorical sum and the categorical product of two objects A and B in a category.

That is enough of the hyperbole; let us look now at the left side version of the arrow theoretic method. We start with a practical example from Category Theory. Figure 60 shows the categorical representation of the categorical sum and the categorical product of objects in a Category. We won't attempt to understand the diagrams for the moment. We simply note that the diagrams look quite neat and clean, quite beautiful in fact.

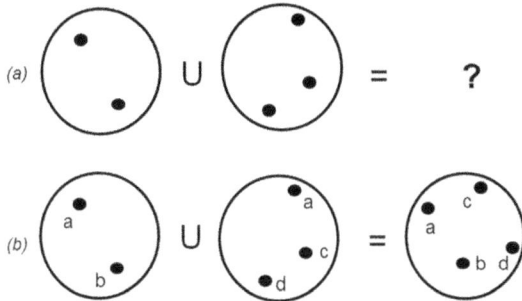

Figure 61 (a) Set theory sum of two sets without labels. (b) With labels.

The notion of the product and sum is fundamental to mathematics and crops up in many guises. Already, the Categorical diagrams reveal something that can be difficult to describe otherwise, In general, product operations are a kind of dual of sum operations. The duality can be seen simply by reversing the direction of the arrows for the sum diagram to produce the dual diagram. The dual of the sum is the product corresponds to the categorical product.

In order to get an understanding of the arrow theoretic methodology involved we will assume that the objects of the category are sets and that the sum of two sets will be set theoretic union and the product will be set intersection. Now consider the set theory problem illustrated in Figure 61(a). The problem is to work out the set theoretic union of two sets without using labels. The answer is that it cannot be done. However, in Figure 61(b) we have invoked the Axiom of Choice, which implies that all the elements of any set can be labelled. The elements of each of the two sets have thus been dutifully labelled and the corresponding result of set theoretic union is shown on the right, problem solved. Of course, if we had chosen a different labelling, we would get a different result. The conclusion here is that the ability to label is very important in Set Theory. That is why Set Theory cannot do without the Axiom of Choice and Zorn's Lemma.

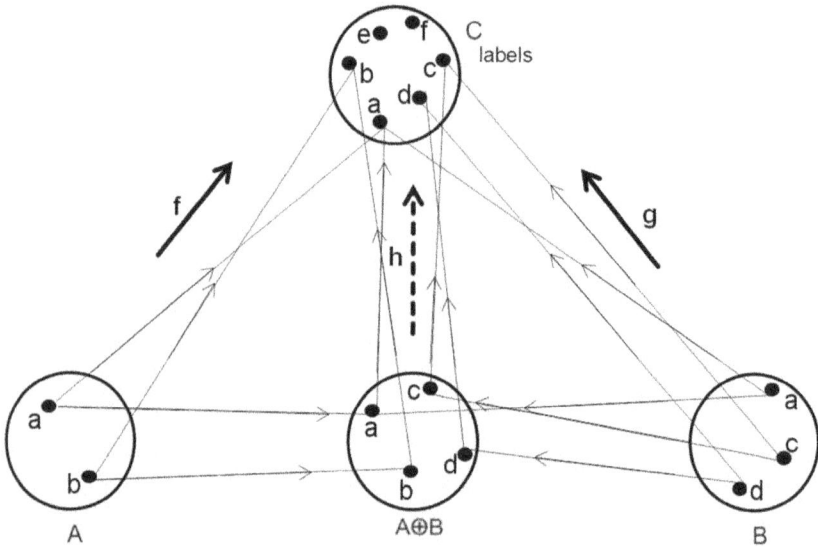

Figure 62 The Category Theory diagram can be thought of as an abstraction of a labelling system. This figure illustrates the implicit labelling underlying the categorical sum operation. In Category Theory all this detail is superfluous and no explicit labelling is necessary.

Let us now start moving towards a categorical approach to the problem and see how Category Theory can get by without any explicit labelling technology whatsoever. The labelling used in Figure 61(b) was quite *ad hoc*. We now interpret this *ad hoc* labelling as a very nuts and bolts model of the Category Theory representation of categorical sum. Using this more elaborate labelling technology, the *ad hoc* labelling is replaced by a system of labelling as shown in Figure 62. In this diagram, the set C consists of a large enough set of labels to do the job. The sets A, B and their sum are each labelled by bunches of connections between each of the sets and the labels in C. The three bunches of lightly drawn arrows are instances of the morphism arrows *f, g*, and *h* of the highly abstract categorical representation of sum shown in Figure 62. The Category Theory represents all the possible indexing instances possible for determining the abstract mathematical notion of sum and does so at the most abstract level.

We will now look at the specificities of this kind of this left side arrow theoretic methodology. Our discussion is motivated by the quest for a sibling right side arrow theoretic methodology. Our aim is to resolve the Kantian problem and so the discussion has a philosophical and even a psychological tone. One of our objectives is to cure mathematics of the hemi-neglect syndrome mentioned elsewhere in this book. Against blatant neglect, we must show that there is another way, another side to it all.

Traditional Arrow Theoretic Methodology

Out immediate task is to characterise the left side arrow theoretic methodology of Category Theory. We will look at how Category Theory respects FC, how and where it violates FC, and its overall architectural characteristics.

The first interesting realisation is that Category Theory, for this example at least, articulates an example of FC, notably that no entity be anterior to any other. What this means is that in this particular Category Theory diagram, no arrow can be said to be before or after any other arrow. There can be no concatenation of arrows. In other words, the Category Theory arrow diagram starts off without any appeal to the composition of arrows. However, there are two caveats. Firstly, we must add the caveat that this FC only applies to the non-dotted arrows. The dotted arrow is added in order to 'make the diagram commute.' The orientation and location of the dotted arrow is such that it always violates FC. Violation of FC is necessary because of the fundamental structural mechanism of Category Theory, the associativity and the composition of arrows.

There is a second caveat that we add in small print. This is to do with identity arrows, which clearly violate FC. In practice, this is the case anyway as identity arrows are usually only explicitly incorporated in the diagram when pertinent and that is usually after adding the dotted arrow, not before. We will sweep this aspect under the carpet by saying that, except for the identity arrows, the rest of the arrows in an arrows diagram, excluding the dotted arrow, all respect FC.

At this point, we will introduce some terminology for describing the two kinds of arrow in an arrow diagram. Mathematicians don't use this terminology but probably would not object. Looking at any Category Theory diagram, such as those shown in Figure 62, all of the arrows will be shown as full arrows except for one shown dotted. We will henceforth refer to the dotted arrow as being *real* and all the other arrows as being *imaginary*. Real and imaginary are used here in a mathematical sense, not in a "real world" sense. And so, every arrow diagram has an imaginary part and a real part. The real part consists of one single arrow, the specificity of which is represented by the geometric configuration of the arrows making up the imaginary part of the diagram. The imaginary arrows, relative to each other, respect FC. They all determine the specificity of the real arrow. We will be using this concept for our right side version of Category Theory, our Anti-Category Theory. Intuitively, the real part of the theory is devoid of determined attribute. The real part is the "answer" to the riddle, the part you don't know directly. The real part is what you don't have. It is non-explicit, the unmanifested, and hence is illustrated with a dotted arrow. The imaginary part of the theory is the explicit part, the part that is known, the part manifested, the part illustrated with full arrows in the diagram.

This arrow theoretic approach with a multiplicity of explicit "imaginary" arrows and one non-explicit "real arrow" is about the nearest that the left side paradigm can get to a monism. The real part corresponds to the non-explicit One and the imaginary part to the many aspects determining the One.

The Category Theorist does not bother to solve the problem and make the dotted arrow explicit. He is merely interested as to whether *there exists* an arrow to replace the dotted arrow that fits into the diagram. If an arrow exists, the diagram is said to commute. Category Theory is all about finding arrow diagrams that commute.

The arrow diagram of Category Theory is a technology for representing the specificity of a real entity, the entity represented by the dotted arrow added to the diagram in order to make it commute. All of the specificity is encoded in the topological configuration of the imaginary arrows relative to each other and to the real. There is no need for labels. In practice, mathematicians label the various arrows making up the imaginary part and real parts of the diagram, but this is only to make it easier to talk about the structure. The labels themselves impart no additional structural specificity to the diagram.

Workers in Computer Science sometimes employ Category Theory. However, they muddy the waters by considering the imaginary arrows as signifying real things such as datatypes. Thus, they treat the imaginary arrows as types and so give arrow labels meaning, a datatype meaning. Even worse, they use the same label for different arrows in the same diagram, declaring that this construct represents a "polymorphic" datatype. The Category Theory of the Computer Scientists is not pure. The approach manually and arbitrarily reintroduces typing or labelling into what was a fundamental self-typed system. The approach is not wrong. It merely means that the approach properly belongs to a Computational Category Theory, not to Category Theory pure. We are concerned only with the latter. Curiously, though, the polymorphic types that the Computer Scientists want to graft onto Category Theory start to take on a different allure when we go over the right side alternative to classical Category Theory. However, the right side types don't arise from the everyday concerns of Computer Science, but from the very demands of FC itself. In other words, the polymorphic types must be based on ontological gender.

The Right Side Alternative

One of the main activities of Category Theorists has been described as "diagram chasing." Ignoring the identity arrows, each diagram has an imaginary part made up of arrows arranged in a configuration that does not violate FC. This arrow configuration is different for each abstract concept so represented. The arrows themselves are ubiquitous, having no specificity except for their relative positions in the overall diagram. All of the arrows are 'meaningless' except for their relative position in the structure. The real part of the diagram consists of a single arrow that "makes the diagram commute' and in so doing violates FC. The violation introduces a concatenation of arrows. The imaginary part of the diagram is a structure devoid of order. Its role is to determine ordering and to do so in accordance with the minimalist demands of associativity. Syntactical structure is the end result. This kind of structure is the essence and the speciality of left side sciences in general.

The paradigm of left side mathematics finds its loftiest expression in the form of the mathematical Category of Category Theory. We now turn to revealing its right side sibling, the Anti-Category. In the process, one should be very conscious of the fact that a very distinct dividing line is being crossed. We are crossing over to what many philosophers see as metaphysics, up until now a science free zone.

Amongst the many detractors of this kind of science was Charles Sanders Peirce, declaring that

> *...almost every proposition of ontological metaphysics is either meaningless gibberish -- one word being defined by other words, and they by still others, without any real conception ever being reached -- or else is downright absurd. (Perice, 1905).*

Notwithstanding Peirce's negative assessment of this kind of endeavour, the relativistic notion of "one word defined by other words" goes along the lines of what we are advocating. It is more a strength than a weakness. To make the relativity tractable however, we will be using a relativistic arrangement of arrows determining other arrows instead of just words determining words. As for Peirce's absurdity accusation, we agree that this may be the appearance but appearances can be deceptive.

Peirce's other accusation is that ontological metaphysics never reaches any real conceptions. Once again, we will take this negative criticism as being a positive characteristic of this kind of science, The fundamental characteristic of the generic is the absolute lack of specificity. Rather than change our ways, our next step is to formalise this art of talking about things that forever lay outside of one's grasp. In this game, one must not fear what initially may apparently appear absurd.

Thus, we come back to the mathematical Category as the formal statement of general mathematical specificity. Our task is to formalise Anti-Categories, the ultimate formal expression of non-specificity, the ultimate expression of the generic. Our task is to make the transition from the abstract world of mathematical Categories on the left side of science to that of the generic Anti-Categories on the right. Both sides employ their own kind of arrow theoretic methodology. We know the left side version; we must now start to understand the right side version.

We note that the following salient points concerning the Category Theory version of an arrow diagram:

1. The arrows in the imaginary part of the diagram (drawn as full arrows) contains no concatenation of arrows. No arrow is before or after any other. In this respect, the structure does not violate FC.

2. The real part of the diagram (the dotted arrow) that makes the diagram commute, always concatenates with at least one other arrow in the diagram. In other words, the real part violates FC.

3. The diagram has no need for labels. Labels, if present, impart no additional formal structure to the diagram.

4. The configuration of arrows is different for each different category. The specificity of the mathematical concept being represented is entirely represented by the relative configuration of the arrows in the diagram.

We now come to the arrow diagram of an Anti-Category. The corresponding salient points are as follows:

1. The same condition applies to Anti-Categories as for Categories. The arrows making up the imaginary part of the diagram contain no concatenations No two arrows are placed with the head of one coincident with the tail of the other. FC prevails.

2. The real part of the diagram (the dotted part) does NOT violate FC and so the diagram, still contains no concatenation of arrows. The difference between the real part and the imaginary arrows is not determined by orientation but by typing, The real part has no specific type whilst the arrows of the other part are all explicitly typed.

3. Unlike the Category diagram, which is devoid of labels having any explicit semantic implications, the Anti-Category arrows are labelled by their types and thus the labelling is semantically meaningful. There are four types MF, FF, FM, and MM, corresponding to the four binary gender typings. Alternatively they can be labelled A, U, G and C respectively, as a shorthand measure. All arrows of the diagram are typed in this way except for the real part that obtains its typing from

the other arrows. Since different arrows can have the same binary gendered type, the Anti-Category can be said to be fundamentally polymorphic. This contrasts with the pure mathematical Category, which is fundamentally non-polymorphic. The real arrow can be called the *indeterminate dyad*, a very ancient term going back in time to The First Ennead (Plotinus, 250 AD), and even much further back to Speusippus, Plato, and finally back to the Pythagoreans. Specifics of the undetermined dyad cannot be known directly but can be known indirectly through the imaginary part of the diagram where the semantics are expressed in the highly relativistic gender typing.

4. Unlike the Category, the Anti-Category does not express specificity via different geometric arrangements of arrows. For Anti-Categories, they all have the same identical generic structure. Only the gender typing of arrows is different. This is the characteristic structural feature of right side science. Being the science of wholes, every whole has the same generic form, the form of a whole, a Three-plus-One structure. There is no departure from the rule. A whole is a whole or it is not. There are no half measures, no half wholes. Each structure is a variant on the same theme, over and over again, no matter which way you look at it. This is the *univocity* aspect of the science. The theme, as we shall shortly investigate in more detail, is that of the Three-plus-One Structure. Despite what so many have thought in modern times, rather than being the most vaporous of intellectual projects, the constraints on this system are becoming truly draconian. There is not much room for the *ad hoc* in this right side science.

The above two sets of matching points put the differences between the left side and right side science paradigms into stark relief. Carrying the bundle for left side sciences was their most abstract discipline, Category Theory. Category Theory is a mature mathematical discipline, the contours of which are reliably well known. On the other side of the divide, we have a science with a much more ancient pedigree, an incredibly ancient pedigree in fact, but which still languishes in almost total immaturity. Here we are dragging this right side science into modernity by confronting it, point by point, with the most abstract pinnacle of the sciences of the day. By qualitatively making this point-by-point comparison and displaying the dialectic of the oppositions at play, we start to get a firmer intuitive handle on this most slippery of sciences. In one take, we start to see the epistemological essence of the left side scientific paradigm, as formalised by mathematical Category Theory. The fundamental essence is that of syntactical structure. Left side science is a specialist in this domain. However, left side sciences have never made much inroads into the semantic side of knowledge. The left side paradigm has not been able to come up with a science of meaning. This is the speciality of right side science.

Having intuitively sketched out the epistemological profile of the right side science paradigm, relative to left side version, we now turn to a more formal understanding of the generic structures involved. We referred to these generic forms in point 4 above. Typed arrows making up the imaginary part of the diagram can represent these generic forms. We call this structure imaginary because it acts as the *de facto* attribute of the real part of the diagram. We know the real part in terms of the attribute part. We cannot know the real directly. The composite attribute must be constructed without recourse to any *a priori* considerations. Our *science without attributes* actually involves attributes by constructing its own. This involves the gender construct, which in turn leads to the possibility of typing arrows with binary gender. We must now look at generic structure composed of a composite of arrows. What is the elementary generic form underlying any generic structure?

A generic structure must be such that the components can be distinguished from each other. In other words, the structure must be self-labelling. In addition, all labelling is relative to the subject, which will be represented by a dotted arrow. This is the indeterminate dyad. The subject provides the polarity convention for the overall structure.

One structural solution might be to concatenate the arrows to form an ordered chain with the first arrow followed by the second and so on and attach the first arrow to one end of the indeterminate dyad. However, such ordering of arrows violates FC and so must be rejected. The solution that does not violate FC is not difficult and can be constructed as follows.

The first imaginary arrow starts from the shaft end of the indeterminate dyad. The problem is that the indeterminate dyad is so indeterminate that we do not know anything about it. Thus, we do the next best thing and start from anywhere whatsoever. Since this science is starting point invariant, we will get the same result wherever we start. In fact, we *must* get the same result as that is the overall system constraint – the theory must be starting point invariant. This is a self-justifying theory. Soldiering on, we start by constructing our first arrow dyad for the imaginary part of the structure, somewhere. By convention, we will call this first dyad the Red dyad. The Red dyad, by its presence, determines the shaft end of the indeterminate dyad as shown in Figure 63(a). Using a terminology fundamental to Peirce, we will say that the Green dyad expresses generic Firstness. The next dyad will be called the Green dyad, which will be said to express generic Secondness. There are two possible locations for the Green dyad as shown in Figure 63(b). The case where the Green dyad shares the same shaft end as the Green is useless as there is no way that one can be distinguished from the other. Also, there is the possibility that only the two Red and Green dyads are necessary to provide the imaginary structure necessary to determine the indeterminate dyad. However, this would violate FC as the indeterminate dyad and the Green dyad would be concatenated.

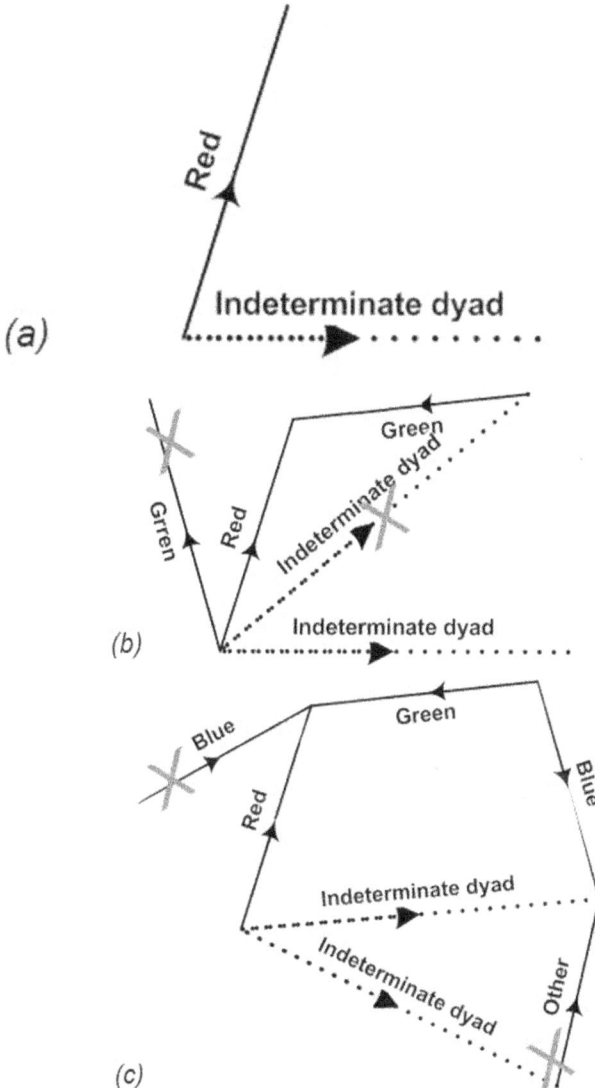

Figure 63 Constructing the RGB triad, the representation of generic Firstness, Secondness, and Thirdness.(b)

The next step will involve the Blue dyad, which expresses generic Thirdness. The only viable position in the diagram is shown in Figure 63(c). There is now the possibility of a fourth coloured dyad to express a generic "Fourthness." This is ruled out, as it would violate FC. In addition, it would add ambiguity to the diagram as this fourth dyad could not structurally be distinguished from the Red dyad. The end result is that there is no such thing as generic "Fourthness," as far as generic attributes go. The same applies to 'Fifthness." When it comes to "Sixthness," this can be assimilated by two of these RGB triads placed end to end. Given such an arrow structure, there is no ambiguity in figuring out how to colour the arrows. This applies to any multiple of RGB triads placed end to end. This is a self-labelling structure that, although free of explicit ordering of arrows end to end, still determines a generic first and last end point. This is essential for determining the underlying real structure of the indeterminate dyad.

The Generic Codon

We have already considered the binary gendered typing leading to four types of arrows. Gender typing is a totally relativistic typing mechanism, free of *a priori* determinations. A single such typed arrow represents a "subatomic" attribute. Three such subatomic attributes are necessary to define the composite attribute necessary to determine a whole, any whole. The elementary determination of any whole demands this Three-plus-One structure, three imaginary parts and one real. The three imaginary parts can be represented by an RGB triad, where each one of the dyads is binary gender typed. Each of these typed triads will be called a *generic codon.* Each codon is thus typed by a string of six genders, six letters taken out of the two-letter alphabet made up of M and F. This is a kind of "Y King" structure, one of the most ancient symbolic systems known to man. However, we will not dwell on that matter here. All in all, there are 64 such elementary codons. These elementary codons can be considered as the elementary building block for the *generic code*, the code that can generically determine anything provided that the demands of FC are respected.

The Generic Code in Nature

If you pour jelly into a square bowl, you end up with square jelly. If you express knowledge in terms of the left side dualist, atomist, reductionist, *a priorist* paradigm, then the kind of reality so articulated will have all of the imprints of the epistemological machinery employed in its production. The right side generic science paradigm removes the bowl and the *a priori* squareness that comes with it. The jelly must provide its own bowl. Rather than developing a dualist argument, the alternative must be based on maintaining a non-dualist monism, free of absolutist dualities. This is what FC is all about. Reality must be understood, viewed, and coded in terms of unified wholes. Each whole is a view of the totality from a determined viewpoint. There are as many wholes as there are viewpoints. These viewpoints and their corresponding wholes can be formalised generically. There are only a finite number of ways of skinning the cat, so to speak.

One of the central claims of this work is that the organisation of living systems is based on FC and as such, capable of being coded and described in terms of the elementary generic structure based on the 64 gender typed codons of a generic code. For biological systems, there is an obvious similarity between the 64 codons of the genetic code.

Science Without Number

This completes our initial adventure into science without number. We ended up with a science based triadic arrow structures called codons. Each triad was made up of three bases taken from a four-letter alphabet. The four letters of the alphabet are generic types that can type the three bases making up each codon. We have worked out this structure from first principles according to the demands of First Classness. The principle of FC is an *a priori* condition for generic systems. It is a self-negation condition because it demands that there be no such *a priori* condition. This is enough to make the head spin. Who cares, as long as it is enough to make the reality that we live in spin along every dy. According to our impeccable reasoning, this must be the case, at least if we live in a FC world.

In the end, we arrive at a science without number based on letters only - an algebra. In fact, there are not even four letters in the alphabet, but just two, the Masculine and the Feminine types. In the next appendix, we look at the generic geometry behind this science without number.

16

Appendix B

Generic Geometry and its Code

Introduction

A common paradigm underlies the modern sciences, including mathematics. These sciences totally dependent on what the philosopher Immanuel Kant called *a priori* knowledge; *a priori* experimental data for the sciences, *a priori* axioms in the case of modern day mathematics. We will refer to these kinds of sciences as *left side* sciences. Kant proposed a diametrically opposite kind of scientific paradigm leading knowledge that did not depend on any *a priori* determinations whatsoever. We will call this kind of scientific knowledge *right side* science. Unlike left side science, which employs a bottom up form of reasoning, right side science must rely on top down reasoning that is free of any *a priori* constructs. Based upon pure reason alone, right side science must be capable of constructing knowledge about the generic shape of the world and the generic forms that abound therein. Unlike the traditional left side sciences, right side science will not be an abstract science. Instead, the science must present a radical alternative to abstract thinking. Instead of trading in abstractions, right side science must trade in the *generic*. It will be a *generic science*. It must describe and proscribe a generic world.

Kant used the term *metaphysics* for what we are calling right side science. He bemoaned the total lack of progress in this domain, a lack of progress that persists to this day. The purpose of this book is to tackle this ancient problem, the metaphysics problem that dates right back to antiquity. We will present the foundations of right side science in a form anticipated before Kant even posed his famous question.

Before Kant, Leibniz proposed an embryonic solution to this kind of problem. Leibniz believed that the world was the most perfect possible. Such a world must be profoundly simple whilst providing a correspondingly maximal richness in phenomena. He proposed a science that would conquer this profound simplicity and equally profound profundity:

If it were completed in the way in which I think of it, one could carry out the description of a machine, no matter how complicated, in characters which would be merely the letters of the alphabet, and so provide the mind with a method of knowing the machine and all its parts, their motion and use, distinctly and easily without the use of any figures or models and without the need of imagination. Yet the figure would inevitably be present to the mind whenever one wishes to interpret the characters. One could also give exact descriptions of natural things by means of it, such, for example, as the structure of plants and animals. (Leibniz)

A Generic and Geometric Algebra

In 1846, to celebrate the bicentenary of Leibniz's birthday, a mathematical competition was organised in Leipzig to solve the problem posed by Leibniz. The problem posed was to develop an algebraic geometry without number. There were no entrants. The terms of the competition were subsequently relaxed to that of a geometry without coordinates. There was only one entry and one winner. Encouraged by Möbius, it was Hermann Grassmann who presented the first version of a coordinate free geometric algebra. Then came Clifford who combined Grassmann's work with Hamilton's quaternions to produce the Clifford Algebras. In modern times, David Hestenes restored the original geometric flavour of Grassmann to the Clifford Algebra approach to promote what is now generally known as Geometric Algebra (GA). Hestenes has applied GA to many different domains ranging from classical dynamics to general relativity and Quantum Mechanics. Hestenes champions GA as the universal language of all physics and mathematics. At a 2009 symposium, celebrating Grassmann's bicentenary, Hestenes presented a paper summarising the development of GA from its historical roots (Hestenes, 2009). This is where we are today, in the shadows of a double bicentenary of a very fundamental idea.

In this chapter, we endorse Hestenes' claim that GA has the ingredients of a universal language for all of physics and mathematics. However, we contend that this is only a possibility, as there is a missing ingredient, as we shall see. As well as championing Hestenes' claim, we intend to go even further. We intend not only to build on Grassmann's visionary perspective but also to go back to that of Leibniz himself. The Leibniz vision went much further than physics and mathematics. For Leibniz, this proposed algebra, this universal geometric language without number, was destined to provide the calculus capable of *describing* even "the structure of plants and animals." We now know that there does appear to be an algebra that codes the structure of all plants and animals. It is called the genetic code. We also know that this code does not change over time, appearing to be impervious to evolutionary change. The genetic code for an organism flourishing a billion years ago was based on the same algebra as that of any organism today.

The biochemists see the generic code as a mere transcription language, coding the building blocks of living substance. To biochemistry, the genetic code has only a single articulation, articulating the observable morphemes of the flesh and bones of the organism. The notion that such a code could have a second articulation that might algebraically code the spatio-temporal semantics of the organism is foreign to biochemistry. However, such was the vision of Leibniz. In his vision, the same code would provide a means for knowing not only the animate machines of nature but also any aspect of nature including the inanimate, right across the board. Leibniz's claim is certainly spectacular, one universe, one Code. In this chapter, we subscribe to the Leibniz vision of a single universal geometric algebra "without numbers." We see GA as a promising start. However, in order to bring the original vision of Leibnitz into the realm of science we must determine the missing ingredient in Hestenes' version of GA. It is only then that we can arrive at the truly universal, generic, geometric algebra of the Cosmos. That is the objective of this chapter.

Because the chapter is more technical than the main body of the book, it appears as an appendix.

The First Principle

The paradigm of right side science forbids any appeal to *a priori* determinations of any kind. One might be tempted to think that such a constraint makes the construction of any science impossible. If there is nothing tangible to start with, how can it be possible even to start? In fact, the opposite is the case. Forbidding the *a priori* is the most draconian constraint that can possibly be placed on any system, the constraint of being totally unconstrained. This means that the system must be free of absolutes. No systemic entity can claim absolute precedence over any other be it spatial, temporal, logical, or in any other modality. What this means is that no systemic relationship can be absolutely determined. Systemic relationships must be relatively determined, all bound up in a mutually self-determining manner. Employing terminology from Computer Science, we will say that any such system will be based on the principle of First Classness (FC). In a system satisfying FC, all entities are First Class and hence free from any Second Class entities suffering from absolute determinations. FC is a kind of democratic principle, but much more subtle. Instead of advocating equal rights, FC places more emphasis on responsibilities. For example, the Democratic Principle would imply that a photon has the right to travel in any direction in space. However, this is only a vacuous possibility. In practice, the photon will travel in the direction that it is obliged to travel. After all, the photon has the responsibility to spring into existence whenever the spectre of a violation of FC is raised. To *be* is never a given.

The First Principle of right side science cannot precede the science because this would make it an *a priori* determination. The *a priori* determination is the stuff of the left side paradigm, not the right. Rather than coming before the science, the First Principle must be synchronically *with* the science. In this way, right side science distinguishes itself from the traditional left side. The traditional sciences all rely on flowing deductive forms of reasoning. This can be summed up by saying that such sciences have a *diachronic* epistemological structure. In contrast, right side science forbids any entities explicitly preceding or preceding other entities as this would violate FC. The *a priori* and the *a posteriori* are both forbidden. Thus, such a science must have the opposite epistemological structure to the diachronic. The epistemological structure of right side science must be *synchronic*. This is a very important observation as it enables us to start to get a tractable conceptual grasp on the nature of this mysterious "metaphysics" that has haunted thinkers for over two millennia.

In order to illustrate the difference between the diachronic and the synchronic, consider the double-barrelled kind of analysis employed in applied physics and engineering. Technically trained students are first exposed to the diachronic take on reality in senior secondary education where they learn how to solve differential equations, and they learn the hard way. Later, they will learn that this is called "working in the time domain," the diachronic domain. Very early into tertiary education the student then encounters a much easier way of solving differential equations using the operational calculus first developed by Heaviside. Solving differential equations can be simply transformed into algebraic problems that are easy to solve. The student learns how to analyse systems by "working in the frequency domain," the synchronic domain. By the time of graduation, the student will be equally fluent in analysing systems either diachronically in the "time domain" or synchronically in the "frequency domain" and think nothing further of it. Few students would think that "working in the frequency domain" was a metaphysical experience, but we may be wrong on that count.

In the bigger epistemological picture, we find the whole gamut of traditional sciences and mathematics making up the epistemologically diachronic world of left side science. In this case, the diachronic does not refer to just temporal flow, but rather to flow in knowledge, knowledge being deduced from a preceding form of knowledge. Even the operational calculus, despite its flirtation with the other side, requires *a priori* constructs and so is still stuck on the left side of the great epistemological divide. When we now look around for the synchronic partner of this epistemological Cosmos, we find, to our surprise, that it is missing. When it comes to the big picture, science in our day only has a single barrelled take on reality. The epistemology of all present day scientific knowledge is diachronically based. There is no scientific discipline satisfying the purely synchronic paradigm. Of course, the aim of this book is to establish such a discipline. Such a science is not some kind of ephemeral metaphysics, but hard-core science. In the light of the great diachronic-synchronic epistemological divide, we start to see that our task is to develop a mega version of an operational calculus, a totally generic operational calculus. If things go according to plan we should start to see highly technical and intricate left side problems being transformed into much more tractable and easily managed algebraic forms, much as Leibniz imagined it. We should end up with a generic algebra capable of coding anything whatsoever. For example, the generic algebra should code the elementary generic forms of the physical world. As Leibniz wrote, the algebra should be capable of coding "the structure of plants and animals." We now know that the structure of plants and animals are indeed all coded by the same code, a genetic code based on a few letters as Leibniz envisaged. The science should not only be applicable in the domain of the inanimate but also the animate. Is this generic code an instance of an even more extensive generic code extending to all aspects of the physical world? Can we work out these generic forms from pure reason alone?

In a nutshell then, our task is to develop a totally generic science and its algebra, which, in order not to violate FC, will be totally synchronic at heart. Any aspect of traditional left side diachronic knowledge will find its synchronic counterpart in right side science in a way analogous to the interplay of functional and operational methods in mathematical physics. Both takes on reality are necessary to arrive at mature, reliable knowledge. The fundamental principle underlying the new science is the non-violation of FC. This is our way of formalising the Leibnitz principle that the world must be the best of all possible worlds.

Our quest is for the operational algebraic calculus of the Cosmos. We should continually keep in mind that genetic code employed by all animate creatures of Nature to code themselves, has all the hallmarks of such a calculus.

Nature herself appears to be in a continuous state of diachronic flux but her code remains eternally the same revealing her synchronic side, the Parmenidean never changing side of Nature. It is only in the Parmenidean perspective of the eternal *now* that FC is not violated. Our objective is to exploit this principle. Our objective is to reverse engineer the operational calculus of Nature.

FC and the Stoics

It is worthwhile noting that Leibnitz probably took his cue from the ancient Stoics who also believed that the world was dominated by perfection. Here we find what appears as a great contradiction. This world dominated by perfection, as a direct consequence of that perfection, is not dominated by anything at all. No world can be perfect whilst dominated by something. The perfect world must be a free soul.

Quite obviously, there is a great Ruse at play here. This great Ruse can be discerned in the physics of antiquity as promoted by the Stoics. There we see some of the ramifications of adopting a science based on FC. One key aspect of Stoic physics is that they rejected the dualism implicit in atomism. An atomistic world is based on atoms floating around in a void and so implying an absolute dichotomy between particle and void. Such a dichotomy violates FC and so is unacceptable to this kind of epistemological paradigm. Instead, the Stoics adopted a non-dualist monist perspective. Voids do not exist. Only bodies exist and together form a unified organism. In this monistic perspective, the Cosmos is an instance of such an organism a human being is another. They all form part of a unified world.

In what follows, we will attempt to resurrect the key principles of the ancient science of the Stoics, as this is central to developing right side science. Before moving on, we note one simple dictum of Stoic physics that clearly illustrates the role of FC in Stoic thought. In classical physics, matter is made up of particles each particle having certain properties. Such a perspective violates FC as it establishes a dichotomy between particles and properties: particles are entities whilst properties are not. The Stoics eliminated this dichotomy by resolutely proclaiming that the property of an entity must be an entity in its own right.

Ontological Gender

We now come to the starting point, the nitty gritty of right side science. To avoid a long-winded discussion, we will skip over all of the fine points and go straight to the heart of the matter. The starting point for right side science is simply *any entity whatsoever*. We will call this entity the *generic entity*. The generic entity is the object of right side science. By defining the starting point in this way, we do not violate FC, as we have not privileged any entity at the expensive of any other.

Our objective now is to develop a science of the generic entity. This science must be the same, independent of which entity might have been chosen (or if to be chosen) as the generic entity. Thus, the science must be starting point invariant. This establishes the invariant of the science. No matter which entity you start with, you must always get the same science. Thus, the science is its own invariant. This is a rather tautological world.

The first step is to examine the generic entity from the point of view of characterising it by any distinctive attribute it may possess. Answering this question is simple as the generic entity is totally devoid of any determined attribute whatsoever. This leads to the conclusion that the only specificity possessed by this entity is that it is devoid of specificity. Because the entity has no specificity, this does not imply that it is devoid of attribute. It just means that it has the attribute of non-specificity. This is a bit like a transparent object not having colour. Transparency becomes a primordial property of such an entity. Of course, the entity devoid of specificity cannot be said to be transparent nor coloured. Nor can it be said that the object is not transparent or coloured. Over the ages, many thinkers have laboured over this kind of totally unqualified entity. Kant was one of them. However, we should not lose too much sleep over this problem as it all comes out in the wash.

Non-specificity articulates the nature of the attribute possessed by the generic entity. This entity is highly specific, distinguishing itself from any other by possessing a totally unqualified attribute. Thus, this entity *has* an attribute, albeit totally undetermined. At this point, there is a risk that this embryonic system might violate FC, as there is the danger of an absolute dichotomy forming between the entity and its attribute. FC demands that the attribute of the entity must be First Class: It must be an entity in its own right.

Thus, the first movement of the argument leads not to one entity but two. One entity *has* an attribute, the second entity *is* that attribute. The difference between these two primordial entities can be formalised using an ancient construct as a difference in *gender*. The entity, which has the attribute, will be said to be of feminine gender; the entity that *is* the attribute is said to be of masculine gender. There are two different entities and only one attribute between them: One *has* it and one is it. The two entities are different but obviously indistinguishable. This marks the beginning of the dialectic of *having* and *being,* the core reasoning process of right side science, the science of the generic.

Left side sciences harvest entity attributes from measurements, or in the case of abstract mathematics, attributes are conjured up by axiomatic proclamation. On the other side of the epistemological divide, right side science is totally devoid of *a priori* attributes. However, armed with the gender construct, the limitations of a total lack of traditional attributes can be overcome. The generic entity, starting point for the science, has one single attribute. Of this, we can be certain. This is an entity in its own right, an entity of masculine gender. The remarkable thing about right side science is that all other attributes will be constructed from this base. Right side science does not harvest attributes; it constructs them from this elementary couple, one feminine and one masculine.

Some Quantum Aspects of the Gender Construct

To our knowledge, the gender construct, as presented here, will not be found in any modern scientific or philosophical literature. The construct provides new ways to interpret familiar principles known to Quantum Mechanics.

The Uncertainty Principle

In Quantum Mechanics Heisenberg's Uncertainty Principle plays a central role. A similar principle can be discerned at the very core of right side science based on the gender construct. The principle dates back to Socrates and his famous confession of ignorance. Socrates is attributed as saying that all he knows with absolute certainty is that he knows nothing with absolute certainty.

The gender construct formalises the Socratic confession of ignorance concept by positing the feminine entity as the bearer of the attribute of absolute ignorance. In order not to violate FC, this attribute must itself be considered as an entity in its own right. As such, this masculine entity can be thought of as ultimate singular statement of certitude. In Heisenberg's Uncertainty Principle, one can know with high certainty the momentum of a particle in which case the position will be equally uncertain, or *vice versa*. This opposition of certainty and uncertainty typing is expressed in its most fundamental and generic form in the gender construct, where we will call it the *Socratic Uncertainty Principle*, the elementary building block of all of right side science. The feminine is the ultimate wildcard of indefinite extent when it comes to knowledge and the masculine as pure singularity is the ultimate expression of certitude. These are the two building blocks of generic knowledge. Unlike Heisenberg's Uncertainty Principle, the generic right side version is anti-symmetric. The uncertainty in knowing the feminine stays that way. The feminine remains forever a wildcard to be carried along in the algebra. The same thing applies to the certitude expressed by the masculine. There are no greys. This is a recurring difference between left side and right side science. Left side sciences are based on symmetries. Right side science is fundamentally based on asymmetry.

Superposition Principle

The gender of the generic entity allows a Quantum Mechanics type interpretation where gender appears as analogous to a quantum state. The coherence of the gender quantum state is only apparent from within the system. Within the primordial system, there are only two entities and each has unambiguously its own gender. Determinism prevails in this scenario. However, viewed by a third party outside the system, it is impossible to distinguish the two entities. The system looks like a single entity possessing a single quantum state, which can be either masculine or feminine. The masculine or feminine, as state values, will appear to be "in superposition."

The only way to get to know the actual value of the state of the system is to be within the system. There are two ways to get within the system:

1. To already be inside and so a part of the system.
2. To enter the system from outside and become part of the (new) system.

In the first case, the subject (the generic self) is already inside the system. The view from inside is totally different from that seen from outside. There is no applicable notion of "quantum state": There are simply two entities, each one of different gender. The subject, because of the determined oneness of its cardinality, is none other than the masculine entity. The feminine entity is the other entity and nothing of any consequence is known about it. In this scenario, there is no applicable QM notion of a "collapse of the wave function." One must keep in mind that right side science is operational rather than functional.

At this point, the subject as the masculine entity may reflect on how the pure unqualified feminine happened ever to get to have an attribute. On reflection, it will realise that it only has itself to blame. Then it has to figure out where it itself came from but that is not our problem.

The second case is more complicated. To begin with, there is the pre-existing system consisting of a feminine and a masculine entity. The story can proceed in a number of ways. A simple story is to say that the outside entity could enter into the system by simply reacting with it. The two separate entities would result in a dichotomy violating the monism of FC. To maintain FC there would have to be four entities with mixed genders MF, FF, FM, and MM. Here there are two masculine entities and two feminine entities, all entangled. The M appearing in the second place of the gender letter doublet would correspond to the original subject. This subject can be thought of as the impersonal subject. Geometrically speaking, the impersonal subject could correspond to the perspective of the "point at infinity." The M appearing in the first place of the gender doublet can be thought of as the personal subject and could correspond to the point at the origin. The notion of the point at infinity and at the origin can be illustrated, and thus demystified, by the example of conformal geometric geometry as will be discussed later.

In a nutshell, the exterior entity insists on barging into the tranquil abode of the primordial masculine and feminine. The immediate result is that the system ends up with four entities rather than just two.

From this simple and rather simplistic analysis, we can say that the composite system is made up of four kinds of entity, each distinguished by binary gender typing:

1. MM the masculine *as* masculine
2. MF the masculine *as* feminine
3. FM the feminine *as* masculine
4. FF the feminine *as* feminine.

In Western culture, this construct dates back to Empedocles of the fifth century BCE. As Aristotle records:

Empedocles, then, in contrast with his predecessors, was the first to introduce the dividing of this cause, not positing one source of movement, but different and contrary sources. Again, he was the first to speak of four material elements; yet he does not use four, but treats them as two only; he treats fire by itself, and its opposite—earth, air, and water—as one kind of thing. We may learn this by study of his verses. (Aristotle)

Empedocles called the four classical elements the four *roots*, associating fire and air with male deities whilst water and earth were associated with female deities.

From a QM point of view, this story of composite genders can be retold. In the new story, the outside entity stays outside of the simple system containing the simply gendered feminine and masculine entities. In this case, to the outside entity the system will appear not to have two potential values of either M or F but rather the four potential values MF, FF, FM, or MM. It will appear that the quantum state of the system will be a superposition of these four compound types. However, this is not (yet) the case. The doubling of possibilities only occurs when the outside entity interacts with the system and hence becomes an integral part of it, and this has not yet happened. Once the outside entity decides to actually intervene and attempt a gender typing measurement on the system, the system will "collapse" to one and only one of these four gender typings. It will not get an M or an F typing but an MF, FF,

FM, or MM typing result, a consequence of its very own presence in the system.

There are many different ways of explaining the origin of the generic entity. One could even advance the argument that the cause of these two gender typed generic entities is merely an integral consequence of being analysed. Thus, one of us could be the cause merely because we are thinking about the generic entity. However, one should not linger too long on that kind of reflection.

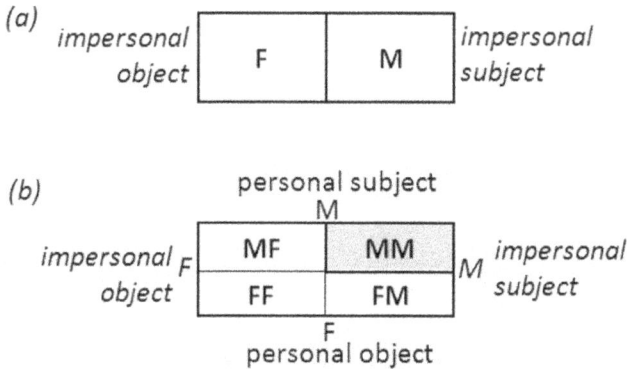

Figure 64 The first determination of the generic entity from an indeterminate point of view. (b) The generic entity determined relative to a determined point of view shown as the personal subject resulting in four substantial parts of the generic entity. The MM part is where the personal and impersonal points of view are effectively identical.

When a second person starts thinking about the result of the first person thinking about it, the second person also gets sucked into the system. We end up with four entities in the system, instead of one, four generic quantum states, so to speak.

However, there is a fine point. The first subject must be totally undetermined relative to the second. George Berkeley would have probably said that the first subject was the God that watches the tree on the hill when no one else is around. The second subject would probably be you when you come to watch the same tree that God is watching. There are four entities in such a scenario. One must keep in mind that God's notion of tree watching is rather more demanding that what you are capable. His tree is different from yours, although indistinguishable from it. He has a taste for generic trees whilst you only watch that tree on that hill over there.

One could undoubtedly have fun with this construct in the school classroom, but we will not dilly dally here. Whatever way it washes out, the generic entity divides into four generic component entities. Using a pseudo quantum theory language is only one way to understanding. Figure 64(a) illustrates the generic entity from a totally undetermined perspective. The two entities are shown as the impersonal object and the impersonal subject. Instead of merely being *any entity whatsoever*, the generic entity in Figure 64(b) illustrates an instance of the generic entity. The instance is determined relative to some determined point of view, which also becomes a part of the systemic scenario, and the entity divides into four as shown.

This four-part structure illustrated in the diagram becomes the point of reference for all of the generic science that follows. More than that, it can work as an oriented reference frame with a left side, a right side, together with frontal lobes and a back region. It can provide a generic ground for the science.

The Gender Exclusion Principle

Another principle in Quantum Mechanics is the Pauli Exclusion Principle. Here we present the gender version of this type of construct.

The four fundamental binary typed entities can be interpreted in a more technical way. Instead of interpreting the subject and object as either impersonal or personal they can be interpreted as placeholder and value respectively. The impersonal subject becomes the subject as a placeholder of value. The personal subject becomes the subject as value. Such a technical interpretation can demystify the construct and provide traction for applying the concept.

Of course, from the generic perspective there are two kinds of placeholder, feminine and masculine. The feminine placeholder is totally catholic and places no specificity requirements on content. Whilst the feminine is always an expression of the pure ignorance side of the Socratic confession, the masculine demands absolute certainty without in any way polluting the generic with any *ad hoc* specificity. How to be specific without being specific that is the question. This can only be achieved by opposing the "anything" mantra of the feminine with the "something" mantra. The feminine can contain anything. The masculine must contain something. This can be formalised by defining the masculine as the Exclusion Principle, something analogous to Pauli's Exclusion Principle in Quantum Mechanics. The Exclusion Principle states that the placeholder can only contain one single entity in any instant; such is the nature of the single tasking masculine placeholder. The feminine can contain any number of entities in any instant: it multi-tasks so to speak. We start to see here that the masculine and feminine in this context, or **as** this context, are primordial expressions of the One and the Many. The masculine, as principle, can be thought of as the Exclusion Principle while the feminine principle can be thought of as the Inclusion Principle. It is the interplay of these two principles that dictate the basic structure of any reality that satisfies the principle of FC.

Geometry with Subject

An organism (as world) or world (as organism) that satisfies the right side scientific paradigm of FC must forbid absolute dichotomies. The most dramatic way of violating FC is to take the left side approach and attempt a science devoid of subject. This approach immediately creates a dichotomy between the "objective" world of objects and the ignored "subjective" world of the subject. To avoid this dichotomy, any take on reality must be a whole with both object and subject present. Because the subject is always present, the very concept of "a take on reality" can be formalised: a take on reality is reality looked at from a particular point of view, the point of view of the subject. Each take on reality becomes reality seen as a whole, from a particular point of view. There are as many wholes as there are points of view.

A geometry that rigorously enforces reality being treated in terms of wholes together with an ever-present subject might seem rather exotic and even esoteric. In fact, the opposite is the case. Advanced computer graphics engines are a very good example of this approach. The geometric science employed is based on the Conformal Geometry Algebra (CGA) of Geometric Algebra presented in its modern form by Hestenes. Unlike the "view from nowhere" perspective of traditional geometry, CGA is highly projective in nature and implements two generic points of view, a generic origin and a generic point at infinity. This is accomplished by adding two extra dimensions to the three dimensional Euclidean base space, whilst retaining the same number of degrees of freedom. The end result is a dramatic increase in First Classness and universality. Computational benefits and conceptual universality of CGA have led to its wide application in Computer Graphics.

It turns out that it is much easier to manipulate geometric objects in 5D representational space than it is in the 3D base space. Instead of manipulating a geometric object by matrix transformations from an "other world" collection M, geometric objects are simply "multiplied" by other geometric objects. The multiplication employed is the geometric product discovered by Grassmann. Unlike the monomorphic multiplication structure of abstract mathematics, the geometric product is polymorphic: geometric objects of different grades can be multiplied to give other geometric objects. Because of the polymorphism, the geometric product does not impose any arbitrary dichotomies between what can and what cannot be multiplied and so does not violate FC. Exploiting the FC characteristics of this generic geometric product, a wide range of geometric transformations becomes possible. The geometric product in CGA encompasses not only rotations, but also translations, reflections, dilations and so on. The surprising aspect about all these transformations, even the translations, is that the generic origin remains unchanged throughout. Of course, this is the way it should be in computer graphics. The scenery changes but the viewing platform of the personal subject always remains stationary.

CGA can be studied via practical examples using freely available graphics engines accompanied by a good practical text (Dorst, et al., 2007). Doran et al in their tutorial paper (Doran, et al., 2002) argue the case that that this kind of approach is conceptually and computationally superior to traditional matrix and coordinate based methods. Indeed, this appears to be the case and explains its wide acceptance in the computer graphics industry. However, we wish to go much further and use CGA as a readily understandable practical example of a non-abstract, polymorphic algebra based on the geometric product. This kind of approach to geometry and algebra is not just "better" than the more literalist traditional approach of linear algebra, it is fundamentally *different*. Rather than being an exercise in abstract algebra, CGA is an exercise in *generic* algebra. The difference between the abstract and the generic is clearly illustrated in geometric algebras of all signatures, not just that of the CGA signature. Abstract algebra is monomorphic whilst generic algebra is highly polymorphic, as demonstrated by Grassmann's geometric product. In addition, abstract geometry is dualistic due to the dichotomy between geometric objects and the matrices that manipulate them. In abstract geometry, a translation of a geometric entity is effectuated by its multiplication by a matrix. A matrix is an array of numbers and is not itself a geometric object. The matrix is thus a second-class object. In generic geometry, the transformations on geometric objects simply involve the geometric product of geometric objects, an example of monism, and an exercise in First Classness.

Geometry without Numbers

In left side geometry, the coordinates of vectors, and the matrices that manipulate them, are all fundamentally based on a predefined concept of number. Numbers in this context are predefined symbolic entities with certain combinatorial properties expressing monomorphic semantics, the semantics of quantity, and strictly quantity only. Our task now is to tackle the Leibniz problem of developing a GA without number. Out inspiration comes from GA based upon the geometric product. The geometric product is one of the key ingredients for constructing geometry that does not violate FC. However, present day GA is a hybrid system and not purely generic. Grafted on to the algebra is a classical n-dimensional vector space structure. Each GA is defined by its signature (p,q,r), where there are p basis vectors which square to 1, q that square to -1, and r that square to 0 with n = p + q + r. Thus, CFA mentioned above has the signature G(1,4,0), 3D Euclidean space G(3,0,0) and Spacetime Algebra (STA) corresponds to G(3,1,0).

Oriented Numbers and Multivectors

Before Grassmann, numbers were considered as magnitudes. Grassmann introduced the idea that not only could numbers have magnitude they could have direction. He discovered *oriented* numbers. In the hands of Gibbs and Heaviside, Grassmann's oriented numbers became what are known today as vectors. This was the left side take on Grassmann's innovation. Any geometric body could be thought of as a set of points in space. Each point could be described by a vector. Thus, for the left side geometric paradigm, a geometric body was nothing more than a set of vectors in a vector space. The left side vector construct was defined as an n-tuple of numbers where the numbers were called the co-ordinates of the vector. The vector with n coordinates was called n-dimensional, leading to the abstract notion of n-dimensional space.

However, the Geometric Algebra initiated by Grassmann relies on algebra rather than coordinate geometry. A geometric object is not determined by a set of points, but as a geometric body with extent and, in particular, with *oriented* extent. The left side notion of dimensionality is replaced by grade. Elementary geometric objects are called *n-blades* where n is the grade. A 1-blade corresponds to an oriented line, a 2-blade to an oriented plane. A multi-blade or multivector is a linear combination of blades of different grade.

The Missing Ingredient in Geometric Algebra

David Hestenes claims that GA provides a universal algebra capable of unifying all of mathematics and physics. However, there is one missing ingredient in the GA universal pudding. The algebra lacks a universal alphabet. GA, as it stands, employs the same sort of labelling technology as traditional mathematics. It uses arbitrary letters taken from the Latin or Greek alphabet where each letter has no intrinsic meaning. This kind of *de Saussure* labelling technology belongs to the left side scientific paradigm. For the right side scientific paradigm, any letters used in the algebra must come with their semantics. The semantics of the symbol can neither be determined *a priori* nor *a posterior* to its employment. Unlike the left side, right side science and mathematics is synchronic. All players of the whole view must be synchronically present. The symbol must come with its semantics built in. An example is the photo shoot. In order not to violate FC, the photographer and even the camera must all be in the photo.

Our analysis has shown that there are only two symbols that come with built-in semantics. These symbols are m for entities of masculine gender type and f for entities of feminine gender type. Entities of f type *have* something. Entities of m type *are* this something. One can see that m and f type entities have a very generic kind of meaning. The two generic semantic building blocks of meaning are *to have* and *to be*. Our immediate task is to interpret this generic semantic of gender in the realm of geometry. What do the masculine and feminine mean geometrically?

We take the feminine first. The only thing known about a geometric body of feminine gender is that you cannot put your finger on it. It lacks localisation. Thus, a geometric body of type f can intuitively be imagined as an undetermined mass of extent. The f type body is totally devoid of any determined singularity. In the absence of any other qualifying agent, nothing more can be said of the f type body. Such a body, in isolation, is mathematically intractable.

We now consider the primordial masculine geometric entity. A geometric entity of pure masculine type m corresponds to pure presence in the form of a pure singularity.

One is tempted to say that the masculine singularity is a body of zero extent. However, this would imply that the two gendered types of geometric object make up an absolute dichotomy with the zero extent masculine entities on one side and the non-zero extent feminie objects on the other. The more formal approach to the subject must take this into account. For the level of rigour of our current exercise, it suffices to say that the masculine is a singularity *relative to* the feminine. In this way, an absolute determination can be avoided.

We now comes our most important geometric innovation. Instead of employing the traditional notion of basis vectors defined in reference to some predefined number system, we propose a much purer solution. We replace the traditional orthonormal basis set e_0, e_1, ... e_n with the basis entities f and m, corresponding to the two fundamental generic entities. The only specificity of the entity f is that it has an undetermined attribute. Of course, the attribute m is that attribute. Moreover, the attribute m is an entity in its own right. Keep in mind at all times that f and m are extremely polymorphic entities. These two gendered entities are required to express extreme polymorphism, as they are the only two players involved in the entire construction of our science.

Making the switch from a traditional vector basis set to one defined by ontological gender is our very radical step. It opens the way forward to a truly generic geometric algebra (GGA). It is by making this radical change that the Hestenes image of a universal algebra and geometry unifying all of the science joins up with a historic dream. It joins up with the prophetic vision of Leibniz for a geometric algebra without number. Instead of traditional numbers, we employ gender.

First Order Logic

The difference between gender and number can be seen more formally from a logical point of view. Left side mathematics is based on what is called Standard Analysis, which relies on abstract reasoning constructs expressible in terms of second order logic. First order logic only allows non-abstract reasoning over individual elements. Second order logic is an extension of first order logic that allows reasoning over the properties of elements or, equivalently, over sets of elements. Thus, the number system of left side geometry is based on Standard Analysis and second order logic.

A system of analysis that only relies on first order logic is called Non-Standard Analysis. In what follows we will see that a system of analysis based on the two *f* and *m* gendered entities only makes sense in terms of first order logic and so is a non- Standard Analysis.

In making our radical break with the prevailing paradigm of present day science, we are revisiting an event that happened in antiquity. The prevailing logic of the time of the Stoics was Aristotle's syllogistic logic. The logic definitely reasons over properties of sets of individuals and so qualifies as a kind of second order logic. The great Stoic logician Chrysippus, as an integral part of the Stoic system, invented a logic reminiscent of modern day propositional calculus. The logic was presented in the form of five undemonstratables where the reasoning was restricted to particulars. Unlike Aristotle, there was no reasoning over species or genus. Stoic logic was definitely first order logic, and proud of it. Defiantly the Stoics proclaimed that Aristotle's syllogistic was "useless." (Chénique, 1974).

In this work, we will not be as brash to say that Standard Analysis, second order logic and all of the corresponding traditional left side science are "useless." More meekly, we will declare that we simply do not need any of that technology for the immediate task in hand. Our task is to develop the foundations for the long missing, high level, top down, operational sibling of present day science. It will not be "better," just totally different and highly complementary. The reasoning involved must be based on first order logic and so qualifies as Non-Standard Analysis. Left side reasoning allows abstract higher order logic but must deal with a never-ending plethora of flat monomorphisms. Right side reasoning is restricted to low level first order logic but deals with tight, highly polymorphic structures. There is the interesting question of which form of reasoning is superior, right side, or left side? This question as to which side is the master and which is the emissary, we leave to the reader to decide. First Classness forbids an absolutist answer to the question.

The Generic Product

The generic product of two generic entities in GGA is similar to the GA geometric product. The main difference is that the generic product applies to the generic entities f and m and the compounds they form whilst the geometric product applies to abstract vectors and the resulting multivectors. GA is considered here as a hybrid half generic, half-abstract form of geometry, whilst GGA must be fully generic.

At this point, we must emphasise the most important difference between abstract left side geometry and generic right side geometry. Traditional left side geometry is based on an absolute dichotomy between space and the geometric objects that inhabit that space. In right side geometry of GGA, any such absolute dichotomy violates FC. Thus, the only acceptable geometry is one where both space and the objects inhabiting the space are all made of the same stuff. In other words, there is no absolute distinction between space and object. There is only stuff. The starting dichotomy is that between masculine and feminine typed stuff. Since this dichotomy is defined relative to each of the participants, it is relative and not absolute and so does not violate FC.

Another important difference concerns nominalism. Left side mathematics and sciences employ a nominalism based on arbitrary labels. The labels have no inherent meaning. Right side geometry is also nominalist but in a radically different way. Every entity must be labelled by its gender type, starting from the two generic generators f and m, and the four binary typed compound elements mf, ff, fm, and mm. The four binary base elements provides the operational bases for the geometry and so form the four "letters" for gender typing geometric entities. The concept goes back to Empedocles. In passing, the left side and right side nominalism discussed here seem to correspond closely to the two radically different kinds of nominalism employed by the Epicureans and the Stoics.

The other crucial difference between abstract geometry and GGA is that, GGA only treats reality as a whole. This means that all pertinent participants are present, including the subject in its various guises. GGA articulates reality in the form of an ensemble of wholes, each whole corresponding to a subjective point of view. As we shall see, there are 64 generic points of view and thus potentially 64 kinds of generic stuff. We will investigate some of the generic geometric properties of these composites. We will then look at two application areas; elementary particle physics from a generic point of view and biological structures coded by the genetic code.

We will now look at the generic product construct using GA as a template. The product of two elements a and b is written here as ab. GA then introduces generic constructs by decomposing the product into the sum of two other kinds of product where **a.b** denotes the *inner* (dot) product and denotes the *outer* (wedge) product of *a* and **b**. Thus

$$ab = a.b + a \wedge b \tag{1}$$

The two products are defined in such a way as to avoid any *a priori* assumptions except that *a* and **b** are generally different and, in general, so will be their products **ab** and **ba**. The two products are defined as follows:

$$a.b = \tfrac{1}{2}(ab + ba) \tag{2}$$

and

$$a \wedge b = \tfrac{1}{2}(ab - ba) \tag{3}$$

If *a* and **b** are of grade p and q, the grade of the dot product part will be p-q whilst the grade of the outer product part will be p+q. These products allow generic geometric interpretations of the relationship between *a* and **b**. For example, if the geometric product commutes, then the outer product is zero and so *a* and **b** can be said to be parallel to each other. If the geometric product anti-commutes the inner product is zero and so *a* and **b** can be said to be orthogonal.

Basis of Traditional GA

Traditional GA is built upon a basis set of n orthonormal vectors in the traditional way. The only feature distinguishing one basis element from another is the square of each basis element. A basis element vector may square to 1, -1, or 0. These are the three possibilities considered possible. Basis vectors with the same norm are distinguished from each other by arbitrary labelling, as in left side mathematics. The end result is a basis set of what is called as being of *mixed signature*. Like the abstract geometries of left side mathematics, all of the geometries are identical except for the metric determined by the signature. The only difference between a Euclidean, a Minkowski, or a de Sitter space is the metric. They are all founded upon the same abstract affine geometry of points and lines. This is an essential aspect of geometry founded on abstraction.

Gender as a Geometric Basis

In our GGA, the basis elements are taken from the generic entities determined by the different products of the two embryonic base elements *f* and **m**. Both *f* and **m** are highly polymorphic, as we will soon find out. It is important to note that gender allows an intuitive geometric interpretation. The feminine entity exists as a body possessing undetermined attribute. The masculine is this attribute. Relative to each other, the feminine as geometric body, can be thought of as an undetermined extent of non-localisation whilst the masculine poses as an entity of relatively null extent, a singularity relative to the feminine. Neither is capable of determined existence without the other. The situation changes with the two combined in the one system. The feminine has extension but no form. The masculine has form (the form of singularity) but no extent. These are the two basis elements, the two primordial generators of GGA.

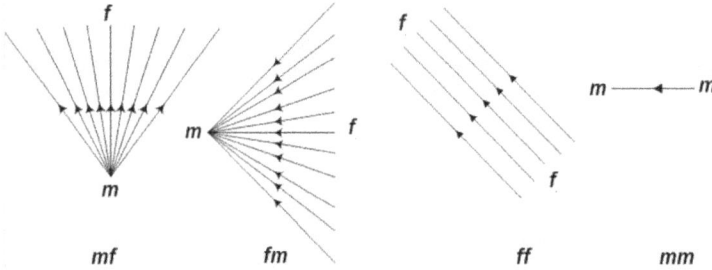

*Figure 65 The four elementary geometric products **mf, ff, fm,** and **mm.***

Thus, instead of n basis grade 1 vectors as in GA, generic geometry only has two basis entities **f** and **m**, each also considered to be of grade 1. We now look at the geometric product of the two geometric genders. There are two apparent possibilities, depending on the order of the geometric product. The two geometric products of **f** and **m** can be written as **mf** and **fm**. The geometry immediately starts to become tractable. An informal sketch of the generic 2-blades **mf** and **fm** is shown in Figure 65.

Note that

$$fm = f.m + f{\wedge}m \tag{4}$$

and since **f** and **m** are orthogonal

$$f.m = 0 \tag{5}$$

Thus, **fm** is simply given by the wedge product

$$fm = f{\wedge}m \tag{5}$$

and so **fm** is a multivector of grade 2. The same applies to **mf**.

Mixed Gender

The two basis elements for generic geometric geometry are the two geometric entities typed **f** and **m**. We will call these two entities the sub-bases of the geometry. The bases of the geometry will be made up of the four possible generic products of the sub-bases. Thus, the bases of the geometry are **mf, ff, fm,** and **mm**. Each base element is a multivector of grade 2. The generic product will be modelled along the lines of the geometric product of GA. Unlike the basis elements for GA and traditional abstract mathematics, the four kinds of basis elements will not only differ metrically, but will also differ by their geometric form and innate geometrical properties. Because of the underlying two gendered sub-base elements, we end up with a finer and richer geometric structure than is possible with Standard Analysis. The standard approach to geometry is literally blind to this finer grade structure. However, it must be admitted that in many cases the generic synchronic structures revealed here will have familiar corresponding interpretations in the traditional diachronic domain of abstract mathematics. It should be kept in mind that GGA is the *operational* version of traditional linear algebra and geometry.

Note that the bases for generic geometry are of grade 2 whilst the bases for traditional geometry, including GA, are grade1. Note also that the bases for traditional geometry are fully determined. In generic geometry the bases are only partly determined. For example, the *mf* and *fm* bases have one end of the base arrow determined and the *ff* base has no determined ends for the base arrow dyad. The only fully determined arrow is the mysterious *mm* dyad, which has no properly defined equivalent in GA.

Generic Dimensionality

Left side geometry takes a completely general, abstract approach to the dimension of its abstract spaces allowing space to be "n-dimensional," where n can be any positive number including infinity. In addition to arbitrariness, there is the "fictionalist" aspect of axiomatic mathematics. (Irvine , 2009) (Leng, 2010) When it comes to determining the basis elements, traditional mathematics has a free reign. Not only are the numbers of dimensions up for invention, but the way they are labelled. The labelling is totally lacking in rapport to any subject, observer, or whatever. The typing of the bases is only determined by the three possibilities of how each base squares. The square of a base can be a scalar 1, -1, or 0. The arbitrariness of labels raises its head again. In traditional geometry, it does not matter how the base elements are labelled. Figure 66(a) illustrates the case of three orthonormal basis elements of the same kind of norm labelled with one of the RGB colours. Clearly, the labelling is non-unique and so has no meaning, no intrinsic semantics.

When it comes to right side geometry, the dictates of FC eliminate fictionalism. There is only one story. In order not to violate FC, the basic requirement is that basis blades must be unambiguously distinguished from each other. The geometry must be capable of answering the question: "Which base?" The question will not be answered by specifying the type of the base. All the bases present may be of the same gender typing, for example. Specifying by gender type obviously does not work. Traditional left side geometry answers this question by simply specifying the axis by label. The axes might be labelled x, y, and z, for example. To the question, "Which axis?"

comes the response, "The y axis?" Right side geometry does not enjoy such a simple life. Right side geometry must be a self-labelling system. The labelling apparatus must form an integral part of the geometry.. Each basis element must possess an inherent identifying label that cannot be arbitrary. The label must be based on a generic geometric.

Without going into details, the only FC compatible self-labelling configuration is in the form of a triad of base elements arranged free of any concatenation of blades, as shown in Figure 66(b). Two adjacent elements must share either the same target or the same source. Although the triad is free of any ordered concatenation of blades, there remains a geometrically expressed notion of Firstness, Secondness, and Thirdness. In in Figure 66(b) these three self-labelling properties are allocated that colour labelling of Red, Blue, and Green, respectively. The structure is similar to that of the biological genetic code, where the triads can themselves be chained end on end potentially producing a non-ambiguous generic self-labelling coding system. There is no structure to learn, as there is only one generic solution to the problem that is compatible with FC.

The way this works is that bases from the alphabet of four generic binary typed base elements can type each of the three dyads in the RGB triad. The RGB colours are themselves not types, but relative positions. No dyad is more important than any other dyad in the triad because of the geometry where none absolutely proceeds or precedes any other dyad in this concatenation free structure. Each triad, when suitably gender typed, will be a concatenation of 6 binary valued entities where the binary values are either m or f. This is all part of the Grand Ruse: the RGB triad is a concatenation free structure that, after gender typing, articulates a concatenation of six genders.

Here, we will jump the gun and make an association with the genetic code in biology, which we are here attempting to reverse engineer based on FC. We will make the association that the four binary typed elements **mf, ff, fm,** and **mm** of GGA can be labelled with the single letters **a, u, g,** and **c** respectively, corresponding to the A, U, G and C bases of the RNA encoding of the genetic code. In this chapter **a, u, g,** and **c** are not only self-labelling letter labels but also 2-blade base elements of a totally generic geometric algebra (GGA).

Each RGB triad, suitably gender typed, encodes the genetic geometry of an organism as a whole as determined from a particular point of view. There are an enormous number of particular points of view, each with an associated whole. The only requirement is that the system not violates FC. This is illustrated for a biological organism in Figure 25. Biological systems employ an the genetic code in the form of DNA and RNA as intermediaries between to the protein world made up of the amino acids coded by the RNA. In the case of the physical universe, there are no intermediaries and the coding RGB triads and the elementary particles they encode are the same thing. Each elementary particle, from a right side perspective provides a generic view the universe as a whole, as we shall see later.

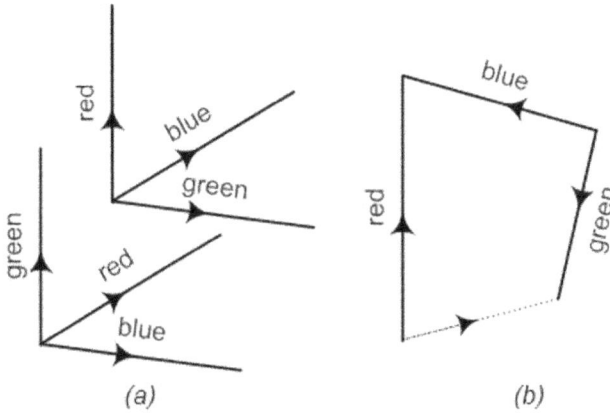

(a)   (b)

Figure 66 (a) Labelling of Cartesian axes is ambiguous. (b) The generic self-labelling RGB triad. The dotted arrow is the indeterminate dyad.

The structure of a generic whole is a Three-in-One structure made up of an "imaginary" part and a "real" part. Hamilton's quaternions are a GA example of such structure. The suitably typed RGB triad makes up the imaginary part and provides the three generic attributes of the whole. The "real" part is the dyad shown dotted in the diagram and can only be known in terms of its imaginary components. This dyad is possibly, what Plato was referring to in *Timaeus* as the indeterminate dyad, a term dating back to the Pythagoreans. The concept was picked up later by the Neo-Platonist Plotinus. Be as it may, we will use Plato's terminology and call it *the indeterminate dyad.*

The Synchronic Product

The difference between an algebraic expression and its evaluated value is an important construct in functional programming languages. In our case, evaluating algebraic expressions can raise the ogre of violating FC. The problem with the geometric product is that it could be seen as a violation of FC. First of all, there will be the algebraic expression expressing the product of two entities. This is the algebraic expression before evaluating the product. Then, after evaluating the product, there will be the result of applying the product. Here we have a before and after. In order to respect FC, there can be no before or after. Consequently, the only way that the geometric product can avoid violating FC is to remain unevaluated. We will call this unevaluated geometric product, the *synchronic product.*

We will use the notation **a**<space>**b** to denote the geometric product of **a** and **b** and use the notation ab to denote the synchronic product.

Thus, the synchronic product **xy** evaluates as

$$xy = x\text{<space>}y = x.y + x\wedge y \tag{6}$$

In what follows, we will ignore the distinction between the geometric product and the synchronic product. Thus, the product the geometric product will often be written simply as **xy**.

The Generic Seed Product and Sum

The most generic form of product could intuitively be imagined as a kind of "ontological product" between the masculine and the feminine, a kind of "seed product." However, such an algebraic operation is undetermined in the sense that **m** neither precedes **f**, as the multiplication **mf** nor proceeds it as **fm**. All that one can say is that **m** is simultaneously present with **f**. As a consequence of this most intangible of algebraic operations, this implicit "seed" product of **f** and **m** leads to the sum of two entities one of which is the singular form without body **m**. The other entity is the pure feminine **f**, the body without form.

In standard GA, one can say that the geometric product of a p-blade **P** and an orthogonal q-blade **Q** is an n-blade **M**, where n=p+q. This is an example of the terminology employed in the abstraction paradigm. The symbols used are arbitrary and have no senses except as monomorphic labels. In addition, the symbols p, q, and n are variables with values ranging over the set of positive integers. Abstraction is a powerful methodology and a powerful way of thinking. However, it is not the only way. There is an alternative, the generic. The generic is an abstraction and generalisation free zone and only deals with the universal.

In passing, one should note that traditional philosophy, science, and mathematics do not make a clear distinction between the general and the universal. In practically all cases, what is referred to as universal is in fact a mere generalisation.

The generic paradigm leads to geometry based on GGA. In GGA, geometric entities are not denoted by arbitrary labels but by their ontological type, that is to say, by their compound gender. The elementary entities are **m** and **f**. The four possible products **mf, ff, fm,** and **mm** are geometric entities in their own right and can sometimes be referred to by their single letter synonyms **a, u, g,** and **c** respectively. Moreover, the grade of blades in GGA is highly restricted and cannot exceed three. Each three blade determines a whole, a holistic way of looking at reality. Highly complex systems exhibit geometric organisation requiring a myriad of such triadic takes on reality.

Elementary aspects of the Algebra of Gender

Here we make our first and undoubtedly rather naïve sojourn into the algebra of gender. We start by considering the four products of the two fundamental gender sub-bases f and m of the gender algebra. Reasoning informally, this should lead to four distinct types of geometric object that are to serve as the basis for all the geometries to follow. The four fundamental basis elements are generated from the four geometric products $mf, ff, fm,$ and mm.

The difference between gender algebra and traditional, left side algebra is in the semantics. Traditional algebra only exhibits first order semantics where the symbols, although rich in syntax, only act as labels for objects. The simple labelling technology of the left side sciences sums up first order semantics. For example, in traditional linguistics like that developed by Chomsky, semantics is expressed in the form of a lexicon. The lexicon formalises pure first order semantics. In the algebra of gender, the semantics is second order. One way of illustrating second order semantics is with the semiotic square, an example of which is shown in Figure 67.

The diagram illustrates two variants of the opposition between subject and object. The two oppositions are then opposed to each other forming the semiotic square, a kind of semiotic binomial. The left-right opposition involves the unqualified or impersonal subject. The front back opposition involves the qualified or personal subject. If preferred, one can use the more geometric term *singularity* instead of *subject*. The subject, or singularity, is gendered as masculine type m and the non-subject, or non-singularity, is gendered as feminine type f.

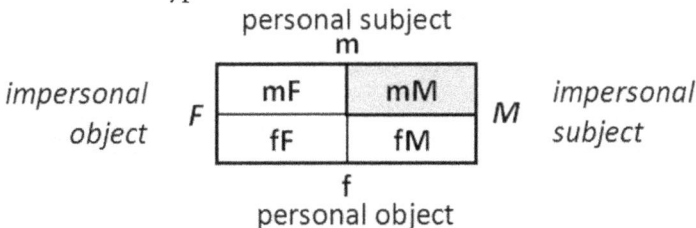

Figure 67 Binomial semiotic square and the four geometric products.

The four terms in the semiotic square shown in Figure 67 are formally constructed by effectively taking the geometric product between the two kinds of subject-object opposition. Instead of using a first order semantics device such as a lexicon, we use the semiotic square. Instead of first order semantics with arbitrary labelling, we construct our labels out of the two elementary gender types f and m. Left side *ad hoc* labelling is replaced by gender *typing*. To illustrate the difference between the impersonal and the personal players in the compound oppositional structure, the impersonal poles of the opposition are shown in majuscule and the personal in miniscule.

Second order semantics naturally leads to a geometric expression of the semantics. A glimpse of this characteristic can already be seen in the fact that the algebraic products under consideration are *geometric* products, not the empty abstract products of traditional algebra. The four geometric products making up the semiotic square form geometric objects. The geometric objects can intuitively be understood in terms of sheaves of arrows as illustrated in Figure 65. The orientation of the arrows is important and can be related to the gender typing in the above semiotic square. We will employ the convention that a gender type shown in miniscule represents the source of the arrows and the gender in majuscule represents the arrow target side. Thus, the four sheaves of arrows corresponding to the four gender types are all oriented from the left to the right relative to the semiotic square as a whole. However, if the entity defined by the geometric product fF corresponds to a sheaf of arrows going from left to right, the geometric product Ff will correspond to a sheaf of arrows going in the opposite direction. The same applies to mM and Mm, with respect to arrow orientation. Interpreting the difference in orientation algebraically, we can write that:

$$fF = -Ff \text{ and } mM = -Mm \tag{7}$$

From this, the inner product of the feminine with the feminine is zero because:

$$f.F = 1/2(f.F + F.f) = 0 = F.f = f.f \tag{8}$$

The same applies to $m.m$. In the case of the other two types of sheaves, we don't need to use the majuscule and miniscule construct as arrow direction is either "outwards" or "inwards" depending on gender typing. We can thus simply write that:

$$mf = -fm \tag{9}$$

From these equalities, we can calculate the four possible inner products between the genders. The inner product is zero in all cases, i.e.,

$$f.f = m.m = f.m = m.f = 0 \tag{10}$$

Hence, the four elementary geometric products of the two genders are equal to the four wedge products and so:

$$mf = m \wedge f, \quad fm = f \wedge m, \quad ff = f \wedge f, \quad mm = m \wedge m \qquad (11)$$

where each wedge product corresponds to a generic 2-blade.

These geometric forms are not unfamiliar to left side mathematics where they can be compared with the building blocks of spacetime algebra, for example. Robert Goldblatt's book (Goldblatt, 1987) is pertinent in this respect. Using the axiomatic approach of abstract mathematics, he explores spacetime geometry from the perspective of orthogonality relationships. From the orthogonality perspective, there are four kinds of lines; timelike, spacelike, and lightlike that make up the usual lines encountered in spacetime geometry. The timelike lines are always orthogonal to spacelike lines and lightlike lines are orthogonal to themselves. There is also a fourth type of line. This is the singular line, which is orthogonal to all lines present, including itself. In what follows we will endeavour to show that the generic versions of timelike, spacelike, lightlike, and optical lines are the **mf, ff, fm,** and **mm** blades respectively of GGA. In the process, we will demonstrate that the generic paradigm reveals a much finer structure than is possible in the abstraction paradigm. Abstraction, by its very nature, is inherently "forgetful" and even "neglectful."

Non-Symmetry

At this point in the development, one might be tempted to think that the gender construct is based on symmetry. The gender algebra considered so far seems to indicate a fundamental symmetry. However, this symmetry is short lived. The symmetry only holds before there is a choice of starting point. One the starting point is determined, non-symmetry becomes the rule of the game. Left side science and mathematics never encounter this fundamental asymmetry as the starting point is not an integral part of such sciences. This is not the case for right side science. In fact, one could say that right side science is the science of the starting point. The starting point becomes a kind of error, a kind of "original sin" which demands continuous atonement. Such is the story of Life, once it starts.

The Start and its Codon

We now move to a more formal approach to the very foundations of right side science and mathematics. We must formalise the starting point. The starting point for this generic science must be *any point whatsoever*, what we will call the *generic entity*. The first problem to raise its ugly head is that there is not one generic entity to deal with but two. In order to gain traction and actually start, we must make a choice and actually choose *any point whatsoever*. Of course, the entity so chosen is not the same entity as the generic entity. Moreover, we have not actually taken the entity into possession, but only the knowable aspect of it. We can only know entities by their attributes, that is to say, via the masculine. We can never know via the feminine, which is inherently unknowable and remains forever the perennial wildcard in the mix. You never pick up a boiling kettle directly, you pick it up by its handle. The handle for the generic entity, for any entity, is the masculine part.

Thus, we will call this first entity so chosen, the personal masculine or the *personal subject*, the subject always being the entity as a singularity. The entity that missed out in the choice process is also masculine as it *could* have been chosen and we will call it the impersonal subject. The impersonal subject should be treated as an entity in its own right.

The impersonal subject is not a stand-alone entity, as it is determined relative to the personal subject. In one sense it is the loser for not have been chosen. However, in another sense it is a winner as it has become less contaminated by subjectivity that in the personal subject. Neither the personal nor the impersonal subjects can claim to be the generic entity, or even to be the generic entity as subject. However, combined together that can make up a structure that represents the generic entity as a whole. This is a fundamental feature of the generic: whenever there is a subject on the scene, it always has company. The personal subject is always accompanied by the impersonal subject. It is this epistemological fact that has long stimulated religious and theological thinking. There are gods, so to speak, in this kind of mathematics.

The essence of generic science is to understand the generic entity in terms of wholes, each whole encompassing a different point of view. The whole that we are considering at the moment is the view of the generic entity at the start. This whole makes up a structure that we will call the start codon. The task before us is the construct the start codon. If we can accomplish this feat, we will have made the first step towards a science that is *starting point invariant*, and thus truly generic.

The Principle of Non-Duality

Our first adventurous step into the generic results in frustration. Instead of one starting point, we have two. The conundrum arises due to the difference between **the** *any-point whatsoever* and **an** any-point-whatsoever. These two entities are not the same entity. The theory instructs us to start with the former, but in practice, we have no alternative but to start with the latter. This is just like our lottery. There are two winning tickets, the ticket winning ticket inside the barrel and the winning ticket drawn from the barrel. Provided the game is not rigged, they are the same ticket.

Thus, we are faced with the enigma. We can start from one point, but how do we start from two? Do we start with the impersonal or the personal subject? The answer to this kind of problem can be found in the thought of the 9th century Vedanta philosopher Sankara in his Principle of Non-Duality. Sankara dealt with the relationship between the impersonal soul, the Brahman and the individual soul, the ātman. Sankara constructed a monism, but avoided naively declaring some ultimate and absolute Oneness. Such naivety would violate FC. Instead, he advocated his Principle of Non-Duality between the impersonal generic and the personal. These two entities are different but do not form a duality. They are indistinguishable parts of a whole. Since Principle of Non-Duality provides the most fundamental underpinning of Hinduism, one might think that this kind of thinking naturally leads into impenetrable mysticism. As we shall see, the opposite is the case.

From our generic point of view, a Principle of Non-Duality is necessary in order to establish a starting point for our science. The first choice of starting point led to the personal subject. This was an error as the generic entity has become contaminated with subjectivity. If we had chosen something else, then we would have got a different result. This is intolerable as our science must be starting point invariant. The only solution is a Principle of Non-Duality where the personal subject and the impersonal are so infinitesimally close to each other as to be indistinguishable. In that way it does not matter where you start, you always get the same result.

We will see that in order for the Sankara Principle of Non-Duality be valid, the world must be constrained to have a particular shape, a particular geometry. Interpreted in the language of physics, it turns out that the Principle of Non-Duality demands a spacetime geometry based on Einstein's Special Theory of Relativity.

Constructing the Starting Point

The problem to be solved is as follows. Our task is to construct a world geometry where the centre of the world is none other than *any point whatsoever*. This means that no matter who or what you are, you are the centre of the world. The principle involved here is the Principle of Non-Duality. The underlying principle is that of FC.

Step 1

The problem will be construed in geometric terms. We must avoid abstract reasoning. Instead, we must argue generically. First of all, we chose a starting point for our science. Because the science must be starting point invariant, we can start anywhere we choose. We will choose the point *m1*. This point *m1* will be any point in the base space *B* of all possibilities, where the geometry of B is yet to be determined. The point *m1* can provisionally be thought of as the origin of the science. As such, this origin is compatible with FC in the sense that it can be any point whatsoever and so no point is more distinguished than any other. However, *m1* is not as advertised. The point *m1* has been chosen as *that-point-there*, which is not the same as any point whatsoever. Thus, **m1** is more distinguished than all the other points. The point **m1** is the winner and all other points are losers. FC has been violated.

We can refer to this winning point **m1** as the personal subject. To overcome this conundrum and remain compatible with FC, a geometric construction is necessary whereby the dimension of the base space is extended to include a point **m2**, which forever stays out of reach of our choosing process. We will call **m2**, the generic origin of the geometry. In order to accommodate the generic origin **m2**,, we must append a one-dimensional space **P** to the base space **B** . We refer to this one-dimensional space **P** as the *Parmenidean dimension*. The total space **T** will be combination of the base space and the Parmenidean dimension. Thus,

$$T = B \oplus P \tag{12}$$

This construct analogous to that used by Hestenes in Conformal Geometric Algebra (CGA) in order to establish a non-distinguished origin (Hestenes, 2001). The basis element **p** defining the Parmenidean dimension will have a zero norm and be orthogonal to all lines in the total space **T**. Without any loss of universality, the generic origin **m2** can be taken as the basis element **p** of **P**. Following the same reasoning as in CGA, the notion of what constitutes a point can be reinterpreted in the total space **T**. Henceforth, any point **m1** in the base space B can be interpreted as a bi-point in the total space **T** determined by the line interval passing from **m1** to the generic origin **m2**. All such line intervals representing points in the base space will be of zero norm.

Wherever our theory starts, we must get the same theory. The theory must be starting point invariant. For this to be the case **m1** and **m2** must form part of the same geometric blade. This blade can be constructed by the geometric product of **m1** and **m2**. We proceed as follows.

The geometric objects **m1** and **m2** are points, or what we call 1-blades. Geometrically, the product of the 1-blade **m1** and the 1-blade **m2** produces the oriented 2-blade **m1m2** shown in diagram Figure 68(a). The blade **m1m2** can be simply labelled (and typed) **mm**, where it is implicit that the first **m** corresponds to **m1** and the second to **m2**.

In order that the blade **mm** avoids FC violation, the two constituent points **m1** and **m2** must be concurrent. Thus, neither point will be *a priori* to the other. Both **m1** and **m2** have to live in the Parmenidean eternal *now*, the personal now of the subject, be it personal or impersonal. The orientation of the **mm** blade is not an *a prior* determination but merely an indication of the degree of relative determination of the two **m** typed constituents.

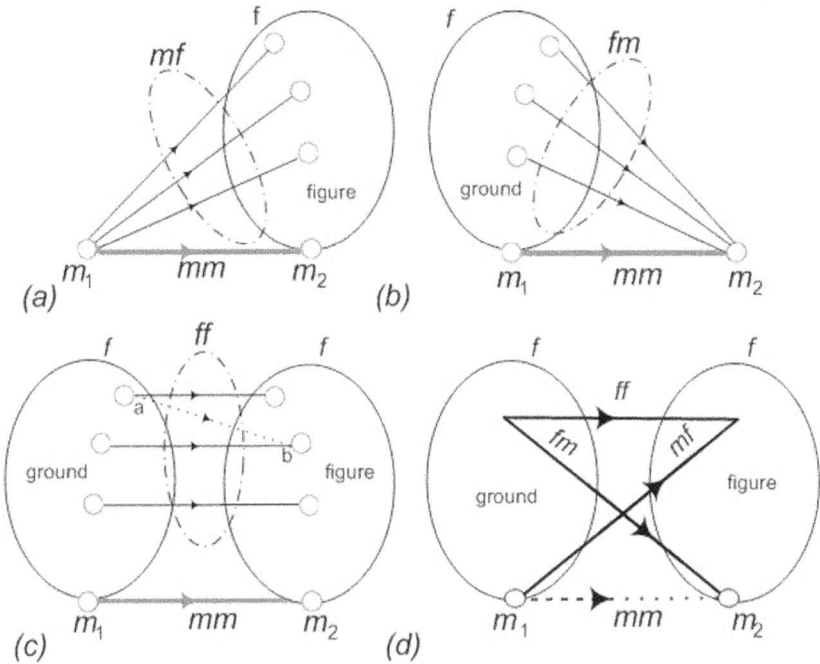

*Figure 68 (a) The **mf** blade constructs the figure (b) The **fm** blade constructs the ground (c) The **ff** blade consists of a sheaf of "parallel" arrows between ground and figure. (d) The complete diagram where the "real" but colourless **mm** dyad is determined by the "imaginary" triad of attributes **mf**, **ff**, and **fm**.*

The purpose of the whole exercise is to construct the shape of reality based on real geometric entities, where real is taken in the sense of Parmenides: what is real is what is *now*. An entity that belongs to the *now* of the subject will be said to be *concurrent* with the subject. What is real, relative to the subject, corresponds to what is concurrent with the subject. Because the blade **mm** is concurrent with the subject, it can be said to be *real*.

In order for the real to exist, there must be entities that are not concurrent with the subject. These entities will be referred to as *imaginary*. The imaginary world is that of the non-concurrent, the world of illusion according to the ontological paradigm of Parmenides, a concept echoed in the Advaita philosophy of Sankara. Of course what is imaginary is, at all times, determined relative to the subject.

It should be kept in mind that the science we are developing is a science *sans attributs*. What this means is that there are no attributes in the Parmenidean real. There are no determined attributes that are concurrent with the subject. Attributes are not real they are imaginary. Attributes are non-concurrent with the Oneness of subject. As the wise sages of former times have claimed, all determined attributes are relativistic illusions.

Our next task is to construct the non-concurrent or imaginary attributes that determine the real *mm* dyad.

Step 2

The origin of the total space *T* is the bi-point < *m1, m2*>, which, from a GGA viewpoint, we know as an *mm* blade, or as simply a dyad typed as *mm*. As a geometric object, *mm* is orthogonal to all lines in *T* and is of zero norm. The source side *m1* of the *mm* dyad is determined relative to the target side, the target being undetermined and totally generic. The *mm* dyad lies in the Parmenidean dimension where, relative to any specific *m1*, it appears to be a single arrow. The *mm* dyad is the only *real* arrow having *m1* as its source. We now consider any others arrows that have *m1* as source, as illustrated in Figure 68(a). The resulting cone of arrows will form a blade of type *mf*. The dyad representation of the *mf* blade will be a single dyad sharing the same source as the *mm* dyad. It will be determined by the geometric product of the 1-blade *m1* and the figure. The 1-blade *m1* will be typed *m*. The figure will be a 1-blade typed *f*. The result will be 2-blade typed *mf* as illustrated.

Step 3

The dyads making up the *mf* blade are not the only dyads that missed out in the choice of starting point. Figure 68 (b) shows another cone of dyads with a common target *m2*. making up the 2-blade *fm*. The sources of the cone of dyads makes up, what we will call, the *ground* relative to *mm*.

Step 4

Not all dyads share a singularity that is concurrent with the origin *mm* of the total space *T*. We now consider the sheaf of dyads with sources in the ground and targets in the figure. The resulting geometric object will be the generic product of the two non-oriented 1-blade *f* entities, the result being a 2-blade entity *ff* as shown in Figure 68(c). The end result, will be the Three-plus-One structure shown in Figure 68d). The orientation of the blades and the very existence of the blades was divined previously in the form of the generic Three-plus-One RGB structure shown in Figure 68(b). Using the RGB triad template, the dyads making up the generic start codon triad are typed *mf, ff, and fm,* respectively.

The 2-blade *ff* can be thought of as a sheaf of arrows going from ground to figure. However, this is not just any sheaf of arrows. The form of *ff* is severely restricted. The first case to be eliminated is the possibility that the blade *ff* consists of only one single arrow ab from ground to figure. This case must be rejected, as it would require an *ad hoc* choice of source in ground and target in figure. We have already spent our one *ad hoc* choice and that was the choice of starting point. Our task is to look at the repercussions of that choice, not to add more mud to the water. The repercussions, we hope, will be that it does not matter what we started with, we will always get the same answer anyway. Thus, the original seemingly *ad hoc* choice will turn out to have been the best choice possible. There is only one way to start, this way.

The next case to reject is that every zero blade in the ground is connected to all the zero blades in the figure. The end result is a simple structure where every point on the left is connected to every point on the right of the diagram. This simple structure, which is supposed to represent the system as a whole, is nothing more than a perfect absolute dichotomy between left and right and so violates FC. Moreover, the loss of any coherence in the gender typing mechanism signals another decent into nihilism.

Other cases can be rejected using similar arguments. The only case that is acceptable to FC and that maintains the integrity of gender typing is where no arrow of the **ff** blade shares a zero blade in either the ground or figure, and that all ground and figure zero blades are end points of the **ff** arrows. In some primitive way, the arrows making up the **ff** blade must be "parallel." This concept can be further formalised once the generic algebra is developed. That is the task tackled in the next section.

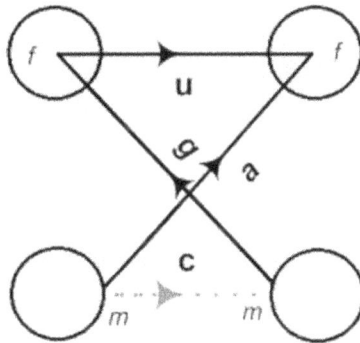

*Figure 69 Schematic diagram of the start codon **aug** where the **mf**, **ff**, **fm**, and **mm** typing have been replaced by single letters **a**, **u**, **g**, and **c** respectively.*

Figure 69 shows the schematic diagram for the start codon. The **mf**, **ff**, **fm**, and **mm**, gender typing has where been replaced by single letter synonyms **a**, **u**, **g**, and **c** respectively with obvious connotations that we will not pursue for the moment. If this all sounds rather extravagant, just keep in mind that we are not dealing with abstract generalisations here. Instead, the structures are generic universals. Instead of a labelling taxonomic hierarchy, the universal is hierarchy free and quite flat in that respect. However, the generic universal makes up by being applicable to literally anything that exists. Consequently, a universal structure like the **aug** start codon will be extremely polymorphic. Physics, biochemistry, foundations of mathematics and so on, it does not matter what the subject, they are all amenable to a universal interpretation, provided that the system involved doesn't have any hidden player behind the scenes pulling the strings. In other words, provided they respect FC.

In summary, determining the generic starting point immediately led to a Kantian style antinomy where the start must the *any entity whatsoever* that needs to be chosen in order to start. This was **m1**. On the other side of the antinomy lies the pure impersonal *any entity whatsoever* **m2**, posing as the correct place to start. Chose **m1** results in immediate error but at least there is a determined start. Do not choose and opt for the impersonal **m2** as starting point and the error is avoided. The downside is that there is no start. The antinomy was resolved where one could have the cake and eat it too. The solution requires a special kind of geometry where **m1** and **m2** can be virtually the same thing. This lead to the generic start codon **aug** providing the three imaginary attributes of the generic real starting point, the zero blade dyad **mm**. This is a generic geometric solution to the generic problem of starting something, starting anything. It is now time to look at the generic algebra of the start.

Category Theory and GGA

Both the traditional, left side mathematics of Category Theory and the right side mathematics of GGA are based on an arrow theoretic methodology. It is useful to compare the semantics and logic of the arrow diagrams in both approaches.

We start with Category Theory, the study of mathematical universals from an abstract, generalised perspective. Category Theory is made up of two kinds of collections, a collection Ω of objects and a collection Σ of morphisms or arrows between the objects. An example is where Ω is a collection of sets and the morphisms Σ are functions between the sets. In this case, for two sets A and B in Ω there will be many morphisms mapping A to B. The arrow diagram for the category should thus have many arrows drawn between A and B. However, in any textbook on the subject, it is most likely that the arrow diagram for the category will show only one arrow between A and B. We can explain this by saying that the multitude of arrows between A and B are *particular* morphisms. Since Category Theory is an abstract science, it is not interested in particulars. It is interested in generalisations. Thus, the arrow diagram showing only one arrow going from A to B is a *general* morphism. Mathematical structure will be expressed in terms of the general morphisms under the supposition that the same structure applies to all of the underlying particulars. Thus, the general morphism from A to B expresses the second order logical property that it is valid *for all* particular morphisms from A to B. Expressed in the quantifiers of the predicate calculus; the single arrow **g** from A to B represents the predicate:

$$g = \forall f\{ f:A \rightarrow B \} \tag{13}$$

The general arrow satisfying this predicate will exist and be drawn in the arrow diagram as a full, non-dotted arrow.

Every Category Theory arrow diagram will have one arrow **h** shown dotted that "makes the diagram commute." This dotted arrow, going between the set X and Y will represent a predicate **h** of the form:

$$h = \exists f\{ f:X \rightarrow Y \} \tag{14}$$

which states that "there exists" a (unique) morphism between X and Y in the context of the full arrows in the diagram. Thus, arrows drawn as full arrows represent universal quantification over the particulars and the dotted arrow represents existential quantification. In this way, one can see that Category Theory fundamentally relies on second order logic.

Turning to the first order logic case of arrow diagrams for GGA, we first note that there is no collection of objects, but merely a mass of arrows. The two ends of an arrow are considered an inseparable moments of the one entity. This expresses the non-dualist, monistic aspect of the mathematics. Instead of a mass of arrows between objects, as in Category Theory, we simply have a mass of arrows. These arrows are all particulars. Certain sheaves of arrows can be distinguished from other sheaves by structure involving the *concurrence* of common source or target ends of the arrows. These sheaves of arrows are called *blades*.

Like in Category Theory, the generic version of the arrow diagram replaces all of the arrows in a blade of particular arrows with one single arrow called a *dyad*. However, this process does not involve any abstract generalisation. Unlike in Category Theory, the dyad is not a general representation of the particular arrows. In fact, the dyad itself is a particular arrow just like any other arrow in the blade. Thus, it must be determined logically by the first order logic of propositions without any recourse to universal or existential qualifiers. What this means is that the blade with its particular gender type is represented by a dyad defined as being **any** arrow of the blade. In this case, the "any" qualifier of the generic has replaced the "for all" universal quantifier of abstract mathematics. The construct involves first order logic, not second order.

It was the Stoic logician Chrysippus that provided the generic form of first order logic. First order logic, viewed from an abstract perspective, is nothing more than the propositional calculus, a logic with only flat first order semantics. However, viewed generically, the first order logic comes along with its own second order semantics. In this book, we are the first of the moderns to comprehend and explain the significance and ramifications of the generic logic of Chrysippus. Further details will be spelt out in the pages to follow.

Like the classical Category Theory arrow diagrams, the generic arrow diagrams are made up of full arrows and one dotted arrow. The right side version is a Three-plus-One structure with three full arrows and one dotted. The full arrows represent the "imaginary" dyads whilst the dotted represents the "real" dyad. By real, we mean the Parmenidean real. What is real is what is concurrent with respect to a very important entity. Any entity whatsoever shares this same concern regarding what is the most important entity in the Cosmos. The answer is unerringly identical. In a nutshell, the answer is *me*. Thus, what is objectively real is that which is concurrent with the *me* in question, the concurrent real. The concurrent real is devoid of any determined attributes in itself. Attributes are considered as imaginary and are, in essence, partially, or fully non-current with the real. Imaginary attribute must be synchronous with the real, but not fully concurrent. An entity can only be known via its "imaginary," noncurrent attributes. An incredible fact about generic knowledge is that the imaginary attributes of the elementary entities that make up a rational universe (any rational universe), these attributes can be calculated from pure reason. We represent this knowledge in the form of Three-plus-One elementary structure described by a corresponding arrow diagram. In what follows, we will call them *Chrysippus diagrams,* a formal amalgam of first order logic with second order semantics. Further on, we will explore how the elementary Chrysippus diagrams correspond to elementary particles in Particle Physics. The Stoics claimed that Physics and Logic were holistically entwined. Using Chrysippus' Stoic logic we can show how, complete with diagrams. This is an incredible story. However, please do not blame the author for such extravagance. He is but a mere reporter.

Generic Algebra of the Semiotic Binomial

As we have mentioned, one important difference between left side and right side science concerns logic. Right side science is limited to first order logic. This may seem like and incredible limitation until we consider in what corresponding way left side science is limited. There we find that traditional left side science and mathematics is limited to first order semantics. This can be seen in its use of monomorphic labels. Algebraic problems typically involve "solving" equations. Given an algebraic equation

$$X(x) = Y \tag{15}$$

then the problem typically becomes that of "solving for x." In this case, x could be a variable with values ranging of some set Ψ. Solving for x will yield a subset of elements of Ψ as the solution. This is an example of first order semantics. The logic however is second order as it allows variables ranging over sets. As we shall see further on, in any analysis based on the right side paradigm, no abstraction is allowed and hence there are no variables. The logic is restricted to first order logic, but the upside will be a second order semantics.

Any algebra must deal with the relationship between symbol and what is symbolised. There are two fundamental possibilities, one left side, and one right side. We have already considered the left side variant. Left side algebra is based on abstraction where the symbol is considered as a variable having possible values ranging over some kind of collection or set Ψ of objects. By its very nature, this kind of algebra must employ second order logic, a key tool of abstraction. Any solution for x will end up with an equation with x on one side and the value of x on the other. In this scenario, the relationship between symbol and symbolised is that between the symbol and a value object which is entirely dependent on circumstance. The circumstance involved is formalised in terms of the specificities of the given algebraic form in (15). Note that the basic problem of "solving for x" leads to an outcome where the solution will always be a set of values belonging to a subset of the set Ψ. This illustrates a very fundamental aspect of abstract algebra. The relationship between symbol and symbolised is strictly monomorphic.

We turn now to the right side paradigm of algebra. In this case, abstraction is forbidden and so there are variables with values ranging over sets. The elementary algebraic form for generic algebra simply becomes

$$x = Y \tag{16}$$

Here we see the dominant feature of algebra based on first order logic. The statement of the algebraic problem is already the solution. "Solving for x" becomes a non-problem. It looks like our right side paradigm for a new kind of mathematics is trivial. Our right side mathematics is still born.

However, the plot thickens if we investigate the limitations of the traditional left side mathematical paradigm. The traditional paradigm might allow the playground of higher order logic but when it comes to the meaning of its abstract symbols, the symbols are strictly limited to first order semantics. The symbols of traditional mathematics are strictly monomorphic. It is here that the right side paradigm starts to shine. In the following, we will explore the right side paradigm not as based on first order logic but second order semantics. We note all the while that the left side paradigm is based on second order logic and only first order semantics. This difference in semantics translates down to a difference in the kind of objects that the left side and right side paradigms deal with. The traditional left side paradigm translates into a science of measurables whilst the right side paradigm, true to the ancient Stoic perspective, becomes a science of bodies. The right side science must be intensely geometrical, through and through. When it comes to polymorphic forms, we must turn to generic geometry. This move from the scalar to the geometrical has already been anticipated in (16) as the scalar x has been replaced by the geometric object $\boldsymbol{x}$.

Thus, the right side interpretation of expression (16) is that the $\boldsymbol{x}$ signifies a geometric entity from a certain point of view. Second order semantics can be introduced by the two forms:

$$x = Y_1, \quad x = Y_2 \tag{17}$$

where $\boldsymbol{x}$ denotes a geometric object whereas Y_1 and Y_2 correspond to the appearance of the object from two different points of view.

From a left side point of view, Y_1 and Y_2 would be seen as mutually exclusive ways at looking at the object $\boldsymbol{x}$, an either-or scenario. The right side paradigm deals with wholes and works understanding that Y_1 and Y_2 are just two different sides of the one thing and so both sides must be synchronically, if not concurrently, present.

Now the above can be written as

$$\boldsymbol{x} - Y_1 = 0, \quad \boldsymbol{x} - Y_2 = 0 \tag{18}$$

the synchrony of the second order semantics demands that

$$(\boldsymbol{x} - Y_1)(\boldsymbol{x} - Y_2) = 0 \tag{19}$$

when expanded leads to

$$\boldsymbol{x}^2 - (Y_1 \boldsymbol{x} + \boldsymbol{x} Y_2) + Y_1 Y_2 = 0 \tag{20}$$

which looks like an everyday quadratic except that the entities involved correspond to geometric objects and the products are geometric products between geometric objects.

Unlike left side mathematics, we are not going to "solve for x." From the perspective of the right side paradigm, this is impossible anyway as there is no usual notion of variables in the right side realm. The right side view is devoid of abstraction and so cannot even "see" such things as variables ranging over abstract sets. Instead, it sees the higher order semantics implicit in geometrical structure. Instead, we are going to look at the necessary shapes involved in the underlying geometry in order for (20) be satisfied. In other words, what are the generic forms that make second order semantic structures possible?

We have already looked qualitatively at this problem and come up with the **aug** start codon structure as a generic geometric form. The task now is to determine the algebraic properties of the generic **a,u,g**, and **c** bases. This is an important step in the process of "cracking" the geometric semantics of the generic code behind it all.

The geometric objects **x**, Y_1, and Y_2 will contain grade one objects and so the generic quadratic (20) will have scalar zero grade components and higher order components. The sum of the components of the same grades must equal zero. Thus, the sum of the zero grade components must be zero. Noting that the inner product commutes we get the form

$$\textbf{x.x} - (Y_1 + Y_2).\textbf{x} + Y_1.Y_2 = 0 \qquad (21)$$

Organising the terms of the quadratic with the even powers of **x** on the one side and the odd on the other, we get

$$\textbf{p.x} = \textbf{x.x} + \textbf{q} \qquad (22)$$

where $\textbf{p} = Y_1 + Y_2$ and $\textbf{q} = Y_1.Y_2$

We will refer to **p** as the *first* generic quality and **q** as the *second* generic quality of the generic quadratic. The problem in hand is to find the *generic* solution to (22). From a left side perspective, equation (22) is seen as a simple scalar binomial

$$px = x^2 + q \qquad (23)$$

which has a scalar solution.

The solution will fall into one of four cases. Three of the cases lead to different kinds of imaginary number where at least one of the coefficients p and q are zero. Since the approach is qualitative, the non-zero values of value of p and q will be assumed to be unity.

Case 1 Universal Imaginary Number: p=0, q=1

In this case, the quadratic simplifies to

$$x^2 = -1 \qquad (24)$$

The value of x will be the traditional "square root of minus one" and denoted by the imaginary number 'i'. Oriented numbers can be formed by combining the real number with the universal imaginary number ib to give an oriented number called a *complex number*. The *complex number* is a linear combination of real and imaginary parts and *so* has the form (a+ib).

Case 2 General Imaginary Number: p=1, q=0

In this case, the quadratic simplifies to

$$x^2 = -1 \tag{25}$$

The solution for x will be the traditional "square root of plus one" and denoted by the general imaginary number j. Linear combinations of real and general imaginary numbers are traditionally called *split-complex numbers.*

Case 3 Particular Imaginary Number: p=0, q=0

In this case, the quadratic simplifies to

$$x^2 = 0 \tag{26}$$

Since the other imaginary numbers square to -1 or +1, this kind of imaginary number can be said to "squares to zero" and we denote it by the imaginary number k.

Case 4 The Real Number: p=1, q=1

For completeness, we have added in a case for a fourth kind of scalar, which we refer to as a real number. For the fourth case, we set both p and q to 1. The quadratic thus becomes

$$x^2 + 1 = x \tag{27}$$

which has a pair of ordinary complex numbers as roots and so notion of determining a "real" number using this approach appears rather strange. Applying a strict left side "solving for x" approach to equation (27) simply yields a pair of ordinary complex numbers. However, looked at qualitatively, there does appear to be a glimmer of hope. The form (27) can yield some sort of hypothesised "real" number if the following holds:

$$x^2 = 0 \quad \text{and} \quad x = 1 \tag{28}$$

We will denote the *real number* specified by (28) the letter l. At this point, to make any headway, we must start to think geometrically rather in left side terms of pure magnitude. Referring back to Goldblatt's approach to spacetime geometry, we see that the three imaginary numbers i, j, and k can be thought of geometrically as oriented spacelike, timelike, and lightlike lines respectively. The difficult geometric entity in the mix, is Goldblatt's fourth kind of line, the *singular line.* We associate the singular line with the "real" number l. Like the imaginary number j, both numbers "square to zero."

The three imaginary numbers j, k, i, together with the "real" number l, form a Three-plus-One structure, and so can be thought of as being constituents of a whole. To understand this structure properly, we are obliged to change paradigms and move over to a right side perspective. Viewed generically, these four scalar entities morph into geometric entities. In what follows, we repeat our four-sided interpretation of the quadratic. In the process, as well as developing our geometric understanding, we enter into the realm of second order semantics and its logic. What follows is beguilingly simple and equally profound. In other words,, what follows is easy to swallow but needs time to digest fully. For the first time since several millennia, we are putting to work a very ancient but little understood gem of wisdom, a gem from the Stoic era.

Before going on to the four case analysis of the generic binomial, we briefly discuss first and second order semantics.

First Order and Second Order Semantics

The left side, scalar approach to solving a quadratic brings into prominence a characteristic feature of left side reasoning. Given a quadratic form, our analysis shows that there are four cases for solving it. From a left side perspective, the cases are mutually exclusive. The solution will be an instance of case 1 OR case 2 OR case 3 OR case 4. This is an example of first order semantics. Solving equations in traditional mathematic is always an exercise in first order semantics.

Second Order Semantics demand Geometry

We now come to second order semantics. In this case, all four solutions must be valid in the one moment. Thus, disjunction is replaced by conjunction. The OR is replaced by the AND. When it comes to the traditional scalar binomial, it is not possible that the four cases be valid in the same moment. The disjunctive approach of first order semantics is the only option. The conjunctive, "all solutions are valid in the same moment" is impossible thus, second order semantics is impossible.

First order semantics only provides a diachronic approach to the quadratic problem where the solution presents as one particular solution amongst many possibilities. We are interested in the synchronic approach where all solutions are valid in the same moment. The synchronic approach becomes possible by considering the quadratic geometrically where all the constituents are geometric objects rather than mere scalar. The geometric approach to the quadratic allows more degrees of freedom so that the four kinds of solution can be considered as Lego building blocks, instead as a list of possible outcomes. By fitting these four fundamental building together to construct a single entity, all four solutions can be considered to be synchronously valid in the same moment.

Sidenote

We use the term "in the same moment" rather than "at the same time." This is because the building blocks involved incorporate fundamental temporality. Thus, the resulting structure built from the four constituents can be said to be synchronous whereas not all parts of the structure are simultaneously concurrent at the same time. The synchrony can be thought of as a moment, not a point in time. The generic structure is not a snapshot in time but can be thought of as a structure in spacetime.

Note also that the approach is generic. In its application, the structures involved may have little to do with the spacetime of physics. It is just that spacetime constructs are quite familiar and provide a practical example of the application of generic concepts.

In what follows, we will investigate second order semantic structure underlying the humble quadratic form. The genus of the quadratic form is that it comes about by the product of two solutions in the one moment. Qualitatively, this kind of structure arises by opposing two oppositions together in the one moment to give rise to the ubiquitous semiotic square. One opposition might feature the impersonal subject, and the other the personal subject. More geometrically, it might involve the opposition between the point at the origin and the point at infinity. Another opposition is between the any-point-whatsoever and an any-point-whatsoever pulled out of a hat. A recurring theme is that the two singular entities in the two oppositions should both be objectively two poles of the one solution.

Treated in an analytical, quantitative manner, this bilinear, "two solutions in one," presents as a scalar quadratic form. Treated synthetically, where the qualitative and the quantitative are both present in the same moment, the problem presents as a highly geometric expression of the quadratic form. All of the scalar constituents become geometric operations and multiplication of objects becomes the geometric product inspired from GA.

Second Order Semantics demand Stoic Logic

Second order semantic demands a new kind of geometry. At its core, this geometry turns out to have a highly logical structure. What we need is a generic form of logic capable of embracing second order semantics. The logic will consist of syllogisms and propositions. In traditional propositional logic, propositions have truth-value and so can be considered true or false, depending on circumstance. The proposition "The door is open" may be true or false. However, the propositional calculus cannot handle the situation where both possibilities are valid in the same moment. The propositional calculus is based on first order semantics where only one of a seemingly contradictory set of propositions can be valid in the same moment. What we need is a synthetic form of logic that can provide an operational form of logic where propositions are Lego building blocks for building a higher truth rather than being mere flickers of the true and false. This synthetic logic will be capable of expressing second order semantics.

An important fine point concerns the difference in nature between first order and second order semantics. First order semantics can be summed up as the semantics of the barcode, a simple labelling technology. The French pioneer of General Linguistics, Ferdinand de Saussure, gave the labelling pair more fancy French names calling them the *signifiant* and the *signifié*. However, the fact remains that this this is merely a barcode labelling technology in action here. As de Saussure remarked, a crucial feature of this labelling technology is the arbitrariness of the *signifiant*, the "arbitrariness of the barcode," so to speak. In traditional left side linguistics, this observation has been turned into a mantra where it is believed that the basic atomic traits of natural language are arbitrary. Thus, the individual phonemes making up speech have no intrinsic meaning. For written language, the spelling of morphemes, the atoms of meaning, is also seen as arbitrary. The word 'cow' for example, has no intrinsic meaning except as a collectively agreed upon audible and symbolic convention to "barcode" cows. Underlying this mantra concerning the arbitrariness of the *significant* in natural languages, is the hypothesis that natural languages are limited to expressing first order semantics only. To the author's knowledge, this hypothesis has never been seriously challenged in modern times, at least not in any formalisable scientific manner.

First Order Semantics			Second Order Semantics	
label (*signifiant*)			signifier (*general*)	signification (*singular*)
labelled (*signifié*)			signified (*particular*)	sign (*universal type*)

Figure 70 First order semantics involves labelling. This is the "semantics of the barcode," one barcode per product. Second order semantics involves understanding an entity as a whole. This requires alignments of the particularity of the object with its abstract form, the signifier and its universal form, its generic type as a sign.

The structure of second order semantics is illustrated in Figure 70. It involves the generic structure of a whole, a Three-plus-One structure. Charles Sanders Peirce homed in on the triadic component of the structure and initiated a discipline he called *semiotics*. His whole philosophy was based on triads. For him, semiotics involved the Three-plus-One structure minus the One.

Despite the contributions of Peirce and the work of many others, what has become known as semiotics is far from being a rigorous science. In particular, semiotics is noted for its rampant excesses of jargon and an absence of any respectable kind of formalisation or rigour. To avoid falling into the morass we will simply illustrate the Three-plus-One structure shown in Figure 70 with our example of developing a generic algebra. In so doing, we must determine the generic solutions to the generic quadratic.

Our problem is one of *alignment*. We must align the key elements of the puzzle in order to arrive at a satisfactory coherent whole. The whole, as a generic entity, will be based on first order logic and second order semantics. In order to advance our science we must provide a formulation of such a logic.

This is an extremely fundamental problem, so fundamental in fact, that it is highly unlikely that nobody has ever thought of it before. Fortunately, the ancient Stoics come to our rescue, and in particular, the great Stoic logician Chrysippus. His key contribution, for our purposes, is his set of five syllogisms, usually referred to as the five undemonstratables. Our task now is to show how this ancient structure can be used as the foundations for a first order logic expressing second order semantics. If we can satisfactorily complete our task, then we will have constructed the cornerstone of a formalism that is diametrically opposed to all the traditional left side science: these left side sciences are based on second order logic and first order semantics. Armed with the two complementary left and right side paradigms, we can then start to move beyond the limitations of the present day sciences that are all effectively half-brained.

Chrysippus Diagrams

Simple constructs in left side logic, together with the simple first order semantics, can be illustrated with Venn diagrams. We need a similar iconic device to illustrate constructs in right side logic and its semantics. Stoic logic is fiercely first order logic and so it might seem that Venn diagrams would be a satisfactory illustrative tool. However, such an approach would mask the non-appreciated and little understood other side of Stoic logic. The logic inherently involves higher order semantics.

The semantic atoms of Stoic logic are expressed in the form of the five syllogisms, the undemonstratables. Four of these syllogisms can be represented by a corresponding Chrysippus diagrams. The exception is the Stoic third syllogism that has no determined diagram but rather expresses the underlying glue that holds the system together. This is the *incompatibility syllogism* shown in Figure 71. Unlike the other four undemonstratables, the third undemonstratable has no Chrysippus diagram representation.

3. Incompatibility Syllogism

> **Either the first or the second but not at the same time
> the first
> hence not the second**

Figure 71 The third undemonstratable expresses the logic of the Shafer stroke. It has no Chrysippus diagram, but demands the black and white coherence of the other four syllogisms. No greys allowed.

In traditional left side logic, the third undemonstratable construct can be recognised as the Shafer stroke, the dual of the *Peirce dagger*. As Peirce showed, all of the primitives of the propositional calculus can be built up from the dagger. Much latter, Shaffer independently proved that the same result for the dual case of the stroke. In right side logic, the incompatibility syllogism underpins the coherence of the whole system. There can be no greys. In the context of a system based on difference uniquely expressed in terms of gender typing, this requirement becomes sacrosanct: Either masculine of feminine type and never both at the same time. There can be no neuter.

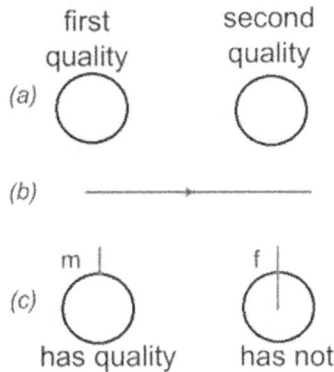

Figure 72 The elementary components of a Chrysippus diagram.(a) Icons for the first and second qualities. (b) The dyad relating the two qualities. (c) Convention for the gender typing: if arrow terminates on circle the entity has the quality and so is of type masculine at this end of the dyad. Otherwise, the arrow end terminates indeterminately inside the circle. The entity has not the quality and so the type at this end is feminine.

Each of the other four syllogisms can be represented by what we are calling Chrysippus diagrams, the right side alternative to Venn diagrams. Each elementary Chrysippus diagram represents the elementary generic characteristics of an entity. The entity can be in the possession or not in the possession of each of two interdependent qualities represented by the two circles. There is also a logical inference direction represented by an arrow dyad. The dyad also represents gender typing specificities by interpreting masculine and feminine according to the convention illustrated in Figure 72.

The four elementary Chrysippus diagrams

Figure 73 shows four schematic diagrams illustrating four of the five Stoic undemonstratable syllogisms. The third undemonstratable, the incompatible syllogism is not included as it has no such schematic diagram. We call these schematic diagrams, *Chrysippus diagrams* and propose them as the right side alternative to Venn diagrams.

To the Stoics, there was only one universal philosophy that embraced and unified all knowledge. The key elements of the science were Physics, Logic, and Ethics. The unification was very tight as can be seen by looking at the schematics in Figure 73, which profile the logical basis. At least, this is the way the author sees the Stoic doctrine. As for the ethical and moral dimension of the science, that will be seen in the critical and normative role that right side science is destined to play with respect to the abstract, open loop rationality, and ethics of left side science.

4. OR exclusive

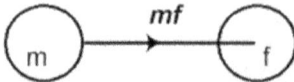

| Either the first or the second |
| the first |
| thus, not the second |

1. Conditional

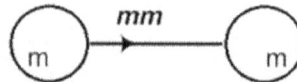

| If the first then the second |
| the first |
| thus, the second |

2. Contraposition of the conditional

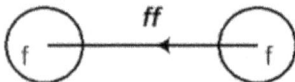

| If the first then the second |
| not the second |
| thus, not the first |

5. OR non-exclusive

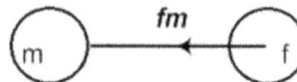

| Either the first or the second |
| not the second |
| thus, the first |

Figure 73 The Chrysippus schematic diagrams for four of the five undemonstratable syllogisms, complete with gender typing.

We have added a gender typing interpretation to the diagrams whereby it becomes quite apparent that the schematics can also be understood as outlining the underlying gender typing of the Stoic theory of the Four Elements, the basis for the physics of antiquity. It is difficult to imagine a closer unification of logic and physics than the author's interpretation of Stoic philosophy as sketched out in the Chrysippus diagrams. With a bit of imagination, one can even get a glimpse of what the Stoics saw as the mysterious fifth element of ancient physics. This appears in the form of the syllogism that has no schematic, the third undemonstratable that demands that the generic qualities involved must not blend but remain mutually incompatible, no matter what. We interpret the third undemonstratable syllogism in terms of the incompatibility of the two genders. For system coherence to prevail, no gender ambiguity can be tolerated. Gender typing cannot permit an entity to be typed as both masculine and feminine at the same time. This the most fundamental organisational and regulatory principle governing any entity organised according to gender, that is to say, according to FC.

We have not the time here to go into a detailed analysis of the Chrysippus diagrams. However, in passing, one cannot but marvel at the simple beauty of the system. Note how the gender typing aligns with our initial semiotic diagram in Figure 64. The handedness of the diagram is important. Just as in Figure 64, the convention adopted in the Stoic logic schematics has MF on the top left side and FF on the bottom left side. This is the convention that we have adopted, whilst we admit that any individual organism may be based on an opposite handedness conventions.

The handedness is important here as, unlike its left side sibling, right side science always carries along a pertinent representation of the left side perspective on reality. In this case, we see that on the top left side of the semiotic square constructed from the undemonstratables we find the Exclusive Disjunction construct. Diagonally opposite, on the right side we find the Non-Exclusive Disjunction. The latter is constructed merely by reversing the flow of inference in the Chrysippus diagram. We have thus two OR operators. We note that the left side OR respects the Law of the Excluded Middle and the right side OR does not. This illustrates a fundamental difference between left side and right side reasoning, a difference long remarked upon when comparing formal mathematics with constructionist mathematics.

It is interesting to note that right side reasoning, as expressed in Stoic logic, includes left side reasoning as a special case. The right side can explain how the left side thinks. The left side cannot even fundamentally explain its own logic let alone that of the right side.

Right side reasoning must come to terms with the dialectical rapport between all of the four generic building blocks of rationality. On the other hand, left side science always ditches the right side of the semiotic and uses only the left side components in its rationality. This produces a very simple epistemology between an abstraction and its version of the real. The missing ingredients can simply be patched over by learnt rules. In the heat of the moment, who needs to be contemplating over ones place in the universe? It is quite amazing that this kind of left side science is quiet sufficient to get a man to the moon and back. Perhaps its main limitation is that it has difficulty in explaining why. The left side has no time for the subjective components with which right side science must contend. In fact the left side can remain totally ignorant in that domain, a blissful case of "hemi neglect."

The advance from our initial binary gendered semiotic square in Figure 64 to the four Chrysippus diagrams underling generic logic is an important advance. The fact that Chrysippus diagrams can be useful in talking about first order logic in one breath, and equally about the primordial elements of physics in another is also worthy of highlighting. In what follows, we will be looking at logical molecules built from these elementary structures. Such molecules will be based on the RGB triadic organisation of dyads. There will be three elements taken from the four Chrysippus logical forms. It is there that one can glimpse the immense power of such a form of reasoning. The reasoning is not deductive or inductive but based on a synchronous presence of all three bases at any one moment. One base might correspond to the fourth undemonstratable and another one or two might be based on the fifth undemonstratable, for example. This means that there will be a single entity consisting of a mixture of two types of disjunction, one respecting the Law of the Excluded Middle, and the other/s violating the same law – in the same moment. Of course, the omnipotent third undemonstratable, that of the inviolate incompatibility of ambiguous typing, that construct is inviolate for all time. First order semantics cannot embrace such synchronic structure.

Genesis of the Imaginary Number

Left side science is based on a single threaded form of rationality. On the other hand, the second order semantic based right side science must be inherently multi-threaded. According to Stoic physics, if we take the five undemonstratables as a template, there are five fundamental threads at play. Vedanta philosophy also talks about the "five breaths." The notion finds echoes in Sankara's theory of the "quintuple dosage." There are five doses, five threads. There are four "elements" or "letters." There are three fundamental qualities. The Hindus call them the three *guna*. This has all been informally known and rummaged upon for millennia.

As for the three qualities in modern times, biochemistry witnesses them in the genetic code as the three bases that make up any codon. Traditional particle physics provides ephemeral glimpses of this fundamental triadic structure in the form of quark triplets. Left side physics then courageously declares that some particles are based on quark triads, and some are just single stand-alone entities with no internal structure at all.

In generic science, these three qualities are arranged in an RGB triadic structure where each of the three qualities are arrows corresponding to any one of the four, *mf, ff, fm* and *mm,* binary typed elements, making up 64 possible combinations in all.

The three qualities are the important ingredient of this five-threaded reality by which entities can be known. Any entity, seen as a whole, will consist of three qualities. However, these qualities are illusory. They are not real but simply a means by which the real can be known. An entity can only be known by what it **has**, not by what it **is**. Any entity **has** attributes but attribute it **is not**.

We will refer to these three illusory elements, as *imaginary*. The generic RGB structure consists of three such imaginary entities that determine a fourth element that we refer to as being *real* (relative to the imaginary).

Left side science tries to harvest accidental attributes as a basis for forming knowledge. Right side science constructs generic attributes based upon gender typing, a totally relativistic mechanism. Even the notion of what is real and what is imaginary, is determined relativistically through the RGB triad construct. The three gender typed arrows in the RGB triad are forcibly imaginary. The triad of three such imaginary arrows determine s fourth arrow, which, relative to the others, is real. This is a fundamentally generic, Three-plus-One structure. There is no absolute restriction on which of the four element **mf, ff, fm**, and even **mm,** can be imaginary. It all depends on context. It is now time to look at the generic triad in the mathematical arena. In this scenario, the three illusory components of the RGB triad will turn out to align with the well know mathematical construct of imaginary numbers.

The universal imaginary entity *fm* (p=0 q=1)

This case aligns with the fifth Stoic undemonstratable with the following Chrysippus diagram:

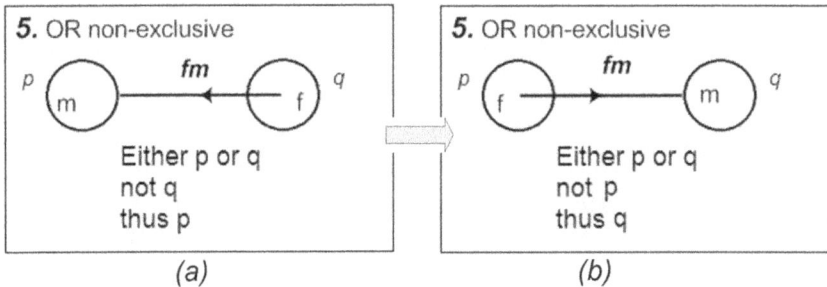

(a)	(b)

The diagram shown in (a) corresponds to the original version of the Chrysippus fifth syllogism. This diagram is the same as for the fourth syllogism except that the direction of the arrow is revered. This transforms the inclusive disjunction of the forth to the exclusive disjunction of the fifth. In what follows, we adopt the convention:

1. The diagram arrow goes from left to right
2. The source side of the arrow is referred to as the p in the syllogism

Adopting this convention transforms the fifth into the equivalent from shown in (b). All the other Chrysippus diagrams need no modification except for the second syllogism, nut that makes no difference, due to symmetry. We can now proceed and consider the four possibilities depending on whether the qualities p and q are present or not.

The signifier is, from expression (28),.the quadratic $p.x = x.x + q$ where the first quality is p, the second quality is q, and $x^2 = x.x$. As if for the scalar case, since we do not have the first quality we set p = 0 but we do have the second so q = 1. The quadratic thus becomes

$$x^2 = -1 \qquad\qquad\qquad (29)$$

which corresponds to the definition of the ordinary imaginary number i of Standard Analysis where "i squares to minus one." However, the difference in (29) is that the imaginary entity is a non-scalar geometric object. This we will write symbolically as a blade i having the property

$$i^2 = (fm)^2 = -1 \qquad\qquad\qquad (30)$$

which illustrates how this imaginary entity i aligns with the generic entity mf, a generic two-blade. Expression (30) illustrates a finer structure than Standard Analysis. The ordinary imaginary number 'i' becomes a geometric object i based on the geometric product of two non-standard demi-entities m and f.

Some would argue that this is not mathematics. We would agree. We call it *anti-mathematics*, the right side, and Non-Standard Analysis version of left side mathematics. In Standard Analysis, the ordinary imaginary number i combined with a real number is called a *complex number*. Combined with the exponential, the imaginary number is useful for defining the trigonometric functions *sin, cost,* and tan, in Standard Analysis.

The general imaginary entity mf (p=1, q=0)

This case aligns with the fourth Stoic undemonstrable with the following Chrysippus diagram:

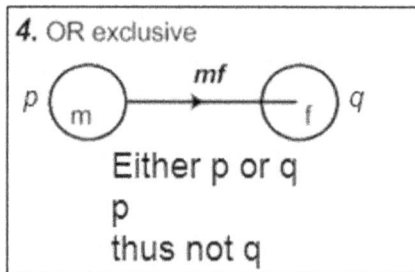

The signifier is once again the quadratic $p.x = x.x + q$ where this time we have the first quality but not the second. Thus, $q = 0$ and p, which cannot be a scalar, has the property that $p.x = 1$. The quadratic thus becomes

$$x^2 = 1 \qquad\qquad\qquad (31)$$

which aligns with the definition of another kind of imaginary number 'j' of Standard Analysis where "j squares to one." Once again, the imaginary entity is a non-scalar geometric object. This we will write symbolically as a blade *j* having the property

$$j^2 = (mf)^2 = 1 \tag{32}$$

Once again, the analysis has been very "non-standard." In Standard Analysis, the imaginary number *j* combined with a real number is called a *split complex numbers or hyperbolic numbers.* Combined with the exponential, the imaginary number is useful for defining the hyperbolic functions *sinh, cosh,* and *tanh.*

The imaginary entity ff (p~0, q~0)

This case aligns with the second Stoic undemonstratable with the Chrysippus diagram below:

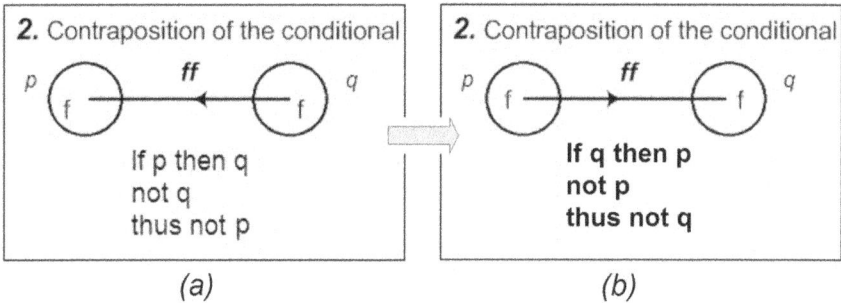

(a)   (b)

Like the case for the fifth syllogism, the original form has to be reversed in accordance with our convention. However, due to the symmetry, the reversion makes no difference.

The signifier is once again the quadratic $p.x = x.x + q$. In this case, p and q will be null. Thus

$$x^2 = 0 \tag{33}$$

which aligns with the definition of the another kind of imaginary number 'k' of Standard Analysis where "k squares to zero." From the right side science perspective, this imaginary entity *k* is a third kind of non-scalar geometric object. This we will write symbolically as a blade *k* having the property

$$k^2 = (ff)^2 = 0 \tag{34}$$

In summary, we have three geometric entities **i, j,** and **k,** that we interpret as synonyms to the three fundamental binary typed elements **fm, ff,** and **mf,** respectively. The three elements concerned here, all exhibit the geometric characteristic of extent. Each of the three elements has the distinctive algebraic quality of squaring to a particular value, either *-1, 1,* or *0.* These values will exhibit the same or similar semantics of the ordinary scalar numbers

The "real" entity *mm* (p=1 q=1)

This case aligns with the first Stoic undemonstratable with the following Chrysippus diagram:

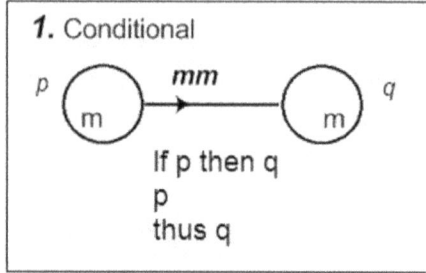

The signifier is once again the quadratic $p.x = x.x + q$. Unlike the previous three "imaginary" cases where the form was broken down into its three generic sub forms, this time the form is taken as a whole. This is why we are terming it the *real* case as distinct from the three fragmentary *imaginary* cases. This time we have the first and the second quality and so both p and q are simply left unaltered. They can be any value. The quadratic to solve is the quadratic as a whole, as is.

With the three imaginary entities *i, j,* and *k,* we were in relatively familiar waters as each entity has a familiar left side counterpart. The essential difference is that the right side version interprets these three imaginary entities as geometric objects whereas traditional left side mathematics treats them as three kinds of scalar numbers differencing only by their respective squares. In this case, the imaginary entities *i, j,* and *k,* become simply the imaginary numbers i, j, and k. GA and the Clifford algebras only deal with these three imaginary numbers. Each space category is defined by its signature (p,q,r) where p is the number of bases that square to one, q the number that square to -1, and r the number that square to 0. From a traditional mathematical perspective, it is generally thought that the Clifford classification of vector spaces based on the quadratic form is complete. However, viewed from a right side perspective, it will become apparent that there is much more structure in play.

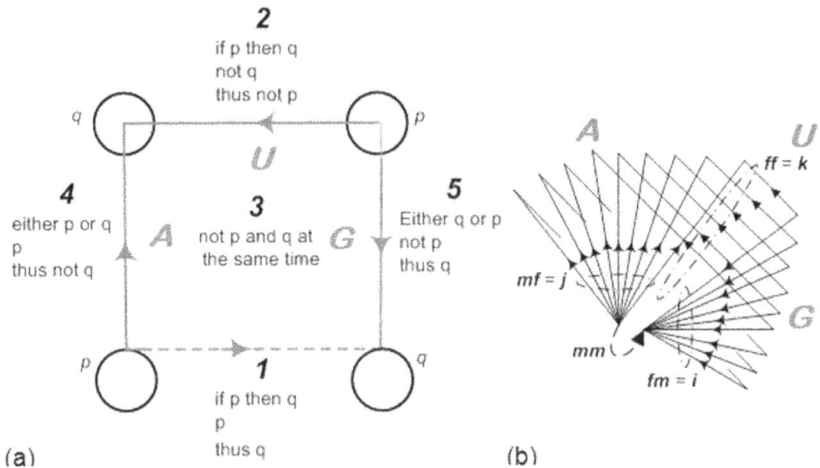

Figure 74 (a) The compound Chrysippus diagram for the AUG start codon. (b) The spacetime version of the AUG start codon. This is the underlying structure of the Special Theory of Relativity, the structure of Minkowski space.

Right Side Geometry is Non-Cartesian

There is an intimate relationship between intuition and visualisation and, in particular, geometric visualisation. The richness of one's intuition is conditioned by the relative richness of one's geometric visualisation capability. In the sciences, the upper limits of intuitive profundity are capped off by the subject's intuitive mastery of geometry. In general, a poor and naïve grasp of geometry will lead to an equally impoverished quality of intuitive understanding. This leads to a central message of this book, notably that the sciences and mathematics of today are impoverished by a geometric science that is quite inadequate for tackling the real and important problems of our age. To overcome this crucial and most debilitating missing ingredient in present day understanding, a new kind of geometry is urgently required.

Unlike other areas of science and mathematics, formal understanding of geometry advances at an incredibly slow pace. For two thousand years, the geometry of Euclid was dominant in Western culture and remained virtually unchanged throughout that period. It seems that geometry does not evolve. Rather, it abruptly emerges from the shadows. Once consolidated, geometry then seems to be relegated to become a tool of the intuition rather than its object.

The move from Euclid towards the kind geometry that dominates the scientific and engineering world of the moderns started to emerge only two centuries ago. It was Grassmann who provided the initial stimulus. He was responsible for initiating two new strands of geometry, one fundamentally based on coordinates and one without. Historically, it was the coordinate geometry version that won out. Kicked off by Gibbs and Heaviside, and formally consolidated by Hilbert, the end result was what is called Cartesian Geometry or Analytic Geometry. This geometry dominates the mathematical sciences of today.

Cartesian Geometry is an extremely simple construct where an n-dimensional space is defined in terms of a set of n orthonormal unit basis vectors. Any vector belonging to the space can be defined as a linear combination of these basis vectors. Each vector can be determined by an ordered set of n numbers called the *coordinates* of the vector. From this, one can see that Cartesian Geometry is very "square shaped," or to be more precise, n-dimensional cube shaped. Moreover, the geometry is always accompanied by a certain, never challenged, attitude of mind. The attitude is that of the Cartesian mind and is summed up as an unquestionable belief in dualism. The geometric expression of Cartesian dualism is the dichotomy between the world of the Cartesian space and the world of things. The doctrinal assumption is that things "live in" the Cartesian space. The space itself is not considered as a thing, or at least not as a thing that lives in something.

Because of the innate characteristics of Cartesian geometry with its flare for squareness, spacetime geometry is very handy for architects drafting up plans for hotels, houses, huts, and even hovels. However, its utility does not stop there. Mathematical physicists have been able to adapt Cartesian geometric technology to encompass such areas as spacetime geometry. Very wide ranges of such problem domains can be covered by simply altering the signature (p,q,r) of the vector space and thus alter the metric. The spacetime geometry of Minkowski space is determined by one such signature and corresponding metric.

The second strand of new geometry initiated by Grassmann was the one that really concerned him, a coordinate free geometry. Through the following of Hamilton, Clifford, and reinvigorated in more recent times by Hestenes, the discipline came to be called Geometric Algebra. GA enjoys some fundamental characteristics that place it apart from classical Cartesian geometry. In our terminology, Cartesian geometry fits into the traditional, left side, dualistic scientific paradigm. It is a left side geometry. GA presents itself as a possible candidate for a right side version of geometry, a non-dualist geometry. In this respect, the fundamental GA notion of the geometric product is very promising. Instead of an algebra involving a dual world mathematical transformations applied to geometric objects, the geometric product eliminates the "other worldly" notion of transformations. Instead of multiplying a geometric object by a mathematical transformations, the geometric object is multiplied by another geometric object. In brief, the geometric product of GA is a first class operation only involving geometric objects.

However, under the hood, GA still potters along on the same old primitive left side motor as traditional Cartesian geometry. Although the algebra of GA can avoid going into the details of coordinates, it is still Cartesian at the core. Just like classical Cartesian analytic geometry, GA is still based on set of n basis vectors. Even the notion of a signature (p,q,r) is still exactly the same as the Cartesian version.

From our perspective, a true right side geometry must correspond to the Leibniz view as being a geometry without number, a geometry based on letters. Moreover, according to Leibnitz, the resulting geometric algebra should be so universal as to "give exact descriptions of natural things by means of it, such, for example, as the structure of plants and animals." The central contribution of our work is to realise the vision of Leibniz, and to do so to the letter. Such a project cannot be achieved via the traditional left side scientific paradigm. In particular, the project cannot be even visualised in term of tradition left side geometry. Cartesian geometry can only aspire to act as a container for natural things. It cannot describe natural things.

When we make the paradigm switch from left side Cartesian geometry to right side geometry, the dichotomy between that geometry as container and the thing contained vanishes. No longer is there a distinction between the thing in space and the space. The two become one and the same; they both become the same stuff. A geometric space becomes a body, each elementary body becomes its own geometric space.

In order to illustrate what we mean and to highlight the limitations of Cartesian geometry, consider the following. We try to interpret Cartesian geometry as a way of describing the plants and animals as mentioned by Leibniz. We first note that all the variants of Cartesian geometry share the same affine skeleton. All have the same underlying bony structure. One could imagine a biological organism based on this technology as some kind of monolithic, fleshless, box-like vertebrate. Cartesian geometry does not allow for much biodiversity. The only biodiversity allowed in Cartesian geometry is the hide that can be pulled over and around these craggy bones. For each signature of the space, there will be a different hide. In this way, Cartesian geometry can be visualised as a bag of bones. The bones provide the common, underlying affine structure and the bag, made of hide, provides the shape. According to this technology, each elementary biological organism can be described by a codon made up of three numbers (p,q,r) called its signature where the numbers can be any non-negative number. All of the possible signatures enumerate all of the possible elementary "biodiversity" possible with this kind of Cartesian technology.

It is nonsense, of course, to talk of Cartesian geometry in this biological way. This Cartesian geometric technology just does not possess the wherewithal to be a player in the Leibniz world of natural things. Right side geometry is the answer. The Geometric Algebra initiated by Grassmann becomes our first candidate. However, GA also suffers from the same limitations as traditional Analytic Geometry; GA is also founded on a Cartesian basis. In our earlier analysis, we showed how the Cartesian basis could be replaced with a totally generic basis, founded on gender. This leads to Generic Geometric Algebra. The elementary entities in GGA can be described by a three letter codons where the letters belong to a four-letter alphabet, just like in the genetic code. We call it the *generic code* as its generic vocation opens up the possibility of describing all of the natural things of Leibniz, not just the plants and animals. It is GGA that realises the dream of Leibniz.

Traditional GA and the Clifford Algebras treat spaces with signatures (p,q,r) but in practically all case, the signature is limited to the cases (p,q,0) . The geometries with non-zero values of rare considered as degenerate. Using basis vectors that "square to zero" can be very misleading. This can be seen in Goldblatt's excellent axiomatic treatment of spacetime geometry. Avoiding using coordinates, and concentrating on orthogonality properties, Goldblatt brings into clear relief, four different kinds of lines. Firstly, there are the usual spacelike, timelike, and lightlike lines of spacetime geometry. From a right side perspective, we treat these lines as $i, j,$ and k geometric entities, respectively.

In STA, GA, and the Clifford Algebras in general, there is no mention of a fourth kind of line. However, there is a fourth kind of line even traditional mathematics recognises, what Goldblatt calls the *singular line*. Like lightlike lines, singular lines also square to zero. Such lines are hence orthogonal to themselves. However, the singular line distinguish itself from the other three types of line in that it is not only orthogonal to itself but to every other line present. From the abstract perspective of left side mathematics, singular lines appear too degenerate to be useful. One should remark an important point here. The directionality along a geometric line describes a certain sort of dimension of space. Of particular interest are the lightlike line and the singular line. Each kind of line describes its own version of a possible spatial dimension. In both case a basis element in this dimension will square to zero. The traditional signature based Cartesian approach to geometry is totally oblivious to this distinction. To avoid conflict and confusion, left side geometry should not meddle with Cartesian dimension that "square to zero." Left side geometry should restrict itself to basis elements that have positive or negative squares.

In the right side geometry of GGA, the elementary basis does not consist of an arbitrary number of orthonormal basis vectors. Each type of elementary space cum entity is made up of exactly three and only three basis elements arranged in the standard generic RGB triad configuration. Each basis element is typed with one of the four generic types. The gender expressions of the types are *mf, ff, fm,* and *mm*. The genetic version uses the four letters A, U, G, and C. Interpreted geometrically, they correspond to the geometry sub-bases *j, k, i,* and *l* respectively. All four versions of the four elementary forms share the same generic semantics and logic

Right side geometry takes a markedly different position to traditional geometry. It considers this "singular" kind of entity as a geometric entity in its own right and of equal geometric status to the other three elementary types . This entity aligns with the generic quadratic in its entirety where the coefficients p and q in the semiotic quadratic be any value whatsoever, except zero. This situation aligns with the first of the Stoic syllogisms, the one that we have gender typed as *mm*. It is this *mm* entity, as a geometric object, that we associate with the singular line that is orthogonal to all other entities present, including itself. This "singular" entity *mm* is indeed a difficult entity to come to grips with.

The Logic of Appearances

Figure 74(b) illustrates how the *mm* entity can be seen as the result if the combination of the three imaginary entities *mf, ff, fm,* and *mm*. The triadic combination of the three dyads respects the RGB structure, previously explained. The RGB orientation convention avoids violating FC as no dyad can be said to precede or precede any other. The triadic RGB structure determines a forth entity, the *mm* dyad. This implicit *mm* dyad is the only dyad in the configuration where both p part and the q part are in possession at the same time.

However, to have both p and q at the same time would contradict the third Stoic syllogism, the most fundamental of them all. This can quickly become a tine of worms. If p and q are different, which they are, then the time of p will be different from the time of q. However, if p and q are indistin-guishable, the times will be indistinguishable. One could say that p and q are in the one moment. One way around the conundrum is to recognize that p and q can be interpreted as gender typings. There, in order to maintain systemic coherence, the gender typing cannot be ambiguous. There can be no greys. The incompatibility syllogism provides such a guarantee.

One way to understand the generic structure in Figure 74 is to interpret the imaginary entities as belonging to the world of appearances. These "attribute dyads" do not correspond to what the entity *is*, but rather to what the entity *has*. What an entity *is*, and in particular, what it is *now*, embraces its whole being. This is the "real" aspect of the entity. The entity can never directly be known for what it is, it can never be known directly by its real, but only in terms of what it *has*. The principle possession of an entity, the most important aspect of what it has, is its appearance.

Compound diagrams can be made up of triads of simple Chrysippus diagrams where, remarkably, all three syllogisms are synchronically true. Each triadic structure provides a logical formalisation of the appearance of an entity. It is via these triadic structures that the entity can be known. What the entity actually and really is, can be formalised in terms of these triadic, imaginary structures of appearance.

Figure 74 is of cardinal importance in the science of the generic. It formalises the starting point and ensuring reference point for all that follows. We call this triadic structure, the *start codon*, and using the letters *a, u, g* as synonyms for *mf, ff,* and *fm,* respectively, we label the *aug* codon. The universe gyrates around the *aug* start codon at least that will be the way it appears to any organism in the universe, including the universe itself.. This might seem to violate FC as FC demands that there be no absolute centre for the universe. The centre of the universe can be any point whatsoever, according to the principle of FC. But, this is the purpose of the exercise. The *aug* codon formalises the very notion of the "any point whatsoever." This notion is codified in the dyad *mm*, which expresses the non-duality between _the_ any-point-whatsoever, and *any* any-point-whatsoever. This Principle of Non-Duality is fundamental to Advaita philosophers in their explanations of the relationship between the Brahman and ātman It is nothing short of astonishing to see it appear here in our formalisation, not only in the guise of the *aug* start codon, with all its biological implications, but in the spacetime guise of the Special Theory of Relativity.

The Parmenidean Dimension

The *aug* codon is literally Ground Zero for our science. As the reader can see, it is rather crowded territory. Literally, anything can be found here. Before venturing on from this crowded terrain, we note the geometric interpretation of the *aug* codon, as illustrated in Figure 74(b). Viewed from a traditional left side mathematical perspective, the geometry can be interpreted as the familiar diagram consisting of a cone of timelike lines and a cone of spacelike lines connected together by a bundle of lightlike lines. This is a diagram of a two dimensional Minkowski space, the fundamental space for spacetime geometry. The direction of the lines involved might appear a bit odd, but we can overlook that fine point. What we cannot overlook is the **mm** part of the diagram. From a left side perspective, this mysterious **mm** line should not even be there. Instead, the two end points of **mm** should be replaced by a single point. The **mm** type line seems to have no place in left side mathematics and physics.

We now move to the right side perspective. Our problem is to understand what the **mm** dyad in the diagram means. After a bit of reflection on the matter, we know that both ends of the **mm** occupy the same point in time and in space. This agrees with the left side interpretation. The difference in the right side interpretation is that although coincident in time and space, there must be another dimension that allows the end point of the dyad to be distinct from each other. We will call this dimension, the *Parmenidean* dimension thereby implying that only entities in this dimension are real. According to Parmenides' doctrine, only that in the eternal present is real and can be said to exist. All else is imaginary. The length between any two points along this Parmenidean will be effectively zero, or infinitesimally small.

The **mm** type dyad will thus be like the **ff** dyad and squares to zero. However, it differs from the **ff** dyad in that its end points are singular, whereas the end points of **ff** are undetermined. This Parmenidean dyad **mm**, plays a role that is similar to what we have already briefly touched upon concerning conformal geometry. A conformal geometry can be constructed from a standard Euclidean space G(3,0,0) by adding two extra dimensions, each with a basis that squares to zero. One dimension adds a "point at infinity" and results in projective geometry. This does not concern us here. The second extra dimension adds a generic "point at the origin." In this case, the extra dimension can be thought of as the Parmenidean dimension, Because it square to zero, standard GA treats it as the same kind of geometric entity as a lightlike line.

Starting Point Invariance

In this perspective, we can intuitively interpret the **mm** entity in our diagram as having two kinds of end point, both of which are a kind of origin. The target of the **mm** dyad will correspond to the determined starting point of the whole enterprise. This starting point is accidental can be any particular starting point whatsoever. On the other side, the source of the **mm** dyad will correspond to the *generic* starting point, the generic origin, the universal origin.

The whole aim of the exercise is to ensure that any particular starting point is indistinguishable from the generic starting point.

The whole aim of the operation is to ensure that any particular starting point is indistinguishable from the generic starting point. This is a key characteristic of right side science. The science must be starting point invariant. When manifested in spacetime algebra, this requirement demands a geometry with the appearance of Minkowski space.

Spacetime geometry is only a particular manifestation of starting point invariance. In that case, starting point invariance translates into the invariance of the speed of light, a property of Minkowski space. However, there can be other manifestations of the principle of FC and that must hence satisfy the same draconian constraints of starting point invariance. We claim that the fundamental organisational principle of any form of autonomous existence must be based on FC and this includes hydrocarbon-based life forms. Biological life forms themselves are subject to their particular manifestation of relativity theory. By replacing the spacetime physics specific form of relativity with its generic counterpart, we can claim that an underlying feature of the science of life forms must be based on a theory that is starting point invariant. It must start with the **aug** codon, so to speak.

The Particular and the Universal aug Entity

Right side science is based on first order logic and so does not allow abstraction. Thus, the relationship between the particular and the universal is not that between the particular and its abstract generalisation. In the latter case, there is an inevitable duality between the particular thing and its abstract representation. Abstractions and generalisations of any sort simply do not exist. From a right side perspective of reality, the universal is always present with the particular. They are both two moments of the same entity. This principle was demonstrated in the case of the starting point invariance of right side science, the generic way of expressing the Special Relativity. The starting point entity as origin and point of reference and the universal generic origin must be indistinguishable.

What this means is that the **aug** entity has two interpretations. In the case of the spacetime manifestation of the **aug** codon, the universal interpretation is in the form of a Minkowski spacetime geometry that facilitates systemic starting point invariance and hence FC. The particular interpretation of the **aug** codon as an entity, will be in the form of an entity, which actually enforces systemic FC. This will be a "boson" messenger kind of particle. In a later section, we will show that the **aug** quantum numbers indicate that it corresponds to the photon.

In the biological manifestation of our generic science, the two-pronged aspect of the **aug** codon is quite apparent. From the universal perspective, **aug** acts as the start codon for the translation of gene sequence to a protein. In this role, it does not code any particular amino acid. Interpreted as a particular, it codes the amino acid Methionine that acts as an intermediary in the biosynthesis many essential biological compounds such as cysteine, carnitine, taurine, lecithin, phosphatidylcholine, and other phospholipids. In plants, it is also used in the synthesis of ethylene in the Yang cycle.

The deep understanding of generic cum genetic code in the biological sphere will only become clear once we have developed the underlying generic geometry to much greater depth than in this present work.

The **aug** codon is one of only two codons in the generic code that only codes one unique amino acid. The other is the **ugg** codon. We will come across this codon in the main text of the book where we will find that it has equally fundamental generic ramifications as the **aug** codon.

Bibliography

Aristotle Metaphysics [Book].

Bäck Allan On Reduplication - Logical Theories of Qualification [Book]. - Amsterdam : Leiden, 1996. - ISBN 90-04-10539-5.

Bäck Allan What is Being Qua Being? [Book Section] // Idealization XI : Historicla Studies on Abstraction and Idealization / book auth. Frncesco Coniglione Robert Poli, Ronin Rollinger. - New York : [s.n.], 2004.

Becker Lawrence C. A New Stoicism [Book]. - [s.l.] : Sabon Princeton University Press, 1998. - ISBN 0-691-01660-7 .

Berlucchi G Mangun GR, Gazzaniga MS Visuospatial attention and the split brain [Journal]. - [s.l.] : News Physiol Sci, 1997. - Vol. 12. - pp. 226-31.

Bobzien Susanne Chrysippus' Theory of Causes [Book Section] // Topics in Stoic Philosophy / book auth. Ierofiakonou Katerina. - Oxford : Clarendon Press, 1999.

Bobzien Susanne Early Stoic Determinism [Online] // Revue de Métaphysique et de Morale . - 2010. - 22 June 2010. - http://www.cairn.info/article_p.php?ID_ARTICLE=RMM_054_0489.

Bono Edward De The Use of Lateral Thinking [Book]. - [s.l.] : Penguin Books , 1967. - ISBN-10: 0140137882, ISBN-13: 978-0140137880.

C.J.Marsolek Abstract visual-form representations in the left cerebral hemisphere [Article] // J Exp Psychol Hum Percept Perform. . - 21 Apr 1995. - pp. 375-86.

Caelli Terry, Hoffman William and Lindman Harold Subjective Lorentz transformations and the perception of motion [Journal] // Journal of the Optical Society of America. - [s.l.] : Optical Society of America, US., 1978. - 3 : Vol. 68. - pp. 402-411.

Calogero Guido, and Lawrence H. Starkey Eleaticism [Online] // Encyclopædia Britannica. - 1 5 2010. - 1 5 2010. - http://www.britannica.com/EBchecked/topic/182279/Eleaticism.

Cartwright Mary Lucy Mathematics and Thinking Mathematically [Book Section] // Musings of the Masters. An Anthology of Mathematical Reflections / ed. Raymond Ayoub. - [s.l.] : Mathematical Association of America, 2004. - ISBN: 9780883855492.

Chalmers David Explaining Consciousness: The Hard Problem [Article] // Journal of Consciousness Studies. - 1995. - Special Issue.

Chénique François Éléments de logique classique : l'art de penser et de juger, l'art de raisonner [Book]. - Paris : Bordas, 1974. - ISBN: 2-04-000511-0.

Cicero The Nature of the Gods [Book] / trans. Walsh P. G.. - [s.l.] : Oxford Press.

Colish Marcia L. The Stoic Tradition from Antiquity to the Early Middle Age [Book] / ed. H.A. Oberman H. Chadwick. - Leiden : E.J. Brill, 1985. - Vol. 1 : 2. - ISBN 90 04 07267 5.

Connes Alain Non-commutative geometry [Book]. - Boston MA : Academic Press, 1994. - ISBN 978-0-12-185860-5.

Corazzon Raul Aristotle and the Science of Being qua Being [Online] // Theory and History of Ontology. - 5 5 2010. - 5 5 2010. - http://ontology.mobi/bqb01.htm.

Couturat Louis La logique de Leibniz: d'après des documents inédits [Book]. - [s.l.] : F. Alcan, 1901.

Dawkins Richard The God Delusion [Book]. - [s.l.] : Mariner Books, 2008. - p. 464. - ISBN-10: 0618918248, ISBN-13: 978-0618918249.

de Lacy Phillip The Stoic Categories as Methodological Principles [Book]. - [s.l.] : The Johns Hopkins University Press, 1945. - Vol. 76 : pp. 246-263.

Doran C. [et al.] Lie groups as spin groups [Journal] // J. Math. Phys.. - August 1993. - 8 : Vol. 34. - pp. 3642-3669.

Doran Chris, Lasenby Anthony and Lasenby Joan Conformal geometry, Euclidean space and geometric algebra [Book Section] // Uncertainty in Geometric Computations, Kluwer International Series in Engineering and Computer Science. - Boston, MA, USA. : Kluwer Academic Press, 2002.

Dorst Leo and Lesenby Joan Guide to Geometric Algebra in Practice [Book Section]. - [s.l.] : Springer 2011, 2011.

Dorst Leo, Fontijne Daniel and Mann Stephen Geometric Algebra for Computer Science, An Object Oriented Approach to Geometry [Book]. - [s.l.] : Elsevier, 2007.

Eccles John C. and Popper Karl The Self and Its Brain: An Argument for Interactionism [Book]. - [s.l.] : Routledge , 1984.

Eddington Arthur Stanley The Nature of the Physical World [Book]. - New York : he Macmillan Company, 1928.

Epictetus Discussions [Book]. - 55-135 AD.

Finey Michele Secrets of the Zodiac [Book]. - [s.l.] : Allen & Unwin, 2009. - ISBN-13: 978-1741757446, ISBN-10: 1741757444.

Furley David [Online].

Furley David Parmenides of Elea (Born c. 515 BCE) [Online] // Encyclopedia of Philosophy. - 2006. - 15 Apr 2011. - http://www.encyclopedia.com.

Genesis The Holy Bible [Book]. - [s.l.] : King James Edition.

Goldblatt Robert Orthogonality and Spacetime Geometry [Book]. - Berlin : Springer-Verlag, 1987. - p. 189. - ISBN-10: 354096519X, ISBN-13: 978-3540965190.

Goodman Nelson and Quine W. V. Steps Toward a Constructive Nominalism [Journal]. - [s.l.] : Journal of Symbolic Logic, 1947. - Vol. 12. - pp. 105-122.

Gould Josiah B The Philosophy of Chrysyppus [Book]. - New York : State University of New York Press, 1970.

Gould Josiah The Philosophy of Chrysippus [Book]. - [s.l.] : SUNY, 1970. - ISBN 087395064X.

Grassman Hermann Günther Die lineale Ausdehnungslehre 1995 English translation [Book] / trans. Kannenberg Lloyd . - Chicago : Open Court, 1844.

Grassman Hermann Günther Geometrische Analyse geknüpft an die von Leibniz erfundene geometrische Charakteristik [Book]. - Leipzig : Weidmann'sche Buchhanlung, 1847.

Greimas Algirdas Julien Maupassant: La Sémiotique Du Texte - Exercices Pratiques [Book]. - [s.l.] : Editions Du Seuil, 1991. - ISBN 2020043653 (2-02-004365-3).

Gul Stephen, Lasenby Anthony and Doran Chris Imaginary Numbers Are Not Real - The Geometric Algebra of Spacetime [Journal]. - [s.l.] : Foundations of Physics, 1993. - 8 : Vol. 23.

Hahm David E. The Origins of Stoic Cosmology [Book]. - [s.l.] : Ohio State University Press, 1977. - ISBN 0-8142-0253-5.

Hameroff Stuart Quantum computation in brain microtubules? The Penrose-Hameroff "Orch OR" model of consciousness [Online] // New Frontier in Brain/Mind Science. - 2010. - http://www.quantumconsciousness.org/penrose-hameroff/quantumcomputation.html.

Heath Jeffrey G. Australian Aboriginal languages [Online] // Encyclopædia Britannica Online. - 2010. - 15 July 2010. - http://www.britannica.com/EBchecked/topic/43873/Australian-Aboriginal-languages.

Heath Jeffrey G. Australian Aboriginal languages [Online] // Encyclopædia Britannica Online. - 15 July 2010. - http://www.britannica.com/EBchecked/topic/43873/Australian-Aboriginal-languages.

Hegel G.W.F Lectures on the History of Philosophy [Book] / trans. Haldane E. S.. - 1892-96.

Hegel G.W.F. Who Thinks Abstractly? [Book Section] // Hegel: Texts and Commentary / book auth. Kaufmann Walter. - New York : Anchor Books, 1966 (1808).

Heller Wendy The Neuropsychology of Emotion : Developmental Patterns and Implications of Psychopathology [Book Section] // Psychological and Biological Approaches To Emotion / book auth. Nancy L. Stein Bennett Leventhal, Thomas R. Trabasso. - [s.l.] : Psychology Press, 1990. - ISBN-10: 0805801502, ISBN-13: 978-0805801507.

Hellman Geoffrey Mathematics without numbers: towards a modal-structural interpretation [Book]. - New York : Oxford University Press, 1989. - ISBN-10: 0198240341, ISBN-13: 978-0198240341.

Hestenes David Grassmann's Legacy [Book Section] // From Past to Future: Graßmann's Work in Context / book auth. Petsche H.-J. [et al.]. - [s.l.] : Springer Basel, 2009.

Hestenes David Old Wine in New Bottles: A new algebraic framework for computational geometry [Book Section] // Geometric Algebra with Applications in Science and Engineering / book auth. Sobczyk E. Bayro-Corrochano and G.. - [s.l.] : Birkhäuser, 2001.

Hick John Philosophy of religion [Book]. - Englewood Cliffs, N.J : Prentice-Hall, 1963. - ISBN: 0715601520 / 0-7156-0152-0. - SBN: 0715601520 / 0-7156-0152-0.

Hohm David E. The Origins of Stoic Cosmology [Book]. - [s.l.] : Ohio State University Press, 1977. - ISBN 0-8142-0253-5.

Irvine A. D. Philosophy of mathematics [Book]. - [s.l.] : North Holland, 2009. - ISBN-10: 0444515550; ISBN-13: 978-0444515551.

J. Lasenby A.N. Lasenby and C.J.L. Doran A unified mathematical language for physics and engineering in the 21st century [Journal] // Phil. Trans. R. Soc. Lond.. - 2000. - 358. - pp. 21-39.

Josephson B. D. Mind-Matter Unification Project video lectures [Online]. - http://sms.cam.ac.uk/collection/664697.

Jung Carl A Psychological Approach to the Dogma of the Trinity [Book Section] // The Collected works of C.G. Jung / trans. Hull R. F. C.. - 1960. - Vol. 11.

Kant Immanuel Critique of Pure Reason [Book] / trans. Meiklejohn J.M.D.. - [s.l.] : Cosimo Classics, 1738. - p. 504. - ISBN-10: 1605204498, ISBN-13: 978-1605204499.

Kant Immanuel Prolegomena to any Future Metaphysics [Book]. - 1783.

Keynes John Maynard Essays and sketches in biography,: Including the complete text of Essays in biography, and Two memoirs [Book]. - New York : Meridian Books, 1956. - p. 347. - NBD1521135.

Kiyosaki Robert T. Cashflow Quadrant: Rich Dad's Guide to Financial Freedom [Book]. - 1998. - ISBN: 0-9643856-2-7-Pbk.

Kleve K. Scurra Atticus. The Epicurean view of Socrates [Book Section] // Suzetesis. studi still ' epicureismo greco e romano offerti a Marcello Gigante / ed. Carratelli G. P.. - Naples : [s.n.], 1983. - 2.

Koestler Arthur Janus: A Summing Up [Book]. - UK : Hutchinson, , 1978. - ISBN 0-09-132100-X.

Lacy Phillip de The Stoic Categories as Methodological Principles [Journal] // Transactions and Proceedings of the American Philological Association. - [s.l.] : The Johns Hopkins University Press, 1945. - Vol. 76. - pp. 246-263.

Laërtius Diogenes Life of Chrysippus [Online] // The Lives and Opinions of Eminent Philosophers / ed. YONGE C.D.. - 2 May 2011. - http://classicpersuasion.org/pw/diogenes/dlchrysippus.htm.

Lasenby J., Lasenby A. N. and Doran C. J. L. A unified mathematical language for physics and engineering in the 21st century [Journal]. - [s.l.] : Phil. Trans. R. Soc. Lond, 2000. - Vol. A 358. - pp. 21-39.

Leibniz G. W. Philosophical Papers and Letters: A Selection [Book] / trans. Loemker L. E. . - [s.l.] : Springer. - 2nd Edition 1975. - ISBN-10: 902770693X ISBN-13: 978-9027706935.

Leng Mary Mathematics and reality [Book]. - [s.l.] : Oxford University Press, 2010. - p. 278. - ISBN-10: 0199280797; ISBN-13: 978-0199280797.

Locke John An Essay Concerning Human Understanding [Online] // Classic Works of Literature, Philosophy, Science, History and Exploration and Travel. - University of Adelaide, 1689. - 13 2 2011. - http://ebooks.adelaide.edu.au/l/locke/john/l81u/.

Lowe E. J. The Four-Category Ontology: A Metaphysical Foundation for Natural Science [Book]. - USA : Oxford University Press, 2007. - ISBN-10: 0199229813; ISBN-13: 978-0199229819.

Lumer E. D., Friston K. J. and Rees G Neural correlates of perceptual rivalry in the human brain [Journal] // Science. - [s.l.] : Science, 1998. - 5371 : Vol. 280. - pp. 1930—34.

Marx Karl Outline of the Critique of Political Economy (Grundrisse) [Book] / trans. Nicolaus Martin. - [s.l.] : Penguin Bioks, 1857–1858.

Marx Karl A Contribution to the Critique of Political Economy [Book]. - 1859.

Marx Karl Grundrisse [Book] / trans. Nicolaus Martin. - [s.l.] : Penguin Bioks.

McGilchrist Iain The Master and His Emissary: The Divided Brain and the Making of the Western World [Book]. - [s.l.] : Yale University Press, 2009. - p. 608. - ISBN-10: 030014878X ISBN-13: 978-0300148787.

Montaigne Michel de OF DEMOCRITUS AND HERACLITUS [Book Section] // Essays of Montaigne Vol. 3. - 1580.

Moore Douglas J. Huntington A metaphysics of the computer: The reality machine and a new science for the holistic age [Book]. - San Francisco, Calif : Mellen Research University Press, 1992.

Murray Gilbert The Stoic Philosophy [Book]. - London : Kessinger Publishing, LLC, 2007. - p. 80. - ISBN-10: 1432644467 ISBN-13: 978-1432644468.

Nagel Thomas The View from Nowhere [Book]. - [s.l.] : Oxford University Press, 1986.

Newman W.R. Dibner Collection MS. 1031B [Online] // The Chymistry of Isaac Newton. - Dibner Library for the History of Science and Technology, Smithsonian Institution. - 31 12 2008. - http://webapp1.dlib.indiana.edu/newton/mss/intro/ALCH00081.

Newman William R. The Chymistry of Isaac [Online] // Dibner Collection MS. 1031B. - Dibner Library for the History of Science and Technology, Smithsonian Institution. - 31 12 2008. - http://webapp1.dlib.indiana.edu/newton/mss/intro/ALCH00081.

Okakura Kakuzo The Book of Tea [Book]. - [s.l.] : Project Gutenberg EBook, 1906.

Ouspensky P. D. Tertium Organum: Or, The Third Canon of Thought and a Key to the Enigmas of the World [Book] = Tertium Organum. - 1990. - THE most difficult thing is to know what we do know, and what we do not know.

Peirce Charles Sanders A Boolian [sic] Algebra with One Constant [Book Section] // Collected Papers of Charles Sanders Peirce / book auth. Hartshorne C and Weiss P. - [s.l.] : Harvard University Press, 1880. - Vol. 4.

Penrose R Shadows of the Mind [Book]. - New York : Oxford University Press, 1994. - p. 457. - ISBN 0-19-853978-9.

Perice Charles Sanders What Pragmatisim is [Journal]. - [s.l.] : The Monist, 1905. - 2 : Vol. 15. - pp. 161-181.

Plato Timaeus [Book] / trans. Jowett Benjamin. - 360 B.C.E..

Plotinus The Seven [Book]. - 250 AD.

Poincaré Henri Last Essays (Dernières Pensées) [Book] / trans. Bolduc John W.. - [s.l.] : Dover Publications, New York, 1969, 1917.

Poincaré Henri Why Space has Three Dimensions [Book Section] // Mathematics and Science: Last Essays / trans. Bolduc John W.. - New York : Dover Publications, 1969.

Poincaré Henri Why Space has Three Dimensions [Book Section] // Mathematics and Science: Last Essays / trans. Bolduc John W.. - New York : Dover Publications, 1969.

Poli Roberto QUA-THEORIES in Shapes of Forms [Book] / ed. Albertazzi L.. - Dordrecht : Kluwer, 1998. - pp. 245-256.

Polysynthetic language - Definition [Online] // Word IQ. - July 2010. - http://www.wordiq.com/definition/Polysynthetic_language.

Popper Karl Conjectures and Refutations [Book]. - New York : Rouledge % Kegan Paul, 1963. - ISB 10 0-415-28593-3.

Reesor Margaret E. The nature of man in early Stoic philosophy [Book]. - [s.l.] : Gerald Duckworth & Co Ltd, 1989. - ISBN-10: 0715622560, ISBN-13: 978-0715622568.

Resnik Michael D. Structural Relativity [Journal]. - [s.l.] : Philosophia Mathematica, 1996. - 2 : Vol. 4. - pp. 83-99.

Robb A. A. A Theory of Space and Time [Book]. - Cambridge : Cambridge University Press, 1914.

Russel Bertrand The philosophy of logical atomism and other essays, 1914-19 [Book] / ed. Slater John G.. - London and New York : Routledge Press, 1986 . - p. 418.

Sapir Edward Language. an Introduction to the Study of Speech [Book]. - Teddington, Middlesex : Echo Library, 2006.

Shapiro Stewart Philosophy and Mathematics: Structure and Ontology [Book]. - [s.l.] : Oxford University Press, 1997. - ISBN-10: 0195139305, ISBN-13: 978-0195139303 .

Sharma Arvind The Philosophy of Religion and Advaita Vedanta: A Comparative Study in Religion and Reason [Book] = The Philosophy of Religion and Advaita. - [s.l.] : The Pennsylvania State University Press, 1995. - ANSA Z39.48-1984.

Sharples R. W. STOICS,EPICUREANS AND SCEPTICS An Introduction to Hellenistic Philosophy [Book]. - London : Routledge, 1996. - ISBN-10: 9780415110358, ISBN-13: 978-0415110358.

Sheffer Henry M. A set of five independent postulates for Boolean algebras, with application to logical constants, [Article] // Transactions of the American Mathematical Society. - 1913. - 14. - pp. 481-488.

Sihvola Juha Happiness in Ancient Philosophy [Online] // Helsinki Collegium for Advanced Studies. - 2008.

Sihvola Juha Sexual Desire and Virtue in Ancient Philosophy [Book Section] // Sex and Ethics / book auth. Halwani Raja. - [s.l.] : Palgrave Macmillan, 2007. - ISBN-10: 1403989842, ISBN-13: 978-1403989840.

Stock St. George William Joseph Guide to Stoicism [Book]. - [s.l.] : Project Gutenberg, 1850.

Stoicism [Online] // Stanford Encyclopedia of Philosophy. - 1996. - 2010. - http://plato.stanford.edu/entries/stoicism/.

Thurot Charles Manuel de Épictète accompagnée d'une introduction at revue par Charles Thurot [Book Section] / book auth. Thurot François. - Paris : Librairie Hachette, 1889.

Whitehead Alfred North An Inquiry Concerning the Principles of [Book]. - [s.l.] : Univ. Press, Cambridge, 1919.

Whitehead Alfred North The Concept of Nature [Book]. - [s.l.] : Univ. Press. Cambridge, 1920.

Yonge C. D. [Online] / ed. YONGE C.D.. - 2 May 2011. - http://classicpersuasion.org/pw/diogenes/dlchrysippus.htm.

Zeeman Erik Christopher Causality implies the Lorentz group [Journal] // Journal of Mathematical Physics. - 1964. - 4 : Vol. 5.

Zinoviev Alexandr The Yawning Heights [Book]. - [s.l.] : Random House, 1979. - ISBN-10: 0394427106, ISBN-13: 978-0394427102.

List of Illustrations

Index

www.ingramcontent.com/pod-product-compliance
Lightning Source LLC
Chambersburg PA
CBHW031047280326
41928CB00047B/85